P9-CFQ-526

RELIABILITY PRINCIPLES AND PRACTICES

RELIABILITY PRINCIPLES AND PRACTICES

S. R. CALABRO

President, Aerospace Technology Corporation, Newark, N.J.

McGRAW-HILL BOOK COMPANY

New York San Francisco Toronto London Sydney

RELIABILITY PRINCIPLES AND PRACTICES

 Library of Congress Catalog Card Number: 61-12956

5 6 7 8 9 MP 9 8

09600

PREFACE

The purpose of this book is to present the fundamental concepts underlying modern reliability theory and to illustrate their application to the solution of practical problems. To achieve this end, I have, in so far as is possible, attempted to avoid abstract statistical and mathematical techniques and substituted in their stead effective reliability formulations and mathematical models whose applications are profusely illustrated throughout the text by means of many examples. It must be emphasized, however, that all the formulations and mathematical models employed have a foundation in precise mathematical theory, and rigorous proofs can be developed for each of them.

To the greatest extent possible, I have also attempted to write in a relatively simple style so that a reader with a minimum of specialized knowledge in the field of statistics and reliability analysis might comprehend the subject matter with relative ease. With this in mind, I have purposely engaged in redundancy and constantly reemphasized the meaning of symbols and terms to facilitate reading and complete comprehension. I have also attempted to present and develop complex topics in an elementary fashion, some of which, to the best of my knowledge, have not been formulated previously.

Chapter 1 is a general survey of the reliability field in which comparisons are drawn with quality control. There has unfortunately been a great deal of confusion between these two areas, and the attempt here is to differentiate between the two and yet show their complementary nature.

Chapters 2 through 6 are intended as a foundation for the reader with limited knowledge of probability theory and statistical distributions. No attempt has been made in these chapters to elaborate; instead, only those concepts are explained which are felt to be necessary for a complete comprehension of subsequent chapters. These chapters have, therefore, been written with the emphasis upon application of techniques to the solution of the reliability problem.

Chapters 7 and 8 are primarily devoted to the method of obtaining reliability data and to their analysis and evaluation. Chapter 7 covers some typical field reporting methods, while Chap. 8 gives precise methods of analyzing such data by the use of the chi-square distribution. It also gives approximate methods for achieving similar results.

Chapter 9 represents a real positive contribution in the field of maintainability and redefines availability in terms of mission and equipment availability. This supplements the commonly accepted practice of defining availability only in terms of up-time ratio. In addition, mathematical relationships are developed which correlate reliability and maintainability so that one can be traded off for the other to achieve a specified availability. This relationship between maintainability and reliability is also expressed in graphical form by means of a nomograph in order to facilitate calculations.

This chapter also presents a new concept: the principle of the *maintenance time constraint*, which is embodied as an integral part of the availability models. The maintenance time constraint permits the reevaluation of failure criteria in terms of total permissible down time before a device is classified as having failed.

This chapter also introduces several mathematical models for redundant systems and summarizes them in a general expression called the n-plex equation, which is useful in evaluating the reliability of complex systems consisting of n redundant elements.

Chapters 10 and 11 discuss reliability sampling and control methods. A new statistic, K^2, developed by the author, facilitates the calculation of time sample sizes to any producer's and consumer's desired reliability risk. The acceptance and rejection numbers are also readily calculated by the use of K^2. In addition the author introduces the r chart and subsample f chart, which can be used as media for the control and assessment of reliability. These charts also act as a medium for the evaluation of product improvement and the determination of whether or not the article conforms to specified reliability requirements. When used as an acceptance medium, the subsample f chart makes possible the assessment of reliability by utilizing a minimum of test time and cost.

Chapter 12 discusses reliability and availability prediction methods and gives illustrative examples to emphasize the techniques used.

Chapter 13 outlines the basic principles to be observed in designing for reliability. It does not attempt to give detailed examples but rather emphasizes basic concepts since this field is considered to be too broad to be covered in one chapter.

The remaining three chapters, namely, 14, 15, and 16, are of primary concern to personnel who may be involved in the management of reliability programs. They provide the background and information necessary for an understanding of the elements necessary for the establishment of a workable reliability organization and program.

In a general sense, the object of this book is to provide the necessary knowledge and perspective needed to plan, organize, and implement a reliability function in a typical organization.

It is the author's sincere hope that statistical, mathematical, engineering, and management personnel may find some area of interest in the pages which follow, enabling them to perform their functions with an increased understanding of this particular subject.

In the preparation of this book, I have incurred numerous obligations—both general and specific. The material in this book has been drawn from the experience of many years, and it is with pleasure that I acknowledge my indebtedness to my coworkers and to the representatives of various government and private agencies who have always been courteous and helpful in cooperating with requests for information. Specifically, I would like to mention M. Barov and V. Selman of my present staff, who offered advice and assisted in the review of material.

I would also like to express my appreciation to many other members of my staff past and present, too numerous to mention individually, who assisted in reading copy and furnishing constructive comments, and to the many secretaries and typists faced with the onerous task of constantly retyping the manuscript as the numerous modifications and changes were developed.

My thanks go as well to Dr. R. G. Pfefferkorn, who provided some material on psychophysical methods of reliability prediction.

I am also indebted to Professor Sir Ronald A. Fisher, Cambridge; to Dr. Frank Yates, Rothamsted; and to Oliver & Boyd, Ltd., Edinburgh, for permission to reprint Table IV from their book "Statistical Tables for Biological, Agricultural and Medical Research."

I wish, too, to thank my many friends who encouraged me throughout this work, particularly M. Dubilier, President of International Electric Corporation, who constantly heartened me by his interest. But most of all my thanks go to my understanding wife and children, who always encouraged me and never complained of neglect while I worked on this text instead of devoting my time to my family.

S. R. Calabro

Dedicated to the memory of
my Father and Mother

CONTENTS

CHAPTER 1

AN INTRODUCTION TO THE RELIABILITY CONCEPT

1-1. Definition of Reliability. The RETMA[1] definition of reliability states: "Reliability is the *probability* of a device performing its purpose *adequately* for the period of *time* intended under the *operating conditions* encountered." This definition is now accepted by most contemporary reliability authorities and hence is considered to be standard.

It should be observed that the definition stresses four elements, namely, *probability, adequate performance, time,* and *operating conditions.* These four factors are very significant since each of them plays an important role. Let us consider each of them in turn.

Probability, the first element of the reliability definition, is a quantitative term because it is expressed as a numeric such as a fraction or a per cent which signifies the number of times we can expect an event to occur out of a total number of trials. Thus, a statement that the probability of survival P_s of a device's operating for 50 hr is 0.65 or 65 per cent indicates that only 65 times out of 100 trials would we expect the device to be functioning after a 50-hr operating period.

Adequate performance, the second element of the reliability definition, indicates that criteria must be established which clearly specify, describe, or define what is considered to be satisfactory operation. As an example, suppose a spark plug from an eight-cylinder automobile engine becomes defective; the motor might become noisy, but it will still operate. In this case, the performance might still be considered adequate if the automobile arrived at its destination in the prescribed time. However, if the engine broke down completely or barely functioned, there is no doubt that this condition would be considered inadequate performance.

The third element of the reliability definition, time, is one of the most important because it represents a measure of the period during which we can expect a certain degree of performance. It is the cornerstone of the reliability concept because without a knowledge of the probability of a device's functioning or surviving for a given time, there is no way of

[1] RETMA, Radio Electronics and Television Manufacturers Association, is now known as EIA, Electronics Industries Association.

assessing the probability of completing a mission or task which is scheduled to last for a given period.

The operating conditions under which we expect a device to function are the fourth element of the reliability definition. Typical of these are such factors as temperature, humidity, shock, and vibration. Experience has shown that each of these has a definite effect on performance. Therefore, it is necessary that they be included as part of the reliability specification because if this is not done, the definition of reliability would be relatively meaningless.

1-2. Definition of Quality. A customary definition of quality is: *The quality of a device is the degree of conformance to applicable specifications and workmanship standards.*

Usually quality specifications do not concern themselves with the element of time, as do reliability specifications. For the most part they define workmanship standards as they relate to a particular industry; or they specify tests which are considered necessary to verify that a device is capable of performance within certain limits without reference to time.

For example, in the electronics industry a number of specifications have been developed which define workmanship standards and classify defects in terms of their relative effects on quality. This is usually referred to as a *classification of defects.* The most common classifications are critical, major, and minor defects. These classifications apply to various categories, such as soldering, wiring and wire wrap, hardware, and finishing defects. For example, in the category of soldering, unsoldered or cold-soldered joints are usually regarded as major defects, while insufficient solder might be called a minor defect. Thus we see that quality is a relative measure because it is assessed on the basis of standards which have been established as applicable to a particular type of equipment.

There are other types of quality specifications which require testing under environmental or ambient conditions to determine whether or not the device functions as specified. These may also be grouped according to major and minor classifications.

Experience has shown that good quality is a major ingredient of high reliability because inferior workmanship may shorten the life of a device and therefore affect its reliability. It is for this reason that in this chapter we shall outline some of the basic factors which affect the quality of the manufactured product.

1-3. Importance of Reliability. In this modern day of science, in which complex devices are utilized for military and scientific purposes, a high degree of reliability is an absolute necessity. There is too much at stake in terms of cost, human life, and national security to take any risks with devices which might not function properly when needed most.

In the old days, weapons of war were relatively simple. A saber consisted essentially of one piece of hardened steel which, even if it were broken during battle, only affected the fate of the warrior involved. Nowadays, if a missile should misfire when needed, the dud might affect the entire outcome of the engagement, particularly if a strategic target were involved. Moreover, the present-day weapon, unlike the saber, consists of thousands of small parts, each interwoven into a complex web which constitutes the weapon. The failure of any one of these parts could, and in most instances does, adversely affect the operation of the weapon. Therefore, if a unit is very complex and must be very reliable, the parts of which it is composed must in turn be of a high order of reliability. Oftentimes, the state of the art does not permit manufacture of these parts to the required reliability standards, and therefore other techniques must be employed. We shall discuss some of these methods in Sec. 1-4.

1-4. Methods of Achieving Reliability. The most basic method of achieving a reliable product is through mature design. In some instances this is easy to achieve, while in other cases it is most difficult. The architect has a relatively simple job in providing a mature building design because he has access to much data on strength of materials to assist him in his efforts. He also has available the details pertaining to structures similar to the one he is planning to design. As a result the architect is confident that the building that will be erected from his plans will be very reliable because it will withstand all the possible loads to which it might be subjected for an extended time period.

The civil engineer who designs a suspension bridge has advantages which are similar to the architect's. In addition, both men utilize the safety factor in those cases where there is a doubt concerning the ability of a particular structural member to withstand a specific load. This means that instead of a marginal structural member, one with a high load rating will be specified. Some of these safety factors are as large as four times that which is theoretically required. Usually cost is the sole restriction on the degree of the safety factor which is used for structural projects.

The electronics engineer, on the other hand, has many other restrictions which he must overcome in his designs, as does the aeronautical engineer. Among these are the limitations of cost, weight, volume, and configuration. Moreover, design requirements are in a constant state of change, and data regarding the reliability of parts and components are not always available. Therefore, the engineer must resort to special techniques as well as exhaustive testing to achieve the required degree of reliability within the constraints imposed upon him.

One of the approaches which has been successfully employed is the method of reliability prediction. This is a method of synthesizing the effects of the reliability of the various parts or components which comprise a device and calculating the over-all reliability by using certain statistical techniques.

The main advantage of making reliability predictions is that the designer obtains a broad estimate of the reliability which is achievable. He also is able to assess those parts or elements of the design which adversely affect reliability and therefore can determine when he must resort to redundancy or other techniques. In Chap. 12 we shall discuss the various types of redundancy in more detail. For the present, redundancy is the technique which employs more than one element to assure better reliability. If one element fails, there is another to take its place. It is not a new idea by any means. The dual wheels on trucks and airplane landing gears are commonplace illustrations. If one tire fails, the chances are good that the other will carry the load until the original can be repaired. In electronic equipment, two resistors could be placed in parallel so that if one burned out, the other would take the load. This is a good technique for the engineer to use, provided that he carefully considers all the factors involved in order to assure that the remaining element that has not yet failed has sufficient capacity and that there are no interactions with other elements in the mechanism which would cause other troubles.

Another method of attaining redundancy is by means of switchover. In this case if the primary unit fails, a standby unit is switched in to take its place. An automatic telephone exchange is a good illustration of this method. If a person dials a number and a particular stepping switch is inoperative, another switch will take the call and complete it. In this manner, the subscriber is satisfied that he has received good service and has no idea that there had been an internal failure of a switch. This is one of the methods used to assure continuous reliability of an over-all system even though a component or subsystem might have failed.

Another technique for assuring the reliability of equipment is to use marginal testing. This is prescribed by the designer as a method of predicting the probability of an imminent failure due to degradation. This involves periodic testing on a programmed basis. In the case of electronic equipment, it may be done by imposing limit voltages on certain elements and observing the resulting symptoms. In this manner, it is possible to isolate degraded components or parts and replace them before the actual failure occurs.

Maintainability is also used to provide a high effective reliability. If parts are readily interchangeable and replaceable, failures may be repaired rapidly by replacing defective parts and components with operating spares. When these replacements can be performed expeditiously, we say we have a high maintenance-action rate and correspondingly high availability of equipment, or simply a high availability.

A mature design will provide for rapid interchange of parts to assure a minimum of down time in case of failure. This means that electrical and mechanical tolerances should be sufficiently liberal to assure that the device will continue to function with a replacement unit. Moreover, the replacement must be capable of being performed expeditiously and without requiring the disassembly of adjacent units in order to make the replacement accessible.

Built-in test equipment is another effective technique to assure reliability. In electronic equipment, it is customary to use the type of test equipment which is simple and of the go–no-go and press-to-test type. In this manner, by the simple expedient of pressing buttons, one can tell whether certain circuits are functioning by observing whether the proper pilot lamp lights. This makes possible routine sequential maintenance checks to assure that everything is functioning in proper working order.

Another method of achieving reliability is called debugging or burn-in. In general, a device goes through three separate and distinct stages. The first is called a break-in or infant stage, during which the equipment is characterized by a relatively high failure rate. The second is called the operating stage, during which we experience a constant failure rate. This is also called the stable or operating period. The third stage is called the wear-out phase, which is the time when the frequency of failure, or the failure rate, increases rapidly. When this happens it is an indication that the equipment has aged or become worn.

Debugging takes advantage of the infant stage. It is a method of accelerating the completion of this infant stage by means of operating the unit night and day, if necessary, until all the early failures are isolated and removed. When this has been done, the unit is considered to be in the operating stage, in which it exhibits a constant failure rate and relatively better reliability.

Testing to destruction is another method of isolating potential failures. It is based on the assumption that if a part is subjected to abnormal stresses, it will fail prematurely. On the other hand, if the part survives this treatment, it is presumed that it will be very reliable when subjected to less stringent conditions in actual use. This technique appears to have merit and probably is very effective for static situations. As a matter of fact, the science of mechanics of materials is based on this very concept. A supporting member such as a beam or a truss is certain to stand a lighter load than that which was used to test it. However, when confronted with dynamic conditions such as are encountered on structures subjected to repeated loads or unusual vibrations, this technique's value is doubtful unless the actual conditions can be simulated in test. This is particularly true in the case of electronics circuits because failures are not always due to the application of overvoltage or other extreme conditions, and therefore precise simulations are difficult. Many of these failures are the result of poor application and lack of compatibility of associated circuits. Therefore, the effectiveness of the test-to-destruction technique is questionable for dynamic applications.

Another reliability technique currently prescribed in many reliability specifications is the selection of parts on the basis of special tests. This is a costly method and not as effective as some would like to believe. For the most part, it consists of 100 per cent testing of these parts under extreme environmental conditions. Usually these conditions prescribe cycling between low and high temperatures and humidity as well as subjecting the part to vibration and shock, the assumption being that those which will withstand the rigors of such tests will be reliable when used in the end item.

There are many disadvantages to the selection of parts. Some typical ones are the following:

1. The method is expensive and of doubtful value.

2. The tests to which the parts are subjected do not always simulate conditions of use and application.

3. There is little knowledge about the effects of these rigid tests on the parts themselves. Therefore, the tests might actually result in part degradation instead of reliability improvement.

1-5. Measures of Reliability. An expression of reliability in terms of an abstract number is not meaningful unless the prevailing physical or environmental conditions under which the reliability was assessed are also included as part of the statement. The reason is that the reliability of a device must be specified with relation to its operating conditions, because if these vary, so will the numeric which is used to express reliability.

For example, it would indeed be misleading for an automobile manufacturer to advertise without reservation that his automobiles were capable of operating for over 100,000 miles without requiring a brake change. Obviously, if a brake reliability test is conducted over toll roads which require infrequent stops, the brake usage will be infrequent. On the other hand, if the test is conducted in city traffic for the same 100,000 miles, we would expect faster brake wear due to more frequent usage.

The most common measures of reliability are failure rate r, probability of survival P_s, and mean time between failures m or mtbf. The failure rate is commonly expressed in terms of failures per hour, 100 hr, 1,000 hr, or per cent failures per 1,000 hr. The probability of survival is expressed as a decimal fraction or per cent which indicates the probable or expected number of devices that will function for a required period of time. For example, if the mission time were 10 hr and a large number of units were tested and all of them equaled or exceeded 10 hr of operation, we would have an example of 100 per cent reliability. On the other hand, if only 80 per cent of the devices equaled or exceeded 10 hr of operation, we would conclude that their reliability for this mission is probably 80 per cent. There are statistical methods, which we will study later, which are used to determine the degree of confidence which we can repose in the reliability test results.

The mean time between failures is expressed in hours. The larger the value of mtbf, the greater the reliability. As its name implies, the mtbf is the ratio of the total test time of a device to the total number of failures.

The failure rate is the reciprocal of the mtbf, and therefore the smaller the numerical value of failure rate, the greater the reliability.

1-6. What Is Considered Satisfactory Reliability? As we have seen in Sec. 1-5, reliability is measured with relation to the mission to be accomplished. Ideally, we would like to accomplish our mission 100 per cent of the time. However, we may find that from a practical point of view the ideal is not always possible to achieve. This may be due to many design problems which must be resolved or to cost limitations or to other factors such as time, weight, space, and maintenance. Therefore, in most instances the engineer is faced with trade-offs between all the factors involved in order to achieve the optimum condition. For example, in order to meet budgetary requirements, he might be satisfied with 95 per cent of the missions being completed; or he might permit a maximum down time for repair and consider any repair performed within this time as indicative of satisfactory performance. This latter concept embodies the principle of *availability*, which we shall discuss in Chap. 9.

In any event, the main purpose of considering reliability is to assure that the mission will be achieved. Therefore, it is necessary that this mission be clearly described so that there is no doubt as to what must be achieved. This description should also list the latitude which is allowed before the mission is considered a failure. Once this has been done, the engineer can design with this knowledge in mind and specify the required reliability in terms of the operational conditions involved.

1-7. General Provisions of a Reliability Specification. A good reliability specification must include means for assuring that the reliability which is required has actually been achieved. Many reliability specifications have been written which are very general and lack the detail necessary to satisfy the reliability requirement. Therefore we will briefly outline here the basic elements which a reliability specification should contain. This subject will be discussed in more detail in Chap. 14.

In general, an effective reliability specification should provide means for *measurement*, *evaluation*, *improvement*, and *prediction*. Each specification in turn should define the general *purpose*, *place*, *methods*, *instruments*, *personnel*, *circumstances*, and *procedures* involved. The details of each specification depend essentially upon the characteristics of the device to be evaluated and their importance in determining the reliability.

The following elements briefly summarize the major provisions of a good reliability specification:

1. Definition of the device or system
2. Information concerning system age, production stage, and modification status
3. Criteria of adequate performance
4. Basis for computing time
5. Description of operating conditions
6. Description of maintenance conditions
7. Definition of malfunction and failure
8. Description of sampling procedures and computations
9. Other considerations

1-8. Manufacturing Practices Which Affect Reliability. There are many manufacturing practices which affect the reliability of the manufactured product. The most important of these is quality control, which covers several areas including workmanship, manufacturing processes, materials, storage and issue of parts and materials, engineering changes and deviations, production and incoming-materials inspection and test, vendors' quality performance, and many other activities. As we can see, this is a rather impressive list of functions. However, it should be understood that, as the name implies, the quality control aspect is one of evaluation or control and not one of implementation or performance. For example, quality control is definitely concerned with evaluating the effectiveness of a process such as plating, but the actual job of determining or planning or developing the plating process is normally a function of industrial engineering.

A major concern of quality control is workmanship, because if this is substandard, it may adversely affect reliability. As an illustration, poor soldering usually causes latent defects which definitely shorten the operational life of electronic equipment. This is particularly true of a cold-soldered or unsoldered joint. In each case poor reliability will be the result if such a condition is not detected and corrected in time.

There are two basic methods which can be used to assure good workmanship. The first is to develop good manufacturing methods and techniques. The second method is thorough inspection of the manufactured product. This inspection can be performed at various times during the manufacturing cycle and is called *in-process inspection;* or it may be performed on the finished product, in which case it is called *final inspection.*

The best method of controlling quality is by means of controlling the process, since this is the most economical and positive method. It is for this reason that quality control engineers use process control charts. Moreover, it is always better to control the method of doing things while the process is being performed than to effect correction after the job has been completed. This is why inspection is often considered by some as an unnecessary overhead expense. However, since most processes are dependent on men and machines, which are not infallible, inspection becomes a necessity to assure that a good product has been made.

In any case, regardless of whether good workmanship is the result of good processing or efficient inspection or both, quality control plays the role of assuring that the job has been done in accordance with high quality standards. Since this is not a text on quality control, we will not at this time describe the techniques which are used; any good text on quality control may be referred to for this purpose.

Another element which affects reliability is the use, storage, and issue of materials. If the design engineer inadvertently specifies two dissimilar metals which are incompatible and these are placed adjacent to each other, corrosion may be the result. A good quality control operation would probably detect this condition in time and have it corrected. There are many illustrations of how materials may affect reliability, since latent defects are usually due to their misuse or misapplication. For example, an acid-base solder produces a good temporary solder connection but is also subject to corrosion at a later time and therefore unquestionably will adversely affect reliability.

Improper storage of materials may cause their deterioration, which will probably result in poor equipment reliability. For example, in one case it was found that metal parts had been stored close to the electroplating department and were contaminated by the corrosive plating fumes. At a later time they were spray-painted. The entrapped acids under the paint caused it to peel off, and this adversely affected the reliability of the product.

Improper handling of materials on the assembly line or at other locations may result in their damage. Any damage, even to the slightest degree, may adversely affect reliability. As an example, in one instance parts of a delicate electronic cavity had been contaminated by metal chips and were not cleaned or placed in a separate container prior to use. The result was a shorter life for the cavity due to abrasion caused by the metal chips, and therefore the reliability of the product in this case was appreciably reduced.

Quality control is also interested in pilot runs and new design control, since these in turn affect quality and reliability. It is well known that a new product fresh off the drawing board must undergo several design changes before it becomes practical to produce. In some instances, because of pressures for delivery or because of lack of foresight, the design engineer writes specifications on the basis of incomplete or insufficient data. If these specifications are permitted to remain unchanged in production, either it becomes impossible to produce the item or else much costly tailoring by the factory is the result. This procedure of selective assembly is very expensive and undesirable since parts have to be carefully matched and mated to achieve a working unit. The pilot run is the place to introduce changes which are based on objective evidence. It is during this pilot run that quality control engineers, armed with their statistical training, can do the most good. The quality control engineer collects data for analysis based on representative samples from the processes which are involved. As a result of his statistical analysis, he decides on whether or not the specifications and the process are compatible. Moreover, the results of his investigation are discussed with interested parties including those in the fields of product or reliability engineering, production, and industrial engineering. A decision is then made either to redesign or to change specifications or the process as required.

There are many other functions performed by quality control which definitely contribute to improved product reliability. The quality control department reviews deviations from specifications and has the authority to disapprove if quality and reliability will be adversely affected. It is also charged with the responsibility of conducting or performing a surveillance of reliability time sampling tests. It accumulates test data and failure information and either performs analysis or furnishes information to reliability engineers to effect design improvements. In general, quality control is responsible for assuring an effective failure reporting system which provides for feedback of failure data and for assuring that corrective action is taken in time.

1-9. Reliability Policy. Most companies consider current concepts of what constitutes an effective reliability program as being unduly expensive and therefore are reluctant to become involved unless the cost is borne by the customer. Therefore, they "talk good reliability" but do very little about implementation of such a program. To offset this attitude, in many of its contracts the Department of Defense now specifies reliability as part of the requirements of the contract. The government has found that in some instances the cost of maintenance for unreliable equipment is in excess of 10 times the original cost and that a good reliability program reduces this expense considerably. It is also felt that current military requirements can be met by specifying equipment of high reliability only.

The subsequent chapters will discuss some of the practical methods and techniques which the reliability engineer can use as tools in the development of good reliable equipment.

CHAPTER 2

MEASURES OF CENTRAL TENDENCY AND DISPERSION

2-1. Introduction. There are many measures of central tendency used by the statistician for particular applications. However, in this chapter we shall discuss only those measures which are germane to the subject matter of this text. In other words, only those concepts which are necessary as tools for a subsequent lucid understanding of reliability will be discussed.

2-2. Central Tendency. The term central tendency implies a value which is central or midway between the extreme limits of a set of data, or a value from which individual values deviate in accordance with some pattern. The most useful, commonplace, and popular measure of central tendency is the arithmetic average or mean.

2-3. The Arithmetic Average or Mean. The average of a group of values is simply computed by dividing the sum of these values by their total number. Thus, the average of the values 5, 10, 8, 6, and 11 is 8. The computation is

$$\frac{5 + 10 + 8 + 6 + 11}{5} = \frac{40}{5} = 8$$

In algebraic notation, the average would be shown as

$$\text{Average} = \frac{X_1 + X_2 + X_3 \cdots X_n}{n} = \bar{X} \qquad (2\text{-}1)$$

In the above formula each of the X's represents an individual value, and the subscripts 1, 2, n represent the first value, second value, up to the nth value. The letter n is used as notation which indicates the number of individual values involved. An X with a bar over it, written as $\bar{X}$, is a symbol customarily employed to represent the average value of all the X values. In this case there are five values of X, which respectively are $X_1 = 5$, $X_2 = 10$, $X_3 = 8$, $X_4 = 6$, and $X_5 = 11$. Thus $n = 5$. If many values had to be dealt with, the symbolism represented by the above formula would be very lengthy. Therefore, an abbreviated notation has been developed which, when understood, represents the same thing as Eq. (2-1). This abbreviated formula is written

$$\text{Average} = \frac{\Sigma X}{n} = \bar{X} \tag{2-2}$$

The Greek upper-case letter Σ (sigma) is used to indicate "sum of all the values," and therefore Eq. (2-2) is equal to Eq. (2-1).

2-4. The Median. The median represents the middle value. It may or may not be equal to the mean, depending on whether the values involved form a symmetrical or nonsymmetrical distribution. Thus the median for the numbers previously listed, namely, 5, 10, 8, 6, and 11, is also 8 because two numbers, 5 and 6, are smaller than 8, and two numbers, 10 and 11, are larger than 8. Therefore, 8 is the middle or median value.

2-5. The Geometric Mean. The geometric mean is the nth root of the product of the n individual values. Thus suppose we wanted to find the geometric mean of 27 and 3. Since there are two values involved, $n = 2$, and therefore the geometric mean would be equal to the square root of the product of 27 and 3. This is shown as

$$G = \sqrt{27(3)} = \sqrt{81} = 9$$

In algebraic notation the equation would be

$$G = \sqrt[n]{(X_1)(X_2)(X_3) \cdots X_n} \tag{2-3}$$

where G is the symbol for the geometric mean and the X's represent the first, second, third, etc. values until n values are accounted for.

2-6. Measures of Dispersion. When we defined central tendency, we implied that there was a dispersion of values about the central tendency. The common methods of describing this dispersion are the range and standard deviation.

The Range. The range as a measure of dispersion is calculated by subtracting the smallest value in a set of data from the largest value. Thus for the numbers 5, 10, 8, 6, and 11, the range is 6. It is computed

$$11 - 5 = 6$$

The algebraic symbol for range[1] is R. Therefore, algebraically, the formula for range is represented as

$$R = X_H - X_L \tag{2-4}$$

where X_H = highest value
X_L = lowest value

[1] In this application R is used to indicate range. It is also used later to indicate the rejection number. (See glossary of terms.)

So far, then, it is seen that knowing the average $\bar{X}$ and range R of a group of values gives us some indication of how far it can be expected that a particular value will stray from the average value of the group. However, the range is not particularly efficient for all applications, and because this is so, the standard deviation is used.

The Standard Deviation. Because of the limitations of the range, a more efficient measure of dispersion, called the standard deviation, has been devised. The symbol for standard deviation is the lower-case Greek letter σ (sigma). This should not be confused with the upper-case Greek letter Σ (sigma), which has previously been indicated to mean "sum of." The standard deviation is defined as the root-mean-square deviation of the observed values from their average. Mathematically the standard deviation σ is expressed by the equation

$$\sigma = \sqrt{\frac{\Sigma(X - \bar{X})^2}{n}} \tag{2-5}$$

This equation simply means that the average of the values is subtracted from each of the individual values, and the difference is then squared. The individual squared results are then added, and this sum is divided by a number n representing the number of differences involved. This result is known as the variance. The square root of this variance is the standard deviation σ. The symbol for variance is σ^2, and its equation is

$$\sigma^2 = \frac{\Sigma(X - \bar{X})^2}{n}$$

An example will be used to demonstrate the methods of computing σ.

Example 2-1. Find the average and the standard deviation for the values listed.

$$X_1 = 20 \qquad X_2 = 22 \qquad X_3 = 20 \qquad X_4 = 24 \qquad X_5 = 20$$

Solution

$$\Sigma X = X_1 + X_2 + X_3 + X_4 + X_5$$
$$\Sigma X = 20 + 22 + 20 + 24 + 20 = 106 = \text{sum of all values}$$
$$\bar{X} = \frac{\Sigma X}{n} = \frac{106}{5} = 21.2 \text{ average}$$

To get the standard deviation, first subtract the average from each of the individual values and square the differences. Then sum each of the results thus obtained, as shown below.

$$
\begin{array}{lr}
 & \textit{Squaring} \\
X_1 - \bar{X} = 20 - 21.2 = -1.2 & (-1.2)^2 = \ 1.44 \\
X_2 - \bar{X} = 22 - 21.2 = \ \ 0.8 & (0.8)^2 = \ 0.64 \\
X_3 - \bar{X} = 20 - 21.2 = -1.2 & (-1.2)^2 = \ 1.44 \\
X_4 - \bar{X} = 24 - 21.2 = \ \ 2.8 & (2.8)^2 = \ 7.84 \\
X_5 - \bar{X} = 20 - 21.2 = -1.2 & (-1.2)^2 = \ 1.44 \\
\hline
 & \Sigma(X - \bar{X})^2 = 12.80
\end{array}
$$

Then divide $\Sigma(X - \bar{X})^2 = 12.80$ by $n = 5$. This gives the variance σ^2. Thus

$$\sigma^2 = \frac{\Sigma(X - \bar{X})^2}{n} = \frac{12.80}{5} = 2.56$$

The standard deviation is the square root of the variance. Therefore

$$\sigma = \sqrt{\frac{\Sigma(X - \bar{X})^2}{n}} = \sqrt{\frac{21.80}{5}} = \sqrt{2.56} = 1.60$$

An alternative formula for standard deviation is

$$\sigma = \sqrt{\frac{\Sigma X^2}{n} - \bar{X}^2} \qquad (2\text{-}6)$$

The calculation of σ using the alternate formula for the same values as in the previous problem is as follows:

Square all individual values of X, add the squares, and divide by n. Thus

$$
\begin{array}{l}
X_1^2 = \ \ 400 \\
X_2^2 = \ \ 484 \\
X_3^2 = \ \ 400 \\
X_4^2 = \ \ 576 \qquad \dfrac{\Sigma X^2}{n} = \dfrac{2{,}260}{5} = 452 \\
X_5^2 = \ \ 400 \\
\hline
\Sigma X^2 = 2{,}260
\end{array}
$$

For the average, take the sum of the individual values and divide this sum by $n = 5$.

$$\text{Average } \bar{X} = \frac{\Sigma X}{n} = \frac{20 + 22 + 20 + 24 + 20}{5} = \frac{106}{5} = 21.2$$

Substituting in (2-6), we get

$$\sigma = \sqrt{\frac{\Sigma X^2}{n} - \bar{X}^2} = \sqrt{452 - (21.2)^2} = \sqrt{452 - 449.44}$$

$$= \sqrt{2.56} = 1.60$$

Thus $\sigma = 1.60$, which is the same value for standard deviation as was obtained using the previous formula.

2-7. Simplifying Calculations of Standard Deviation. The calculations for σ may be simplified by subtracting a constant value from each of the individual values. This is possible because the subtraction or addition of an arbitrary constant to each of the individual readings does not change the standard deviation. For example, using the same values as before, the method would be as follows. From each individual value subtract 20. Thus

$$X_1 - 20 = 20 - 20 = 0$$
$$X_2 - 20 = 22 - 20 = 2$$
$$X_3 - 20 = 20 - 20 = 0$$
$$X_4 - 20 = 24 - 20 = 4$$
$$X_5 - 20 = 20 - 20 = 0$$

The average of the differences is the average above the origin of 20, since 20 was subtracted from each of the original values. Thus

$$\bar{X} \text{ (above origin 20)} = \frac{0 + 2 + 0 + 4 + 0}{5} = \frac{6}{5} = 1.2$$

To get standard deviation, substitute in the formula $\sigma = \sqrt{\Sigma X^2/n - \bar{X}^2}$ using the differences obtained above as the values for X.

$$\sigma = \sqrt{\frac{0^2 + 2^2 + 0^2 + 4^2 + 0^2}{5} - 1.2^2}$$

$$\sigma = \sqrt{\frac{4 + 16}{5} - 1.44} = \sqrt{4 - 1.44} = \sqrt{2.56} = 1.60$$

If the true average $\bar{X}$ is desired, add 20 to value of $\bar{X}$ (above the origin of 20), because 20 was the value which was subtracted originally. Thus $\bar{X} = 1.2 + 20 = 21.2$. It is obvious from the above calculations that the subtraction of a constant from each value greatly simplifies the work of calculating σ. In this example, as an instance, only small numbers required squaring instead of the relatively larger ones used in the previous method.

2-8. Computation of the Average and Standard Deviation by Grouping. During the previous discussion the basic formulas for calculating the standard deviation were used with a limited amount of data. In this case no apparent advantage of one formula over another would be very obvious. However, there is a very definite advantage to using alternate forms of the basic formula and other devices to reduce the work of calculations when the data are voluminous. One method of reducing work is to group all like values of data together in order to determine their frequency of occurrence. It is then possible to handle them as a group instead of as individual values. We shall use the same data as before to illustrate the method. It is again emphasized that the only reason for using a limited amount of data is to avoid boring the reader with voluminous figures or deluging him with calculations when all we are interested in doing at this time is to present a method of application. For this illustration we shall use a modified version of formula (2-6), namely,

$$\sigma = \sqrt{\frac{\Sigma(fX^2)}{n} - \left(\frac{\Sigma fX}{n}\right)^2} \qquad (2\text{-}7)$$

Again we have

$$X_1 = 20 \qquad X_2 = 22 \qquad X_3 = 20 \qquad X_4 = 24 \qquad X_5 = 20$$

An examination of the data indicates that we have three values of 20, one value of 22, and one value of 24. Thus, 20 has a frequency of 3, 22 has a frequency of 1, and 24 also has a frequency of 1. The symbol[1] for frequency in this application is f. These data can then be organized in a tabular form as shown below.

X	Frequency, f	fX	X^2	fX^2
20	3	60	400	1,200
22	1	22	484	484
24	1	24	576	576
Totals.....	5	106	...	2,260

Substituting in (2-7),

$$\sigma = \sqrt{\frac{2{,}260}{5} - \left(\frac{106}{5}\right)^2} = \sqrt{452 - 449.44} = \sqrt{2.56} = 1.60$$

which is the same solution as was obtained using the previous methods. Note also that Eq. (2-7) is the same as (2-6), since $\Sigma fX/n = \bar{X}$ and $\Sigma(fX^2)/n$ is the same as $\Sigma X^2/n$.

2-9. Computation of the Average $\bar{X}$ and Standard Deviation σ by Means of Grouping and Simplification. To further simplify the method of computation of the standard deviation, a method of grouping data into subgroups or cells and of using an arbitrary origin has been developed. Usually, it is advantageous and desirable to select the median as this arbitrary origin and call it zero, although any particular value may be selected and called zero without affecting the end results of the computations. The steps involved follow.

Arrange the data in either ascending or descending order in an array as shown below.

[1] In this application f indicates frequency. It is also used later to indicate failures.

X	f	ρ	$f\rho$	$f\rho^2$
24	1	1	1	1
22	1	0	0	0
20	3	−1	−3	3
Totals.....	$\Sigma f = n = 5$	...	$\Sigma f\rho = -2$	$\Sigma f\rho^2 = 4$

The formulas for $\bar{X}$ and σ to be used here will be

$$\bar{X} = \text{assumed origin} + \frac{\Sigma f\rho}{n}\theta$$
$$\sigma = \theta\sqrt{\frac{\Sigma(f\rho^2)}{n} - \left(\frac{\Sigma f\rho}{n}\right)^2} \tag{2-8}$$

In the above formulas θ is the cell interval. In this case $\theta = 2$, since the difference between adjacent values of X is 2. Note that in the tabulation, the column labeled ρ represents intervals from the assumed origin. These intervals are positive in the ascending order of values of X and negative in the descending order of values of X. Thus, substituting in the above formulas,

$$\bar{X} = 22 + \left(\frac{-2}{5}\right)(2) = 22 - \frac{4}{5} = 21.2$$
$$\sigma = 2\sqrt{\frac{4}{5} - \left(\frac{-2}{5}\right)^2} = 2\sqrt{0.8 - 0.16} = 2\sqrt{0.64} = 2(0.8) = 1.60$$

Of all the methods discussed in the preceding paragraphs, the simplified method just presented is the best. Of course, its advantage is not obvious when only three values are involved, as in the example. However, its advantages should become apparent during the solution of the following example.

Example 2-2. The torque in inch-pounds required to lock a control box in place in a radio transceiver was measured on a series of radio units. The results are tabulated below. Calculate $\bar{X}$ and σ.

40	67	73	55	60
58	53	58	57	55
60	60	65	50	65
66	51	60	70	80
67	58	75	60	60
60	60	60	43	55
60	70	73	60	60
80	45	60	55	42
75	57	59	50	60
52	70	70	60	50

Solution. The first step is to examine the data to determine how they should be grouped into cells. A rule of thumb to follow as a guide is to have between 10 and 20 cells with each cell boundary halfway between two possible observations. It is important to do this so that there will be no doubt about the cell in which each of the readings belongs.

The range of the data is found by subtracting the smallest reading from the largest. Thus, $R = 80 - 40 = 40$. In keeping with the desire to have between 10 and 20 cells, we divide the range by 3 to get the nearest estimate of the number of cells. The following arrangement of 14 cells is the result.

Cell boundaries	Cell mid-point	Frequency of occurrence f	ρ	$f\rho$	$f\rho^2$
38.5–41.5	40	1	−7	−7	49
41.5–44.5	43	2	−6	−12	72
44.5–47.5	46	1	−5	−5	25
47.5–50.5	49	3	−4	−12	48
50.5–53.5	52	3	−3	−9	27
53.5–56.5	55	4	−2	−8	16
56.5–59.5	58	6	−1	−6	6
59.5–62.5	61	15	0	0	0
62.5–65.5	64	2	1	2	2
65.5–68.5	67	3	2	6	12
68.5–71.5	70	4	3	12	36
71.5–74.5	73	2	4	8	32
74.5–77.5	76	2	5	10	50
77.5–80.5	79	2	6	12	72
		$\Sigma f = n = 50$		$\Sigma f\rho = -9$	$\Sigma f\rho^2 = 447$

$$\bar{X} = \text{assumed origin} + \frac{\Sigma f\rho}{n}\,\theta$$

$$\bar{X} = 61 + \left(\frac{-9}{50}\right)(3) = 61 - 0.54 = 60.46 = \text{average}$$

$$\sigma = \theta\sqrt{\frac{\Sigma f\rho^2}{n} - \left(\frac{\Sigma f\rho}{n}\right)^2} = 3\sqrt{\frac{447}{50} - \left(\frac{-9}{50}\right)^2} = 3\sqrt{8.94 - 0.0324}$$

$$\sigma = 3\sqrt{8.928} = 3(2.99) = 8.97 = \text{standard deviation}$$

Note: Any value that falls between the limits of the cell boundaries is arbitrarily assigned to the cell mid-point shown and the frequency for each mid-point is determined accordingly.

2-10. Calculation of Standard Deviation by Using the Average Range $\bar{R}$. In addition to the methods described for calculating standard deviation σ, there is another method which depends upon the average value of the range $\bar{R}$. This is the method that is used when making calculations of 3-sigma control limits to be used on control charts for variables.

The method is relatively easy. The first thing that is done is to subdivide the data into subgroups 2, 3, 4, 5, or greater. Then the range is obtained for each subgroup. The average range $\bar{R}$ of all the subgroups is calculated as the ratio of the sum of the subgroup ranges and the number of subgroups. To find σ, divide $\bar{R}$ by a value of d_2, which is found by reference to Table 5, App. 7. The relationship is

$$\sigma = \frac{\bar{R}}{d_2} \tag{2-9}$$

In order to illustrate the method we will use the data of Example 2-2.

Example 2-3. Using the data of Example 2-2, determine the standard deviation σ by the use of Eq. (2-9).

Solution. Tabulate the data into subgroups of 5.

Subgroups					*Range of each subgroup of 5*
40	67	73	55	60	33
58	53	58	57	55	5
60	60	65	50	65	15
66	51	60	70	80	29
67	58	75	60	60	17
60	60	60	43	55	17
60	70	73	60	60	13
80	45	60	55	42	38
75	57	59	50	60	25
52	70	70	60	50	20
					$\Sigma R = 212$

$$\bar{R} = \frac{\Sigma R}{n} = \frac{212}{10} = 21.2$$

From Table 5, App. 7, the value of d_2 for subgroups of 5 is equal to 2.326. Substituting in (2-9),

$$\sigma = \frac{21.2}{2.326} = 9.1$$

which is essentially equal to the value of σ found in the solution of Example 2-2.

PRACTICE PROBLEMS

2-1. What is the average $\bar{X}$ and standard deviation σ of the following data?

8	7	10	1	15
9	6	9	2	13
6	3	8	5	11
15	2	7	7	9
14	16	5	9	7
3	1	3	11	5
10	9	2	13	2

2-2. The following data represent measurements in mils of a sample of stainless-steel shafts. What is the average and standard deviation for these data?

1,548	1,547	1,549	1,547	1,546	1,546	1,547
1,546	1,549	1,548	1,548	1,548	1,548	1,548
1,549	1,540	1,547	1,547	1,548	1,547	
1,546	1,548	1,546	1,546	1,547	1,549	
1,548	1,547	1,547	1,549	1,542	1,546	

2-3. The 100-hr reliabilities of the three major components of a radio receiver are as follows: r-f stage = 0.97, mixer and i-f stage = 0.95, output stage = 0.98. What is the geometric mean value of these reliabilities?

2-4. Ten incandescent lamps were tested to failure. The times to failure in hours were respectively as follows: 900, 221, 630, 682, 700, 672, 715, 501, 385, and 875. Calculate the mean time between failures and the standard deviation.

CHAPTER 3

PROBABILITY AND ITS APPLICATION TO SERIES AND PARALLEL RELIABILITY (REDUNDANCY)

3-1. Necessary Background. In Chap. 1, we defined reliability in terms of probability and as a function of time. In order to understand properly the mathematical techniques associated with the calculations of reliability, it is necessary to understand some basic concepts of probability. This chapter considers basic axioms of probability and also the laws of combination and permutation and their application to reliability analysis.

3-2. General Comments on the Philosophy of Probability. "Probable" is one of the most commonly used words in everyday conversation. People are constantly making predictions based on what they believe is probable. Predictions which are based on hearsay or on personal opinion with no statistical evidence cannot normally be lent credence. For example, predictions are periodically made that the world will come to an end on some specific date. Since these predictions are not made within a framework of objective statistics, the probability of the event actually occurring is considered to approach zero. At the other extreme, the statement that "everybody must die someday" has a probability of occurrence approaching 100 per cent because this is a statement of fact which has been demonstrated to be true for centuries on the basis of available data.

These two examples represent two extremes. It should be noted that we have stated the probabilities as approaching zero and 100 per cent. Theoretically, we can approach but never reach the extremes since all the data in either case can never be available. For actual calculations, however, we ignore infinitesimal values, and when decisions are necessary, we consider these values as the limiting value. From a philosophical point of view, it is possible to consider all causal relationships in terms of probability. For any given cause, we can know only the event which is most likely to occur, but we can never have complete and absolute certainty of the result until it occurs. Because of the mass of previous data, we can normally act upon the basis of a high degree of certainty. In many cases, however, the probability of occurrence of an event is not obvious, and it becomes necessary to establish the relative probability of an event. The statistician has provided himself with a graduated yard-stick which has end points of zero and one. The zero point is a point where there is no possibility of the event occurring, while unity indicates absolute certainty. Any graduation between these two points represents a relative measure of probability. This leads to our discussion of the types and definitions of probability.

3-3. A Priori Probabilities. A priori probabilities are the type that can be specified from the very nature of the event. For example, if we toss a coin over a flat surface, we know, because of our prior knowledge, that the coin must rest on either heads or tails. For the purposes of our calculation, we consider the case of standing on edge to be nonexistent, as the probability of this occurring is infinitesimal. Therefore, since there is an equal chance for a head or a tail, we can say that the probability of obtaining a head is one-half and the probability of obtaining a tail is one-half. Another example of a priori probability is the case of a die. A die has six faces, each with an equal probability of occurrence. Thus, the probability of occurrence of a face for any specific toss is one-sixth. A priori probability is, then, the predicted probability of an event's occurring or not occurring based on prior knowledge of the nature of the event.

3-4. Empirical Probability. Suppose we took the die from the previous illustration and rolled it 1,000 times. Then, according to a priori probability, any given face is expected one-sixth of the tosses, or we would expect that each face would occur 166.66 times. Now we know it is impossible to have a fraction of an occurrence, and therefore we might tend to suspect the accuracy of our a priori probability prediction. However, even if we rolled the die 996 times, we could not expect to get each number exactly 166 times. We would expect to get some distribution of occurrences of faces which would group around 166 times for each face of the die. Moreover, as we increased the number of trials, we would expect each face to occur closer to the a priori estimate of one-sixth of the number of trials. This all leads to a definition of empirical probability.

$$\text{Empirical probability} = \frac{\text{total number of occurrences of an event}}{\text{total number of trials}}$$

As the total number of trials is increased indefinitely, this ratio approaches some limit. This limit is then considered the probability that the event will occur.

As can be readily recognized, empirical probability is based upon actual measured data concerning an event rather than the intrinsic properties of the event.

3-5. Addition Law. The addition law is used to calculate the total probability when the event can occur in different ways, each of which is mutually exclusive. By mutually exclusive we mean that given two occurrences A and B, if A occurs, then B cannot occur. The probability that an event will occur in any one of many possible ways when each way is mutually exclusive is the sum of the probability of occurrences of several different possible ways.

For example, in our discussion of a priori probability, we noted that the probability of obtaining a head is one-half and the probability of obtaining a tail is one-half. By the addition law, the probability of obtaining a head or a tail equals the sum of the individual probabilities, and the total of all cases equals 1. (We have of course neglected the infinitesimal probability of the coin's landing on edge.) Likewise, the probability of obtaining any of the six numbers on a die is equal to the sum of individual values, which would be $\frac{1}{6} + \frac{1}{6} + \frac{1}{6} + \frac{1}{6} + \frac{1}{6} + \frac{1}{6} = 1$.

To illustrate further the principle involved, suppose two players A and B decide to play a game with a die. The rule is established that if an even number appears after each roll, A will win, and conversely if an odd number comes up, B will win. On the basis of a priori probabilities, it would appear that A and B's chances are equal, i.e., each has a 50 per cent chance of winning. Let us verify this conclusion by means of the addition theorem.

It was previously established that the probability of any face of the die appearing after each roll was $\frac{1}{6}$, or one time out of six. On this basis, A's chances of winning will depend on the probability of occurrence of a 2, 4, or 6. Since each of these has a probability of $\frac{1}{6}$, then the probability of an even number's coming up is the sum of the probabilities of the three different ways that an even number can appear, i.e., $\frac{1}{6} + \frac{1}{6} + \frac{1}{6} = \frac{1}{2}$. Thus, the logical conclusion of the a priori theorem has been verified by the addition theorem. Likewise, the probability of B's winning the game is the probability of the appearance of an odd number such as 1, 3, and 5, and again, since each has a probability of occurrence of $\frac{1}{6}$, the total probability of occurrence of an odd number is $\frac{1}{6} + \frac{1}{6} + \frac{1}{6} = \frac{1}{2}$. On this basis, the logical conclusion that each player has a 50 per cent chance of winning is established by the addition law.

3-6. Complementary Law. This rule can be used when it is desired to know the probability of an event's not occurring if we know the probability of an event's occurring. The probability of an event's occurring plus the probability of an event's not occurring is always equal to 1. Or stated mathematically, we can say for some event E:

$$P(E \text{ occurring}) + P(E \text{ not occurring}) = 1$$

or

$$P(E \text{ not occurring}) = 1 - P(E \text{ occurring})$$

For example, the probability of not getting a head with the coin-tossing case equals 1 minus the probability of obtaining a head.

3-7. Multiplication Law. If an event is made up of a number of separate and independent subevents which must occur simultaneously or in series, the probability of occurrence is the product of the probabilities that each will happen. If we assume two independent events E and F, the $P(E \text{ and } F \text{ both occurring}) = P(E)P(F)$. This basic rule can be illustrated by the following example.

Example 3-1. A box contains 10 balls. Four of the balls are black and 6 balls are white. If a person selects 2 balls from the box at random, what is the probability of the sample's containing both colors?

Solution. By use of a priori probabilities, the probability of selecting a black ball is $4/10$. If this happens, since there will be only 9 balls left in the box, the probability of selecting a white ball is $6/9$. Hence, in accordance with the multiplication law, the probability of choosing white after black is

$$4/10 \times 6/9 = 24/90$$

However, the conditions can also be satisfied in another manner, i.e., selecting black after white. In this case, the probability of selecting a white ball is $6/10$. If this occurs, there are only 9 balls left in the box, and hence the probability of selecting a black ball is $4/9$. Therefore, in accordance with the multiplication law, the probability of choosing black after white is

$$6/10 \times 4/9 = 24/90$$

In either case, the sample contains a ball of each color. Hence, according to the addition law, the probability of the event of getting a ball of each color in the sample of two is the sum of the probabilities of the separate events, which is

$$24/90 + 24/90 = 48/90 = 0.533$$

Example 3-2. As an illustrative example of the above laws, let us use all three rules to find the probabilities associated with three coins.

$P(\text{getting 3 tails}) = 1/2 \times 1/2 \times 1/2 = 1/8$ by multiplication law.

$P(\text{getting 3 heads}) = 1/8$ by the same rule.

$P(\text{getting 3 like faces, either heads or tails}) = 1/8 + 1/8 = 2/8$ by the addition law.

$P(\text{getting at least 1 head and 1 tail}) = 1 - P(\text{getting 3 like faces either heads or tails}) = 1 - 2/8 = 3/4$ by complementary law.

3-8. Permutations. Let us suppose that you were asked to determine how many different arrangements could be made of the letters

A, B, and C so that each arrangement would be different from any other arrangement and contain all letters. After some thought, the arrangements would appear as in Table 3-1.

TABLE 3-1. PERMUTATIONS OF A, B, C

ABC	BAC
ACB	CAB
BCA	CBA

The six arrangements of Table 3-1 are called permutations of A, B, and C. Therefore, a permutation of a number of things involves the number of different orders of arrangement of these things which can be made.

It was relatively simple, with some thought, to calculate the permutation of A, B, and C. However, the permutation of A, B, C, and D would be more difficult since twenty-four different arrangements can be developed for four things. However, there is a simple method for determining the number of permutations. This method involves the progressive reduction by 1 of the figure representing the number of things. This procedure results in a number of factors arranged in a descending order such that the succeeding factor is always 1 less than its immediate predecessor. The result so obtained is the number of possible arrangements or permutations. Thus, the number of permutations for the three things A, B, and C shown in Table 3-1 is 6. This is calculated as $3(2)(1) = 6$.

This result checks with Table 3-1. Moreover, note the reduction of 3 by 1, giving 2, and the subsequent reduction of 2 by 1, giving the final number 1, which agrees with our prior statements. Similarly, the permutation for A, B, C, and D consists of twenty-four arrangements, and since four different things are involved, it is calculated as $4(3)(2)(1) = 24$. This principle of progressively reducing subsequent numbers by 1 and multiplying the results by each other is called the *factorial* of the number. Thus, the factorial of 5 is $5(4)(3)(2)(1) = 120$, and the factorial of 6 is $6(5)(4)(3)(2)(1) = 720$. It is seen, then, that the factorial of the number representing the number of different things which are involved will give the number of different arrangements or permutations. The recognized mathematical notation for a factorial is an exclamation point following the number. Thus, factorial four is shown as 4!, and, as was seen, $4! = 4(3)(2)(1)$. Moreover, in general algebraic notation, the factorial of any number N is denoted by $N!$ (N followed by an exclamation point).

To understand why the factorial of a number gives the number of different arrangements, i.e., a permutation, let us again consider the permutation for A, B, C, D. However, this time we will develop the logic as follows. In the first position we have a choice of 4 letters. Having filled the first position, we now have a choice of only 3 letters. Similarly, for the third position we now have a choice of only 2 letters left, and finally for the fourth position, only the single remaining choice is left. Thus, progressively we have 4(3)(2)(1), which is the same as saying that the permutation is equal to factorial four (4!).

The mathematician has a simple shorthand for indicating permutations. Thus, if he wanted to indicate the number of the permutation of 4 things taken 4 at a time, such as A, B, C, D, he would write it as ${}_4P_4$. The number 4 before the P tells us we have 4 different things to choose from, while the number 4 after the P tells us how many different things must be in each arrangement. Thus, ${}_4P_4 = 4(3)(2)(1)$. Suppose we wanted to determine how many different arrangements could be made out of 4 different things with only 2 things appearing in each arrangement; we would denote this as ${}_4P_2$. Moreover, ${}_4P_2 = 4(3) = 12$. Therefore, the permutation of 4 things taken 2 at a time is 12. A simple rule to follow is that there should be as many terms in the factorial as the number which appears after P. Thus, in our illustration, the number 2 appears after P in ${}_4P_2$, and therefore there were only two terms, namely 4 and 3, which resulted in a product of 12.

The general algebraic notation for denoting a permutation is[1]

$$ {}_nP_i \tag{3-1} $$

where n = number of things to choose from

i = number of different things appearing in each permutation

3-9. Combinations. In the previous discussion, it was emphasized that in permutations order is important. However, in combinations, we are not interested in order. In general, the concept of combinations is more important in statistical problems than are permutations. As long as each specific arrangement has a number of different things involved, it is called a combination, regardless of their order. A glance at Table 3-1 will show that although there are six permutations, each permutation has the letters ABC, and since order is of no consequence, there exists only one combination.

Since we can compute the number of combinations from the number of permutations and by the elimination of ordering, it becomes apparent that a mathematical relationship must exist between the two concepts. This relationship is expressed as follows:

$$ \text{Total number of combinations} = \frac{\text{total number of permutations}}{\text{number of permutations of the number of things appearing in each permutation}} $$

[1] The symbol P, when used in conjunction with n and i as shown in Eq. (3-1), indicates a permutation; when used alone, it symbolizes reliability or probability in accordance with the applicable text material.

This relationship can be expressed in the same algebraic notation as used previously:

$$_{n}C_i = \frac{_{n}P_i}{i!} \tag{3-2}$$

where $_{n}C_i$ is the number of combinations of n events taken i at a time.

To illustrate the relationship between permutations and combinations, let us examine a few cases.

Example 3-3. (*a*) How many combinations can be formed of 4 things taken 3 at a time?

Solution. Substituting in Eq. (3-2), we get

$$_{4}C_3 = \frac{_{4}P_3}{3!} = \frac{(4)(3)(2)}{(3)(2)(1)} = 4$$

It may be noted that twenty-four permutations exist, but because we have eliminated ordering, there are a total of four combinations.

(*b*) Evaluate the probabilities associated with obtaining heads and tails in the three-coin problem of Example 3-2.

Solution. There are four cases which we can examine. They are

(1) Probability of obtaining all heads
(2) Probability of obtaining all tails
(3) Probability of obtaining 1 head and 2 tails
(4) Probability of obtaining 2 heads and 1 tail

We can apply the laws of combination to each case separately. In the first case, we are taking 3 things 3 at a time, and so

$$_{n}C_i = \frac{_{3}P_3}{3!} = \frac{(3)(2)(1)}{(3)(2)(1)} = 1$$

This combination can occur in only one way. The same formulation holds for the second case.

For cases 3 and 4, we are considering 3 things taken 2 at a time. Therefore

$$_{n}C_i = \frac{_{3}P_2}{2!} = \frac{(3)(2)}{(2)(1)} = 3$$

For each of these cases, there are three combinations which can occur. It may be seen that there are a total of eight events which can occur. Let us now tabulate our results (Table 3-2).

TABLE 3-2. TABULATION OF PROBABILITY OF OCCURRENCE OF HEADS OR TAILS

Category	Type of arrangement which occurs	Probability of occurrence
3 heads	$H_AH_BH_C$	$(\frac{1}{2})(\frac{1}{2})(\frac{1}{2}) = \frac{1}{8}$
2 heads, 1 tail	$H_AH_BT_C$ $H_AH_CT_B$ $H_BH_CT_A$	$(\frac{1}{2})(\frac{1}{2})(\frac{1}{2}) = \frac{1}{8}$ $(\frac{1}{2})(\frac{1}{2})(\frac{1}{2}) = \frac{1}{8}$ $(\frac{1}{2})(\frac{1}{2})(\frac{1}{2}) = \frac{1}{8}$
2 tails, 1 head	$H_AT_BT_C$ $H_AT_CT_B$ $H_CT_AT_B$	$(\frac{1}{2})(\frac{1}{2})(\frac{1}{2}) = \frac{1}{8}$ $(\frac{1}{2})(\frac{1}{2})(\frac{1}{2}) = \frac{1}{8}$ $(\frac{1}{2})(\frac{1}{2})(\frac{1}{2}) = \frac{1}{8}$
3 tails	$T_AT_BT_C$	$(\frac{1}{2})(\frac{1}{2})(\frac{1}{2}) = \frac{1}{8}$

The subscripts A, B, C are used to differentiate between coins. It should be noted that for each combination, a number of permutations can be derived. As a check, if all values in the probability-of-occurrence column are summed, we arrive at the value of 1. This is in accordance with the addition law.

3-10. Definition of 100-hr Reliability. The definition for reliability which was given in Chap. 1 expressed reliability in terms of probability and as a function of time. In discussing probability we have been concerned primarily with the occurrence or nonoccurrence of events without reference to time. When discussing reliability we always refer to a period of time, and for the purposes of the ensuing discussion we have arbitrarily selected a time of 100 hr. All illustrative examples for series and parallel reliability will be based on this unit of measure.

In terms of our definition of reliability, we now are concerned with the probability that a device will operate satisfactorily under specified environmental conditions for a period of 100 hr. The measure of reliability is expressed in terms of a decimal fraction. Thus, if a series of tests were conducted on a number of parts of the same type for 100 hr and 93 per cent of them survived the test, the reliability would be indicated as 0.93. Likewise, if a number of equipments were tested for 100 hr and 90 per cent of them survived, the reliability for the equipment would be recorded as $P_e = 0.90$, where P_e is the symbol representing the reliability of equipment. In this case, the decimal fractions representing reliability refer to 100-hr reliability. This is an important consideration. For example, it is meaningless to talk about 99 per cent reliability unless it is defined. Does it mean initial reliability which requires that 99 per cent of the time an equipment taken out of a shipping container will function;

or does it mean that 99 per cent of the equipment will function after a stipulated number of hours? Thus, when 100 hr is specified, we talk about 100-hr reliability.

The method of calculating reliability for any given number of hours is accomplished by the exponential failure law, which is discussed in Chap. 6.

3-11. Series Reliability—The Product Rule. If a number of elements of a system are connected in such a way that the failure of any one element causes a failure of the system, then these elements are considered to be functionally in series. In terms of survival, this means that each element must survive if the system is to survive. The system can be no better than the element with the lowest probability of survival. This is merely another way of saying "a chain is no better than its weakest link."

In translating this concept to reliability, we are concerned with the fact that for a given equipment a number of elements are functionally connected in series. This does not mean merely physical connection, but implies that the failure of any part will cause an equipment failure. When these series elements are found to be independent, then our system satisfies the requirements of the multiplication law, and we can find the reliability of the equipment by multiplying the individual reliability of each element together. This is known as the *product rule.* The mathematical equation for product rule is

$$P_e = (P_1)(P_2)(P_3) \cdots (P_n) \qquad (3\text{-}3)$$

which states that the probability of satisfactory operation for some period of time of a system composed of n independent elements in series is equal to the product of the individual values of reliability for each of the n elements.

3-12. Variations of the Product Rule. When the relationships of values between elements in series are known, it is possible to restate the equation for the product rule in various forms which may be more convenient for the user.

When the reliabilities of the several series components are equal, Eq. (3-3) reduces to

$$P_e = P_c^n \qquad (3\text{-}4)$$

where P_c = reliability of the several components when their reliabilities are equal to each other
n = number of components involved

When the reliability of the several components are not equal to each other, then we may use the geometric mean value G for P_c, which is

$$G = P_c = \sqrt[n]{(P_1)(P_2)(P_3) \cdots (P_n)} \qquad (3\text{-}5)$$

and the equipment reliability can then be calculated as before by use of Eq. (3-4).

This can be done because, by using the geometric mean value, we created, in effect, an arbitrary mean reliability value for each component. This method essentially provides a solution equivalent to the prior one, where the reliabilities of the several components were equal to each other.

To illustrate the use of Eqs. (3-3) to (3-5), we shall do an example in which we shall calculate the reliability of an equipment consisting of only three components. We selected such a small number of components merely to reduce the labor of calculations, since we can demonstrate the method with a small number of components just as effectively as with many components.

Example 3-4. An equipment consists of three components D, E, and F, whose respective reliabilities are $P_D = 0.92$, $P_E = 0.95$, $P_F = 0.96$. Calculate the equipment reliability first by using Eq. (3-3) and then compare the result with that obtained by using Eqs. (3-4) and (3-5).

Solution. Substituting in Eq. (3-3),

$$P_e = (0.92)(0.95)(0.96) = 0.8390$$

Substituting in Eq. (3-5), the resulting geometric mean reliability of the three components P_c is

$$P_c = \sqrt[3]{(0.92)(0.95)(0.96)} = \sqrt[3]{0.8390} = 0.9440$$

This indicates that the geometric mean value which can be allocated to each of the three components is 0.916, but since there are three of these components in the equipment, by substituting in Eq. (3-4), we get

$$P_e = 0.9440^3 = (0.916)(0.916)(0.916) = 0.8390$$

Thus, the values of reliability using Eqs. (3-3) and (3-4) agree.

Discussion of Example 3-4. The calculation of the geometric mean reliability results in the establishment of an arbitrary value of reliability which can be assigned to each of n components comprising an equipment. This results in an equivalent component reliability P_c. Thus, Eq. (3-4) for calculating the reliability of an equipment consisting of n components of equal reliability is applicable. The same equations and relationships are also applicable to the parts which comprise a component.

Equation (3-4) is very useful because usually a component consists of several different parts which may be grouped into families of the same reliability. Thus, there may be n_1 parts of P_1 reliability and n_2 parts of P_2 reliability, etc. In this case, the over-all reliability may be expressed as

$$P_e = (P_1{}^{n_1})(P_2{}^{n_2})(P_3{}^{n_3}) \text{ etc.} \tag{3-6}$$

3-13. Effect of the Number of Parts on Series Reliability. The larger the number of series parts comprising a component, the poorer its reliability. This should be obvious, since the reliability of each of the parts is expressed as a decimal fraction, and the product of these decimal fractions gets smaller and smaller as the number of parts increases. Further, it can be seen that reliability must always be less than the value of the lowest element in the series. Therefore, in order to attain a high order of series reliability, either the number of parts must be kept to a minimum or else the respective reliabilities of each of the parts must be improved.

3-14. Series Reliability Theorems. These theorems can be stated as follows:

Theorem 3-1. *The series reliability P of a device comprised of various parts functionally in series and reliabilities expressed as decimal fractions is the product of the respective reliabilities of the parts.*

Theorem 3-2. *The series unreliability U of a device comprised of several parts functionally in series and of reliabilities expressed as decimal fractions is calculated by subtracting the reliability of the device from unity.*

3-15. Parallel Reliability. Parallel reliability is often referred to as redundancy. This is because elements are paralleled functionally so that if one fails, the redundant or parallel unit will continue to do the job. This type of arrangement is usually not necessary if one component is of sufficient reliability to meet specified requirements. In most instances, redundancy is justified when necessary, because the increase in cost due to additional parts or components is usually compensated for by the increased reliability obtained.

In the past discussion on reliability, we used the symbol P to denote reliability. We also indicated that basically reliability is a probability, so that sometimes the symbol P_s is used to indicate the probability of survival. We shall use P_s when we discuss the exponential failure law and shall then refer to it as the probability of survival. Thus, although the two terms might be considered synonymous, P is used to indicate a decimal measure of reliability with respect to a fixed time such as 100-hr reliability, whereas P_s is used to indicate a *probability* as a function of time. The distinction will become clearer as we proceed with the text. The symbol we shall use for probability of failure or unreliability will be the capital letter U. Thus, it is apparent that there must be a relationship between P and U for any component or part. By use of the complementary rule, we can obtain the relationship between P and U, which is expressed as

$$P + U = 1 \tag{3-7}$$

Equation (3-7) stems from the fact that since the sum of all probabilities must be equal to unity, then the sum of the reliability and unreliability for any part must be equal to 1 since these are the only two probable occurrences of the event. Thus, we see that $P = 1 - U$ and $U = 1 - P$. Now let us see whether we can establish the methods of calculating parallel reliability.

We can consider a system which has a set of parallel elements, each with a reliability P and an unreliability U. We know that each unit can separately and independently provide for proper operation. Therefore, the only way the system can fail is if all parallel elements fail. Since each element is independent, we can now apply the product rule to the unreliability of the system, or the probability of failure, which we have designated as U. If we had, for example, n units in parallel, all with an unreliability U_1,

$$U = (U_1)(U_1)(U_1) \cdots \text{etc.}$$

$$\text{Probability of failure } U = U_1^n \tag{3-8}$$

But we know from Eq. (3-7) that

$$P = 1 - U$$

Therefore

$$P = 1 - U_1^n \tag{3-9}$$

Conversely

$$U = 1 - P_1^n$$

We can now state the basic theorems of parallel reliability.

3-16. Parallel Reliability Theorems. From the foregoing discussion, we can state the following theorems:

Theorem 3-3. *The parallel unreliability of a device comprised of several parts functionally in parallel and of unreliabilities expressed as decimal fractions is the product of the respective unreliabilities of the parts.*

Theorem 3-4. *The parallel reliability of a device comprised of several parts functionally in parallel and of reliabilities expressed as decimal fractions is calculated by subtracting the unreliability of the device from unity.*

3-17. Application of Combinations and Permutations to Reliability Calculations. To gain further insight into the significance of these theorems, we can examine the various arrangements of success and failure in terms of permutations and combinations. This can be done best by the solution of a problem involving parallel elements, which will lead to the same theorems.

Example 3-5. An equipment consists of three components D, E, and F, which are connected functionally in parallel and whose respective reliabilities are $P_D = 0.92$, $P_E = 0.95$, $P_F = 0.96$. Calculate the equipment reliability P_e.

Solution. This problem is analogous to the three-coin example which has been previously discussed. In this instance, we have two probabilities for each component D, E, F: the probability of success, or the reliability of the components, and the probability of failure, or the unreliability of the components. Once again, there are four possible cases to be considered.

1. All components will operate simultaneously.
2. All components will fail simultaneously.
3. Two components will operate and one will fail.
4. One component will continue to operate when the other two fail.

We can apply the laws of combinations to each of these four cases, as we have previously done, and tabulate results. Once again, we determine that a total of eight combinations exists. Moreover, the sum of the probabilities of all occurrences must equal unity. This constitutes a check on our answer, since the sum of all probabilities of success and failure must total 1.

With a little thought we can tabulate the various arrangements. The same approach is applicable to any number of components in parallel, although the effort would be extremely tedious. The example shown is representative of the more complex models, and we can use this example to develop the parallel reliability theorems.

TABLE 3-3. TABULATION OF PROBABILITY OF OCCURRENCE OF FAILURES

Category	Type of combination or arrangements in which P and U can occur	Probability of occurrence
No failures	$P_D P_E P_F$	$(0.92)(0.95)(0.96) = 0.839040$
1 failure	$P_D P_E U_F$	$(0.92)(0.95)(0.04) = 0.034960$
	$P_E P_F U_D$	$(0.95)(0.96)(0.08) = 0.072960$
	$P_F P_D U_E$	$(0.96)(0.92)(0.05) = 0.044160$
2 failures	$P_F U_D U_E$	$(0.96)(0.08)(0.05) = 0.003840$
	$P_D U_E U_F$	$(0.92)(0.05)(0.04) = 0.001840$
	$P_E U_F U_D$	$(0.95)(0.04)(0.08) = 0.003040$
All failures	$U_D U_E U_F$	$(0.08)(0.05)(0.04) = 0.000160$
		Sum of probabilities $= 1.000000$

Table 3-3 demonstrates that the equipment will fail only when all three components have failed, since in all other instances there are one or more of the redundant elements still operative. Moreover, it demonstrates that there are a number of ways or arrangements in which P and U can occur. Thus, there is only 1 way in which no failures occur, 3 ways in which 1 failure can occur, 3 ways in which 2 failures can occur, and only 1 way in which all components fail. This last way then represents the situation which results in equipment failure. It is called the unreliability of the equipment and is calculated as the product of the unreliability of the three different components, which product is 0.000160 (as shown in Table 3-3).

The reliability is the sum of the products of all the other remaining arrangements. According to Eq. (3-7), the sum of the reliability and unreliability must equal unity, and we know that this is in accordance with our probability requirements that the sum of the probabilities must equal unity. To calculate the equipment reliability, we subtract the equipment unreliability from unity. This is in accordance with Eq. (3-7). Thus

$$P = 1 - U$$

Substituting,

$$P = 1 - 0.000160$$
$$= 0.999840$$

In other words, the 100-hr reliability of the equipment for the parallel arrangement is that it will function 99.98 per cent of the time. This is an appreciable improvement over the series result of 84 per cent of Example 3-4.

3-18. Series-Parallel Reliability. From the foregoing discussion, it should be obvious that problems of series-parallel reliability can now be simply solved by the use of our theorems. An example will be solved to illustrate the method.

Example 3-6. An equipment consists of 100 parts, of which 20 parts are tubes connected functionally in series (branch A). This branch is in turn connected in series to two parallel branches of 60 and 20 parts (branches B and C). The parts which comprise each of these branches are connected functionally in series. The reliability of each tube is $P_A = 0.95$, and the geometric mean reliability of branch B is $0.93 = P_B$ and of branch C is $0.96 = P_C$.

Solution. The equipment diagram is

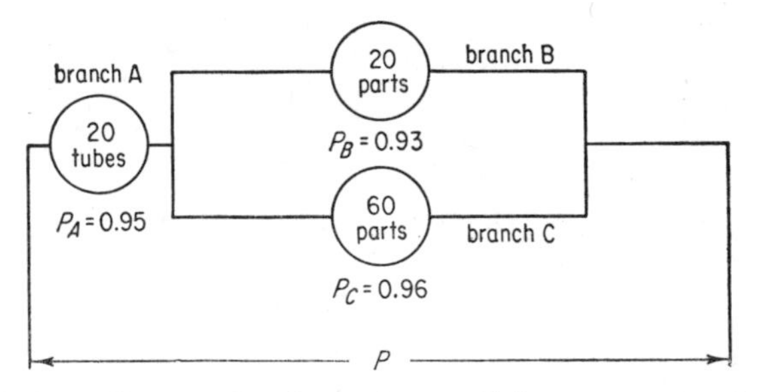

Equivalent diagram showing series-parallel arrangement of parts.

Using Eq. (3-4), for

$$\text{Branch } A \qquad P_A = 0.95^{20} = 0.358$$
$$\text{Branch } B \qquad P_B = 0.93^{20} = 0.240$$

and using Eq. (3-9),

$$U_B = 1 - 0.93^{20} = 1 - 0.240 = 0.760$$

Using the same equations, for

Branch C
$$P_C = 0.96^{60} = 0.0863$$
$$U_C = 1 - 0.96^{60} = 1 - 0.0863 = 0.914$$

Using Theorem 3-3, the unreliability of the two parallel branches U_{BC} is the product of the unreliabilities of branches B and C.

$$U_{BC} = U_B U_C = 0.760(0.914) = 0.695$$

and the reliability of the parallel branches P_{BC} by Theorem 3-4 is 1 minus the unreliability.

$$P_{BC} = 1 - U_{BC} = 1 - 0.695 = 0.305$$

The over-all reliability of the equipment P_e is then given by Eq. (3-3) or Theorem 3-1.

$$P_e = P_A P_{BC} = 0.358(0.305) = 0.109 = 10.9 \text{ per cent}$$

This implies that the probability of the equipment's functioning for 100 hr is very poor because it will only perform this long for approximately 11 per cent of the time.

3-19. Comparison of Series and Parallel Reliability. As was previously stated, parallel reliability always provides higher values than does series reliability. The reason is that parallel reliability provides more

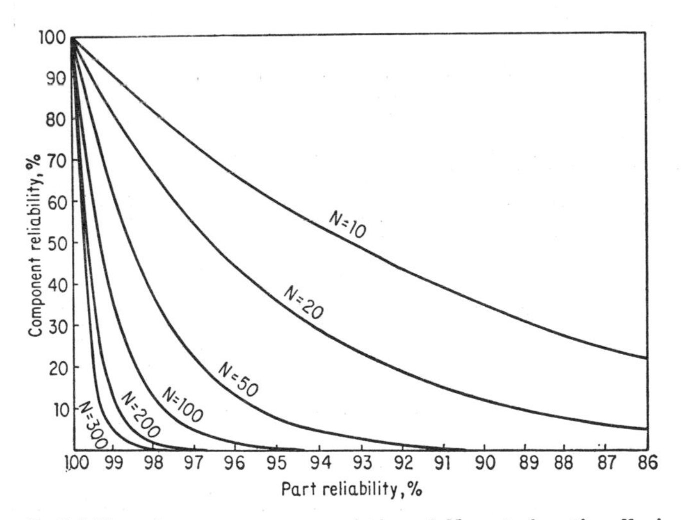

FIG. 3-1. Reliability of a component consisting of N parts functionally in series.

than one work horse to do the job. Thus, if one component fails, there is another to take its place. This is the technique of redundancy, which we mentioned previously and for which we will provide some general applicable theorems (see Sec. 3-20). Figures 3-1 and 3-2 show the relationship between reliability and the number of components involved for both the series and parallel cases, respectively. If the reader will examine both figures carefully, it should be obvious to him that for the same number of parts, a higher order of reliability is obtained when they are used functionally in parallel than is possible if they are in series. The parallel

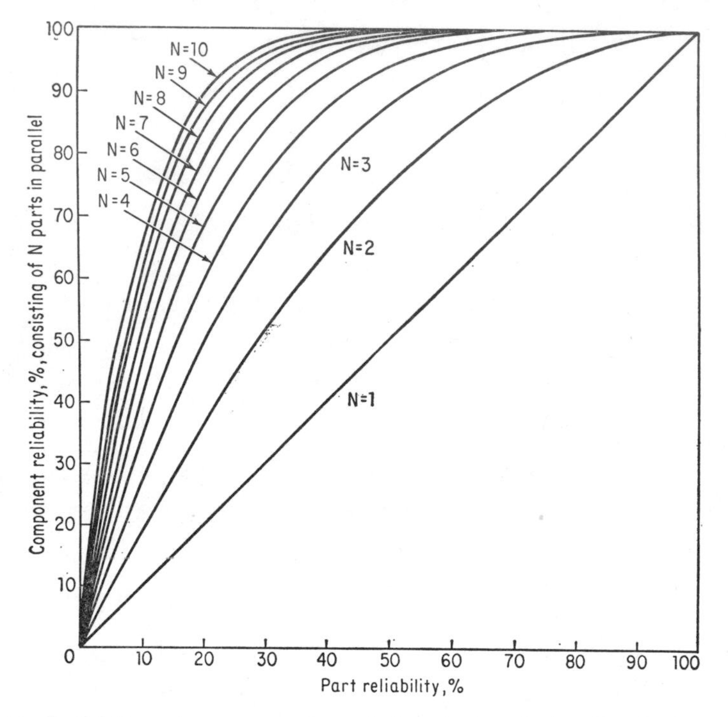

FIG. 3-2. Reliability of a component consisting of N parts functionally in parallel.

functional connection is known as the redundancy technique and is used when better reliability is required.

3-20. Redundancy Considerations. Although redundancy is a common expression among reliability engineers, the term cannot be used loosely. Redundancy is not a cure-all. Indeed, it should not be used at all if the cure is not necessary. To add a component is a relatively simple matter, but before this is done, all factors must be considered. Does the increase in reliability justify the addition of the component? How reliable is the component itself? Is the circuit self-sustaining, i.e., are provisions made to prevent the failure of one component from affecting the other? If one component fails, will the other be able to carry the load completely? How is this switching from one component to the other accomplished? Further, is it possible that redundancy will increase the probability of getting a spurious message through when it is not wanted? These and a list of other questions must be answered by the design engineer before he can employ redundancy successfully. However, as a general rule, the following theorems are submitted as guides.

Theorem 3-5. *When the expected part failures are of the shorting type, add the extra part or parts in series.*

Theorem 3-6. *When the expected part failures are of the open type, add the extra part or parts in parallel.*

PRACTICE PROBLEMS

3-1. What is the factorial of 4? How many combinations can be made of 4 things taken 2 at a time?

3-2. An equipment consists of 100 components connected functionally in series. Each component has a 100-hr reliability of 0.99. Calculate the reliability of the equipment.

3-3. An equipment consists of 3 components. Component 1 has 50 parts, each of reliability $P_1 = 0.98$. Component 2 has 75 parts of reliability $P_2 = 0.95$, and component 3 has 100 parts of reliability $P_3 = 0.999$. All the parts for each component are functionally connected in series. Calculate the equipment reliability if P_2 and P_3 are functionally in parallel and P_1 is in series with this combination.

3-4. By how many times would the reliability of an equipment consisting of 100 parts in series be increased if the parts were reconnected so that 50 parts were in a branch in series with each other and in parallel with another branch in which the other 50 were also in series with each other? The reliability of each part is 0.95.

3-5. An equipment consists of four components, the reliabilities of which are respectively 0.99, 0.93, 0.96, and 0.97. All these components are connected functionally in series. What is the geometric mean reliability? How would you calculate the reliability by means of the product rule? Compare the two results.

3-6. Write Theorems 3-1 through 3-6.

3-7. Calculate the reliability for each of the two connections and compare the results.

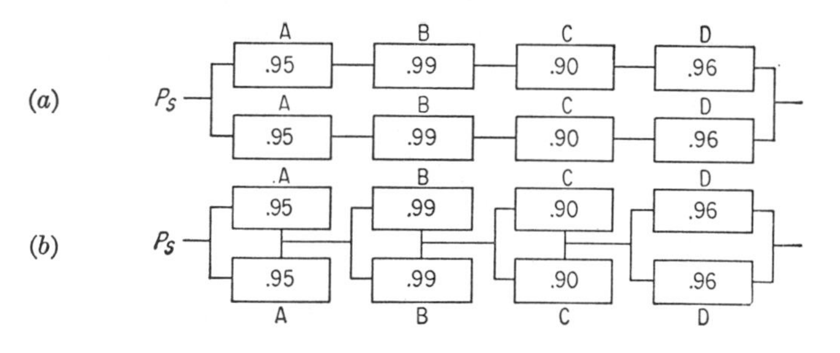

(c) Which arrangement is preferable? Why?

CHAPTER 4

THE NORMAL DISTRIBUTION

4-1. Introduction. The normal distribution is applicable to quantities which vary from some mean or central value in accordance with a particular frequency pattern. The curve which outlines this behavior has a bell shape and hence is often referred to as the bell-shaped curve or the normal curve. The variation which this curve describes is due to a constant-cause chance system. A constant-cause chance system implies that the variation involved is due entirely to the effects of chance and that the causes are not readily capable of being identified or isolated. In statistical language the term used is *unassignable causes of variation.* The extent or range of the variation is a function of the constant-cause chance system operating to produce it. It is also characterized by the fact that it remains within specific limits, and quantities within these limits occur in accordance with a specific relative frequency.

A properly adjusted lathe is a good example of a constant-cause chance system. Therefore, the quality of the item it produces can be controlled by statistical methods; because of this we can label this a controlled process, which is a manufacturing operation which produces an article whose dimensional variations are in accordance with the normal distribution. Thus, if this lathe were turning shafts to a specified nominal dimension, we would find that if we measured the diameters of several shafts, the measurements would distribute themselves in a pattern which approximates the normal curve; the characteristics of this pattern will be discussed in the ensuing paragraphs. In general, the mean diameter would occur most frequently, and dimensions differing from the mean would respectively have a lesser frequency of occurrence as the deviation from the mean increased. This will become clearer to us as we study the normal curve.

Another example of a constant-cause chance system is to be found in gunnery. Let us assume a rifle is trained on the bull's-eye of a particular target. If this rifle is fired a number of times, it will not always hit the bull. However, most of the times it will find its mark. The rest of the time it will deviate by some particular distance. Theoretically, it should strike the bull every time because the rifle is firmly placed in position and all firing conditions are apparently the same; but this does not occur. Instead the shots fall about the target within a specific range. A study of where the shots fall with relation to the bull will reveal that they follow the normal distribution. Therefore, we can conclude that the variation is due to unassignable causes and to a constant-cause chance system.

The characteristics of the normal curve provide a very valuable tool to quality control engineers. They are also of value to reliability engineers when studying wear-out characteristics, as we shall see in Chap. 6.

4-2. The Histogram and the Shape of the Normal Curve. The normal distribution is represented by a bell-shaped curve that describes the behavior of a variable Y with respect to another variable X. This curve

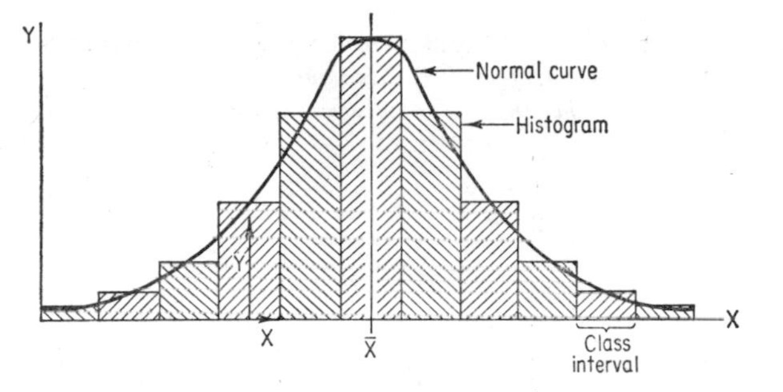

FIG. 4-1. Relationship of bell-shaped symmetrical histogram and the normal curve.

is known by several names. Among these are the *curve of error*, *probability curve*, *normal law*, *Gaussian curve*, and *Laplacian curve*. Although the curve as shown in Fig. 4-1 can be expressed mathematically, it is obvious that the outline of a symmetrical bell-shaped histogram approximates the form of a normal curve. A histogram is a bar chart which represents the relative frequency of occurrence of measurements, values, or classes of items. If the width of each bar is constant, then the height of the respective bars represents the relative frequency. Figure 4-1 shows an example of a symmetrical histogram comprised of several bars under the outline of the normal curve. The width of the bars equals the size of the *class interval.* If the number of class intervals is increased, their width will proportionately decrease until their outline approaches the shape of a normal curve, as shown in Fig. 4-1.

The mathematical equation that describes the normal curve is

$$Y = \frac{1}{\sigma \sqrt{2\pi}} e^{-(X_i - \bar{X})^2/2\sigma^2} \tag{4-1}$$

where σ = standard deviation
X_i = individual values
$\bar{X}$ = average of all values
Y = height of the curve at any point along the scale of X. It is also known as the probability density or frequency of a particular value of X
e = base of Napierian logarithms which equals 2.718
π = 3.1416

The derivation of this equation can be found in any standard statistical text and, therefore, will not be discussed here. However, we shall describe a practical experiment which should prove of interest to the reader. The procedure involves selecting a group of 18 pennies, tossing the entire group 360 times, and counting and recording the number of heads after each toss. Table 4-1 summarizes the data for 360 tosses.

TABLE 4-1. FREQUENCY OF OCCURRENCE OF HEADS IN COIN-TOSSING EXPERIMENT

No. of heads	*Frequency of occurrence*
0	0
1	0
2	1
3	1
4	5
5	11
6	22
7	47
8	58
9	70
10	62
11	40
12	27
13	12
14	3
15	0
16	1
17	0
18	0
Total frequencies =	360 tosses

The same data, plotted in the form of a histogram, are shown in Fig. 4-2. It should be noted that the outline of this histogram, shown as a dashed-line curve, is the bell-shaped curve or normal curve under discussion. This is the type of curve which describes the frequency of occurrence of events which are due purely to a constant-cause chance system. The coin-tossing experiment fits the curve very well, and, therefore, we can conclude that the frequency of occurrence of heads is due to a constant-cause chance system. If we had influenced the results by using weighted or biased coins, we would not have obtained a normal curve. In other words, the normal curve is a probability distribution which describes the probability of occurrence of an event if all factors involved are due to a constant-cause system. As a point of interest, at this time, it might be in order to indicate that the binomial distribution, which we shall study in the next chapter, will also give us essentially the same shape of curve as Fig. 4-2 under specific circumstances. This holds true only for a particular application such as this, since under certain conditions the binomial and normal distribution are practically equivalent. The explanation for

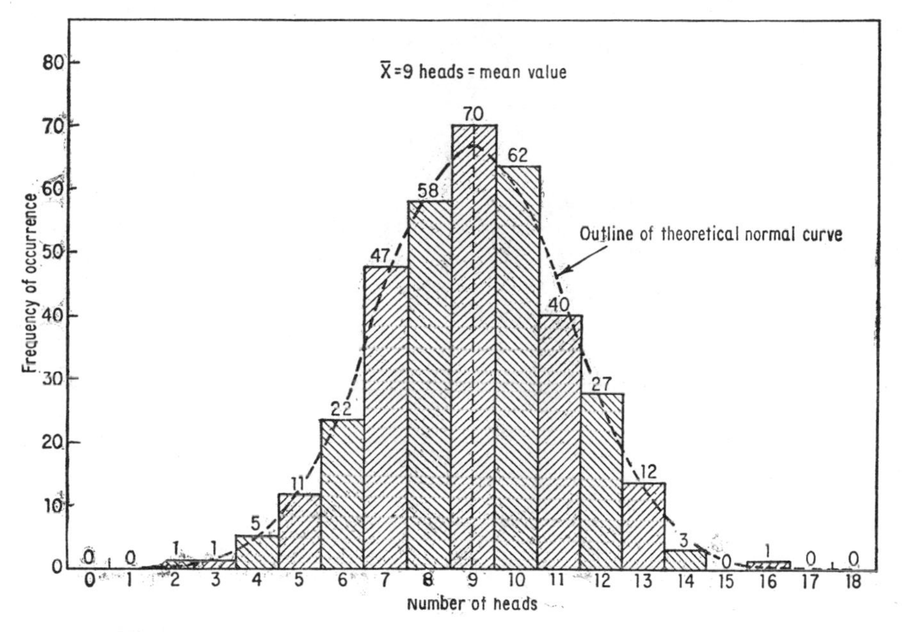

FIG. 4-2. Histogram showing frequency of occurrence of heads in 360 random tosses of 18 pennies.

their behavior is outside the present scope of this book and therefore will not be considered at this time.

Because of its characteristics the normal curve is useful in controlling industrial processes and forms the basis for what are known as *variables control charts.* These charts use upper and lower limits which are actually the extreme or 3-sigma limits of the normal curve. If a measurement exceeds these limits, it may be an indication that the process is out of control and that unassignable causes of variation might be present. A thorough explanation of variables control charts can be found in any good text on quality control.

The histogram of our coin experiment shown in Fig. 4-2 indicates that on the average 9 heads will appear more often than any other number.

This appears to be logical, since there are 18 heads in total and the probability that no heads will show up is very remote. Also, the probability that all 18 heads will appear after a toss is equally improbable. Between these two extremes there must be some value which is most likely to occur, and this is the mean value 9. Values greater and smaller than 9 will occur at a frequency which is less than that of the mean. The theoretical frequencies of occurrence are shown by the dashed outline, which represents the normal curve for the coin-tossing experiment. Thus, from

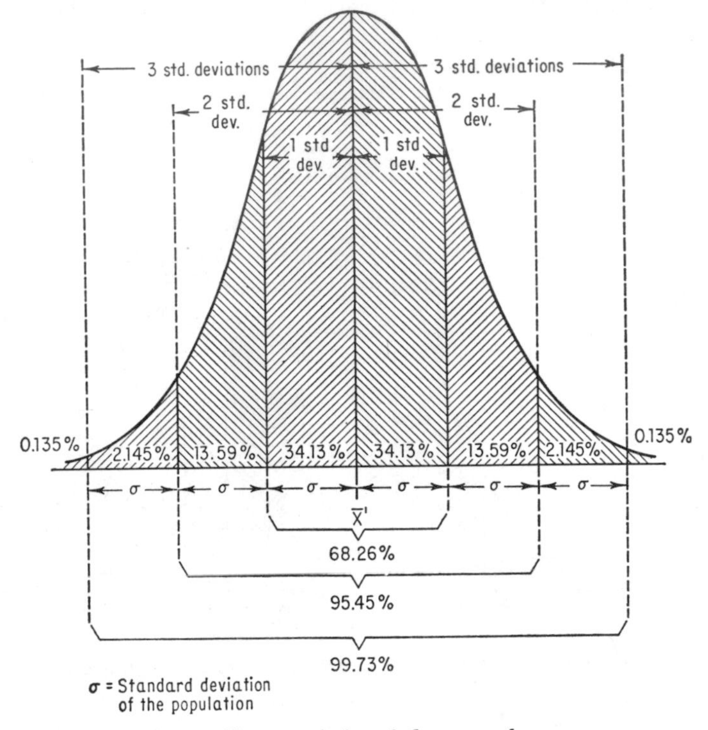

FIG. 4-3. Characteristics of the normal curve.

the curve it is seen that theoretically for 360 tosses 9 heads should occur 70 times, and 8 heads will probably occur 58 times, etc. In other words, the probability curve clearly predicts the number of times an occurrence can be expected to happen. In practice, however, it is very rare for an experiment such as the one we have conducted with the 18 coins to result in the actual theoretical frequencies. However, the agreement is usually close, as can be seen from Fig. 4-2.

4-3. Characteristics of the Normal Curve. A study of the normal curve indicates that it possesses certain basic inherent characteristics, which are summarized in Fig. 4-3. From this figure it is seen that 99.73

per cent of the area of the curve lies within 6 standard deviations, i.e., 3 standard deviations on either side of the average. Also, 95.45 per cent of the area lies between 4 standard deviations, i.e., 2 standard deviations on either side of the average; and 68.26 per cent of the area lies between 2 standard deviations, i.e., 1 standard deviation on either side of the average. It is also apparent that for the *normal curve only*, the standard deviation is approximately equal to $\frac{1}{6}$ of the range. This is a very good estimate since, as can be seen from the curve, only 0.27 per cent of the area is outside of plus or minus 3 standard deviations.

The characteristics of the normal curve are also important because they provide a means for making rapid estimates of probabilities of occurrence. The method of making these estimates will be illustrated by Example 4-1.

Example 4-1. The specification for a certain type of resistor requires that its resistance value be 100 ohms plus or minus 5 per cent. A sample of 50 resistors was selected at random from a batch which had just been received. The average resistance of these 50 resistors was found to be 100 ohms, and the standard deviation was 5.0 ohms. What per cent of the resistors in the batch would probably be rejected as a result of 100 per cent inspection?

Solution. If a normal distribution is assumed, the curve representing the sample is shown in Fig. 4-4. It was drawn by measuring 3 standard deviations on either side of the mean of 100 ohms. The lower and upper specification limits are also shown. Coincidentally, these limits happen to be exactly at plus or minus 1 standard deviation.

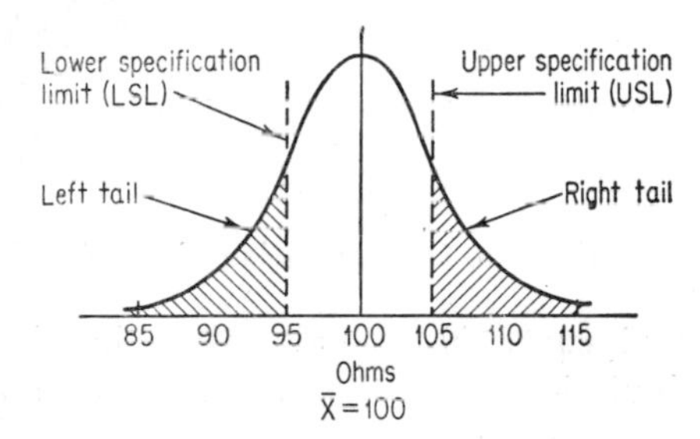

Fig. 4-4. Normal curve of sample of 50 resistors of Example 4-1.

The normal curve representing the sample of 50 resistors indicates that some of the resistors in the batch from which the sample was taken will be beyond the specification limits and therefore rejectable. The specification limits are shown as dashed lines on the normal curve. The upper specification limit (USL) is 105 ohms, and the lower specification limit (LSL) is 95 ohms. Any resistors between these two limits are acceptable. The area of the tails which is crosshatched is a measure of the percentage of the sample beyond the specification limits. This is indicative of the percentage which would be rejected if the entire batch were inspected 100 per cent. Thus, the per cent defective indicated by the sample is also a measure of the per cent defective that can be expected in the batch. This characteristic of the normal curve is a handy device which makes possible predictions (on the basis of representative samples) of the per cent defective that can be expected in the lot itself.

To determine the percentage of resistors which we can expect to reject, we observe that the good resistors lie within 1 standard deviation on either side of the average. Therefore, 68.26 per cent of the resistors meet the specification requirements, and the quantity of 31.74 per cent reposing in the tails of the distribution represents defectives.

Discussion of Example 4-1. This solution was very simple because the specification limits were exactly on the 1 standard deviation point. But suppose they were not so located. Suppose the USL was 106 ohms and the LSL was 94 ohms. Then what percentage would be rejected?

We can determine this percentage of rejections by the use of Table 3 of App. 7. This table shows the proportion of the total area under the nor-

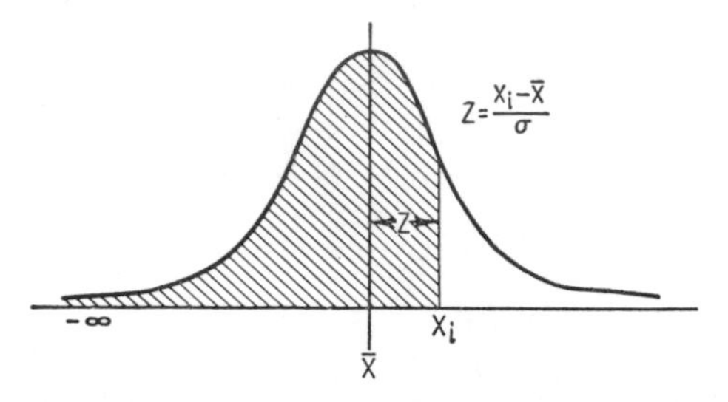

FIG. 4-5. Demonstration of areas under the normal curve from $-\infty$ to Z.

mal curve from $-\infty$ (minus infinity) to a point which is Z standard deviations from the mean. The value of Z is calculated by subtracting the mean $\bar{X}$ from the individual value X_i and dividing the result by the standard deviation σ. This relationship is shown as

$$Z = \frac{X_i - \bar{X}}{\sigma} \tag{4-2}$$

In theory $-\infty$ represents a point at an infinite distance to the left (see Fig. 4-5) at which the tail of the normal curve converges or closes on the abscissa. Actually, for all practical purposes, this occurs at the -3σ point. Moreover, Eq. (4-2) indicates that when X_i is plus, Z has a positive value, and when it is minus, Z has a negative value. The value of Z so determined with reference to Table 3 will give the area under the curve from the left tail extreme, i.e., minus infinity, to X_i.

Now that we have learned about Z, we are in a good position to discuss the effect when USL $=$ 106 ohms and LSL $=$ 94 ohms.

When USL is 106 ohms, $X_i = 106$, and $\bar{X}$ is still 100.

$$Z = \frac{106 - 100}{5} = \frac{6}{5} = 1.2$$

Looking up $Z = 1.2$ in Table 3, we get an area from the left tail to $X_i = 106$ of 0.8849. Therefore, the percentage of resistors expected to have a resistance value of greater than 106 ohms is $1 - 0.8849 = 0.1151$, which is 11.51 per cent.

Similarly, the area from the left tail to $X_i = 94$ is

$$Z = \frac{94 - 100}{5} = \frac{-6}{5} = -1.2$$

In this case the area is directly read as $0.1151 = 11.51$ per cent. This represents the percentage of resistors which are below the lower limit of 94 ohms. Therefore the total percentage of resistors outside the limits is

$$11.51 + 11.51 = 23.02 \text{ per cent}$$

Thus, by widening the specification limits, the number of rejects was reduced from 31.74 to 23.02 per cent.

As we have seen from Example 4-1, the normal curve has some useful properties, but we have not yet learned them all. All normal curves have the same basic standard form. They differ only with respect to two parameters, their average $\bar{X}$ and their standard deviation σ. Figure 4-6 represents a situation in which the two curves have the same mean but different standard deviations. Figure 4-7, on the other hand, shows the two curves with the same standard deviation but different means. Note that the curve with the smaller standard deviation in Fig. 4-6 is taller than the other curve. The reason is that the area of each of the curves must be equal because they are similar probability distributions. Thus, since the two areas must be the same, as the base of one shrinks, its height must increase to assure that its area remains the same.

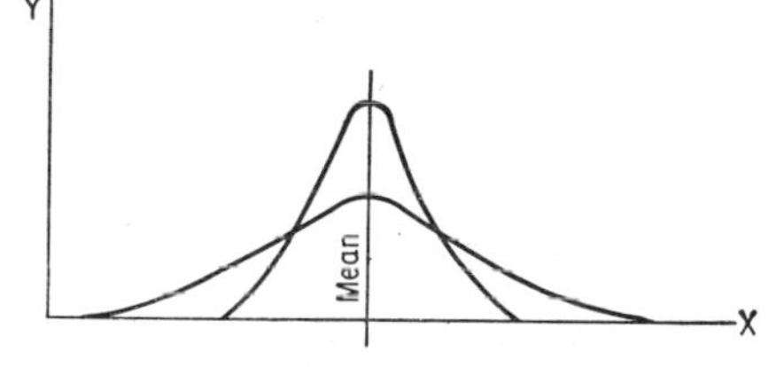

FIG. 4-6. Comparison of two normal distributions having the same mean but different standard deviations.

An examination of Figs. 4-6 and 4-7 indicates that it would be certainly advantageous if there were some method for comparing one distribution with another. Statisticians have accommodated us by developing the desired technique. This is a method of making all distributions comparable to each other by reducing each of them to its basic nature. This is done by considering the mean of each distribution as zero and measuring all deviations from this mean, not in terms of the original units, but in terms of the standard deviation of the distribution. To accomplish this conversion, all that is necessary is to subtract the mean of the distribution from each unit of measure and divide the result by the standard deviation. When this is done, it is equivalent to converting the mean to zero and measuring all deviations from it in terms of Z. [See Eq. (4-2).] The value of Z is very important, since it represents a new code value which can be used in various applications. It is important to remember that it always represents the number of standard deviations from the mean, and hence the area under the curve may be determined from Table 3.

In some textbooks this code value is called t. However, there is another value, also used in tests for significance, which is called "students' t." Moreover, in this text the symbol t is used to represent time, and therefore to avoid all these conflicts, Z is used.

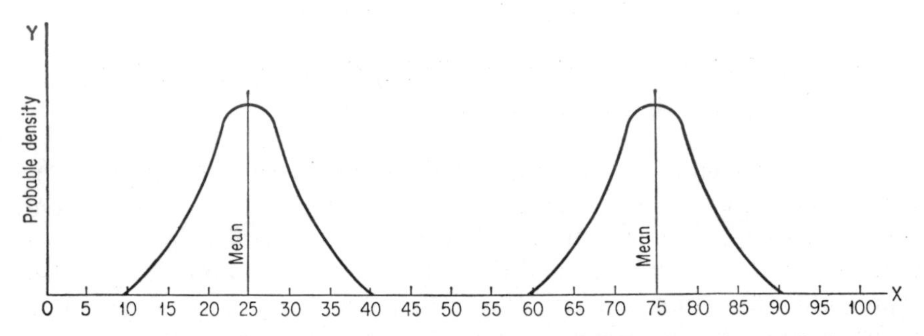

FIG. 4-7. Comparison of two normal distributions of different means and the same standard deviations.

To illustrate the use of the Z table, we shall consider the data of Example 4-1. What is the value of Z when $X_i = 110$ ohms?

$$Z = \frac{110 - 100}{5} = \frac{10}{5} = 2$$

This means that the value of 110 ohms is 2 standard deviations away from the mean value of 100 ohms.

Example 4-2. Using the data of Example 4-1, find the expected percentage of resistors which will exceed 107.5 ohms and the percentage less than 93.75 ohms.

Solution

$$Z = \frac{107.5 - 100}{5} = \frac{7.5}{5} = 1.5$$

Referring to Table 3, we calculate that for $Z = 1.5$ the area of the right-hand tail is 0.067. Therefore, 6.7 per cent of the resistors would be expected to have values greater than 107.5 ohms.

Similarly, to find the percentage of resistors which are less than 93.75 ohms,

$$Z = \frac{93.75 - 100}{5} = \frac{-6.25}{5} = -1.25$$

The minus sign indicates that the 93.75 value is less than the mean by -1.25σ. Therefore when we refer to Table 3, we must be certain to refer to -1.25. This gives us the area in the left tail of the distribution, which is what is required. The value of this left-tail area is found to be 0.106, or 10.6 per cent. Therefore 10.6 per cent of the resistors would be expected to have values of less than 93.75 ohms.

4-4. Construction of the Theoretical Normal Curve. To fit a theoretical normal curve over a histogram which has a known mean and standard deviation, we must first become acquainted with certain basic facts. The general equation for the normal curve is expressed by Eq. (4-1). If we make $\sigma = 1$ and substitute the value of Z as expressed in Eq. (4-2), we can rewrite the equation which expresses the probability density as

$$Y = \frac{1}{\sqrt{2\pi}} e^{-Z^2/2} \tag{4-3}$$

In Sec. 4-3 we mentioned that we could refer all normal curves to each other by considering their means equal to zero, in which case each of the curves would have a standard deviation of unity. Moreover when this is true, it can be shown that the area under the curve expressed by Eq. (4-3) is also unity. Therefore this is the basic probability curve, since, as we have shown in the past, a characteristic of a probability distribution is that the sum of all the probabilities equals unity.

Table 4-2 shows a tabulation of the probability density for various values of Z. These Z values are for the right tail only; however, since the normal curve is symmetrical, the minus values of Z will have the same probability density as their plus counterparts. Actually these probability densities are proportional to Y, which represents the ordinate of the normal curve for any value of Z.

The value of Eq. (4-3) lies in the fact that since it gives relative frequency, i.e., the probability density, it can be used to establish the theoretical curve corresponding to a histogram of n values and of σ standard deviation. Let us use Example 4-1 again to determine how we would construct the normal curve. In this case the total frequency equals $n = 50$ and $\sigma = 5$. Therefore if we multiply Eq. (4-3) by n/σ, we should get the absolute frequency. Thus, substituting $n = 50$ and $\sigma = 5$,

$$Y = \frac{50}{5}\left(\frac{1}{\sqrt{2\pi}} e^{-Z^2/2}\right)$$

but the value in the parentheses can be determined for various values of Z by referring to Table 4-2. When this is done, we get Table 4-3, which gives the absolute values of Y for various values of Z. The normal curve can then be plotted.

TABLE 4-2. TABULATION OF PROBABILITY DENSITY FOR ONE TAIL OF NORMAL CURVE

Value of Z	Y = *ordinates of normal curve (probability density), relative frequency*
0	0.399
0.25	0.387
0.50	0.352
0.75	0.301
1.00	0.242
1.25	0.187
1.50	0.130
1.75	0.086
2.00	0.054
2.25	0.032
2.50	0.018
2.75	0.009
3.00	0.004

It should be realized that we cannot have fractional values for the number of resistors, as is indicated by Table 4-3. Therefore a histogram from actual data would approximate the shape of the normal curve indicated by this table but obviously could never follow its exact outline. It

TABLE 4-3. ABSOLUTE VALUES OF ORDINATES OF NORMAL CURVE FOR EXAMPLE 4-1

Z values	Y ordinates of normal curve	$-Z$ values	Y ordinates of normal curve
0	3.99	0	3.99
0.50	3.52	0.50	3.52
1.00	2.42	1.00	2.42
1.50	1.30	1.50	1.30
2.00	0.54	2.00	0.54
2.50	0.18	2.50	0.18
3.00	0.04	3.00	0.04

should also be recognized that Table 4-3 shows only ordinates for specific values of Z. This is likely to be confusing to the reader because there are other intermediate points which would appear in a histogram to account for the total population of 50 resistors.

For example, if we wanted to find the theoretical or estimated number of resistors which have a value of 101 ohms, we would substitute in Eq. (4-2) and calculate the value of Z corresponding to 101 ohms, i.e.,

$(101 - 100)/5 = 0.2$. Then we would find the corresponding ordinate equal to 0.3910 from Table 4, App. 7, since this is more complete than Table 4-2. Multiplying the value found by 50/5, we would get a value of 3.9. Similarly, we could find the expected frequency for resistors of 102, 103, etc., ohms. If we added all our results, we would find that they would total 50, the number of resistors in the sample. It is left as an exercise for the reader to calculate the remaining values and verify the above statement.

PRACTICE PROBLEMS

4-1. Select 12 pennies and toss them randomly for 40 times. After each toss record the number of tails. Plot the resulting histogram. Estimate the mean from the histogram and make a rough visual estimate of σ.

4-2. From the data of Prob. 4-1 calculate the mean and standard deviation. How do you account for the fact that the estimate of the mean from Prob. 4-1 is very close to the calculation of the mean of Prob. 4-2 but that the two standard deviations do not agree as closely?

4-3. Measurements on a particular vacuum tube resulted in data which gave an estimated mean life of 2,000 hr and a σ of 100 hr. On the assumption that the mean life varies in accordance with the normal distribution in this case, what percentage of tubes do we expect to exceed 2,100 hr of life? What percentage do we predict will last for 1,900 hr or less? What percentage will operate between the limits of 1,900 and 2,100 hr?

4-4. Construct the theoretical normal curve for Prob. 4-2 by the use of Eq. (4-3) and Table 4-2. Fit it over the histogram of Prob. 4-1. Is it a good fit? Explain.

CHAPTER 5

BINOMIAL DISTRIBUTION

5-1. The Nature of the Binomial Distribution. The binomial distribution is a probability distribution and as such conforms to the criterion that the sum of its probability components must equal unity. In its basic form the binomial consists of two parts, as its name implies, since the prefix *bi* indicates two. These two parts may be any two attributes, such as good or bad, or black or white, expressed in decimal fractions. Thus a box containing 100 parts, 90 of which are good and 10 of which are bad, would, in terms of decimal fractions, contain 0.90 good parts and 0.10 bad ones. Moreover the sum of the good and bad equals unity. Suppose we decide to take random samples of 10 from the box (making sure we replace the prior sample before each drawing so that the population remains intact); what proportion of good and bad parts would each sample contain? Obviously we could deduce that there would be some variation between each sample due to the laws of chance. However, whenever we talk about chance, we are dealing with probability. This means that it is probable that in some samples of 10 we would find no bad parts, in some 1 bad, in others 2 bad, etc. In order to determine the probability of getting either 0, 1, 2, 3 or more bad parts in a sample, we have to break the binomial down into its component probability parts. This can be done by means of a mathematical expression which is called the binomial expansion.

5-2. The Binomial Expansion. In order to understand the binomial expansion, we shall discuss it in general terms. Thus, suppose a population is q effective and p defective in decimal fractions. Then the first equation which we can write is that the fraction effective plus the fraction defective equals unity. In algebraic terms the equation is

$$q + p = 1 \tag{5-1}$$

Moreover, if we were to take samples of 2 from a population consisting of q per cent effectives and p per cent defectives, this would result in a number of combinations for samples of 2 which are as shown in Table 5-1.

From an inspection of columns 1 and 3 of Table 5-1, it should be obvious that in samples of 2 the probability of obtaining 2 effectives is qq; of 1 effective and 1 defective, qp; of 1 defective and 1 effective, pq; and of 2 defectives, pp. Column 4 shows the product of all the probabilities for

TABLE 5-1. TYPES OF COMBINATION AND WAYS OF OCCURRING FOR SAMPLES OF 2

(1) Type of combination	(2) Ways of occurring	(3) Probability of each way	(4) Probability of type of arrangement
2 effectives	good, good	qq	q^2
1 effective 1 defective	good, bad bad, good	qp pq	$2qp$
2 defectives	bad, bad	pp	p^2

the respective rows. If we add all the entries of column 4, we get the total probability, which is

$$q^2 + 2qp + p^2 \tag{5-2}$$

If we refer to Table 5-1, we see that the first term of Eq. (5-2) indicates the probability of no defectives; the second term, the probability of 1 defective; and the last term, the probability of 2 defectives in samples of 2. Moreover from elementary algebra we know that $(q + p)^2 = q^2 + 2qp + p^2$. But from Eq. (5-1) we see that $q + p = 1$, and therefore $(q + p)^2$ and $q^2 + 2qp + p^2$ must also equal unity, since these expressions are all equal to each other. If we examine the binomial $(q + p)^2$, we note that its exponent 2 is equal to our original sample size $n = 2$. Further study reveals that if we desire to find the probability of occurrence of defects for any sample size n, we may expand the binomial in a manner similar to that used for $n = 2$. However as n gets larger and larger, it is not practical to expand the binomial by making a table such as Table 5-1 because this method is too unwieldy for large sample sizes. It is also impractical to raise the binomial to high powers of n by multiplying the binomial by itself n times because the computations involved are laborious. Fortunately there is a simple method for expanding the binomial which is based on certain rules of thumb. We shall illustrate the steps involved by using $(q + p)^2$ as a typical example.

Terms of the Expansion. Step 1. To Get First Term. We must raise the term q to the value of the exponent of the binomial. In this case it equals 2. This gives the first term of the expansion, which equals the probability of zero defectives.

First term $= q^2$ (probability of no defectives)

Step 2. To Get Second Term. Use the exponent of the first term as the coefficient of the second term; decrease the value of the exponent of q of the first term by 1 to get the exponent of q for the second term and introduce p in the second term.

$$\text{Second term} = 2qp \text{ (probability of 1 defective)}$$

Step 3. To Get Third Term. Multiply the coefficient of the second term, namely, 2, by the exponent of q in the second term, which is 1; this results in (2)(1). Divide this result by a number which is 1 greater than the exponent of p of the second term. This results in dividing 2 by 2, the quotient of which will be the coefficient for the last term. Then to get the exponent of q for the third term, reduce the exponent of q of the second term by 1, which gives q^0, and increase the exponent of p of the second term by 1; this gives p^2. Thus the third term will be $1q^0p^2$; but $q^0 = 1$, and therefore

$$\text{Third term} = p^2 \text{ (probability of 2 defectives)}$$

Summing all three terms, we get

$$(q + p)^2 = q^2 + 2qp + p^2$$

The rules stated above can be used to expand a binomial for any sample size. A fast check for accuracy of the expansion is that the sum of the exponents for each term of the expansion should equal the sample size, which is the same as the exponent of the binomial. Thus the exponent of the first term q^2 is 2; for the second term $2pq$ it is 1 for q plus 1 for p, which sum is also 2; for the third term p^2 the exponent is also 2; therefore the condition is satisfied. To illustrate the method more thoroughly, we shall do an example.

Example 5-1. Expand the binomial $(q + p)^4$.

Step 1. Raising the first term q to the fourth power, we get q^4 for the first term of the expansion.

Step 2. The exponent of q^4 is 4, and therefore it will be the coefficient of the second term. Decrease the exponent of q^4 by 1 to obtain q^3, which becomes one of the factors of the second term. Introduce p in the second term. Summing all these steps, we find the second term is $4q^3p$.

Step 3. Multiply the coefficient 4 of the second term by the exponent of q of the second term, which is 3. This product of 4 and 3 is 12. Divide this product by 2, which is a number that is 1 greater than the exponent of p of the second term. The quotient of 12 and 2 is 6, which number becomes the coefficient of the third term. Next decrease the exponent of q^3 by 1, which gives q^2; increase the exponent of p by 1, which gives p^2. Therefore the third term is $6q^2p^2$.

The fourth and fifth terms are obtained in a similar manner and the final expansion is written as

$$(q + p)^4 = q^4 + 4q^3p + 6q^2p^2 + 4qp^3 + p^4$$

As a check worthy of note, it should be observed that for each term of the expansion the sum of the exponents of q and p equals 4. A further simplification is to observe that the exponent of p indicates the probability of that many defects for the expression of which p is a factor. For example the expression for the probability of 3 defects is $4qp^3$, since the exponent of p in this expression is 3. In summary the following terms represent the probability of defects as shown.

q^4 = probability of 0 defects
$4q^3p$ = " " 1 defect
$6q^2p^2$ = " " 2 defects
$4qp^3$ = " " 3 defects
p^4 = " " 4 defects

To demonstrate the actual method of making calculations we shall do another example.

Example 5-2. Suppose that samples of 4 were selected from a large lot which is 10 per cent defective. What is the probability of 0, 1, 2, 3, or 4 defects in samples?

Solution. Using the expansion of Example 5-1,

Probability of 0 defectives $= q^4 = (0.9)^4 = 0.6561$
Probability of 1 defective $= 4q^3p = 4(0.9)^3(0.1) = 0.2916$
Probability of 2 defectives $= 6q^2p^2 = 6(0.9)^2(0.1)^2 = 0.0486$
Probability of 3 defectives $= 4qp^3 = 4(0.9)(0.1)^3 = 0.0036$
Probability of 4 defectives $= p^4 = (0.1)^4 = 0.0001$
Total probability $= 1.0000$

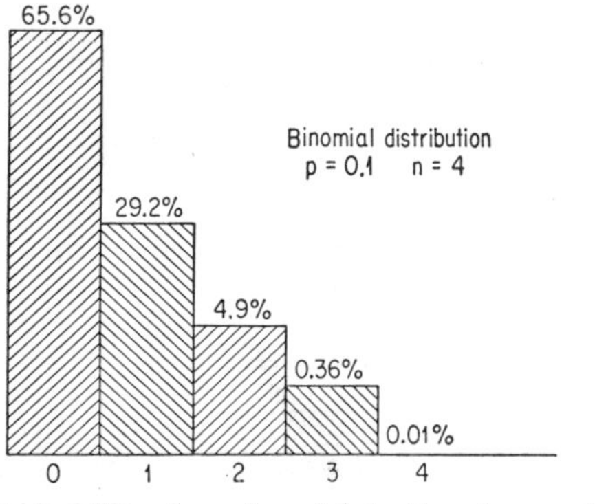

FIG. 5-1. Probability of number of defectives in samples of 4.

5-3. Other Methods of Determining the Coefficients of the Terms of the Binomial Expansion. The solution of Example 5-2 indicated that the terms for the probability of 0, 1, 2, 3, and 4 defectives were respectively q^4, $4q^3p$, $6q^2p^2$, $4qp^3$, and p^4. This suggests that the coefficient of each term could also be found by finding the number of possible combinations for each term by the use of Eq. (3-2), namely, ${}_nC_i = \frac{{}_nP_i}{i!}$. The calculations and tabulations are shown in Table 5-2.

TABLE 5-2. CALCULATIONS OF COEFFICIENTS OF THE TERMS OF THE BINOMIAL BY COMBINATIONS

(1) Term without coefficient	(2) Condition	(3) Formula	(4) Calculation	(5) Coefficient
(1) q^4	Number of combinations of 4 things taken 0 at a time	${}_4C_0 = {}_4C_{(4-0)}$	$\frac{4 \times 3 \times 2 \times 1}{4 \times 3 \times 2 \times 1} = 1$	1
(2) q^3p	Number of combinations of 4 things taken 1 at a time	${}_4C_1$	$\frac{4}{1} = 4$	4
(3) q^2p^2	Number of combinations of 4 things taken 2 at a time	${}_4C_2$	$\frac{4 \times 3}{2 \times 1} = 6$	6
(4) qp^3	Number of combinations of 4 things taken 3 at a time	${}_4C_3$	$\frac{4 \times 3 \times 2}{3 \times 2 \times 1} = 4$	4
(5) p^4	Number of combinations of 4 things taken 4 at a time	${}_4C_4$	$\frac{4 \times 3 \times 2 \times 1}{4 \times 3 \times 2 \times 1} = 1$	1

We can see from the tabulation that the numbers in column 5 represent the coefficients of the terms in column 1. Also the calculation in row 1, namely ${}_4C_0$, is shown as ${}_4C_{(4-0)}$ or ${}_4C_4$. This comes from the fact that whenever we make a choice of the items to include in a combination, we automatically exclude the remaining items. Thus for row 1, q^4 represents 4 effectives, and this is equivalent to excluding 4 defectives; but the combination of 4 things taken 4 at a time is equal to the combination of 4 things taken 4 − 4 or 0 at a time. Thus the combinations of what was included should be the same as what was excluded. Therefore Eq. (3-2) could also be written

$$ {}_nC_i = \frac{{}_nP_i}{i!} = {}_nC_{(n-1)} \qquad (5\text{-}3) $$

From Table 5-2 it appears that a certain symmetry exists in the number of combinations that are possible for various sample sizes. Pascal found this to be true and developed what has since been known as Pascal's triangle. This triangle, shown in Table 5-3, is a simple means of determining the coefficients of the various terms of the binomial expansion for different values of n.

TABLE 5-3. PASCAL'S TRIANGLE

Sample size, n	*Coefficients*
1	1
2	1 2 1
3	1 3 3 1
4	1 4 6 4 1
5	1 5 10 10 5 1
6	1 6 15 20 15 6 1

Thus for $n = 6$ the expansion for $(q + p)^6$ can be immediately written as $(q + p)^6 = q^6 + 6q^5p + 15q^4p^2 + 20q^3p^3 + 15q^2p^4 + 6qp^5 + p^6$. It should be noted that the coefficients of Pascal's triangle for $n = 6$ apply. The symmetry on either side of the median, whose value is 20, is another characteristic which it is important to bear in mind. Pascal's triangle is easy to determine for any sample size. Thus for $n = 2$ the middle term is the sum of the two adjacent terms $1 + 1 = 2$. For $n = 3$ the number 3 is obtained as the sum of 1 and 2 from the row above, and, again, the other 3 is obtained from $2 + 1$ of the row above. Similarly for $n = 4$, the first 4 is the sum of $1 + 3$ from the row above; the 6 is the sum of $3 + 3$ from the row above; the last 4 is the sum of $3 + 1$ of the row above. After each row is thus calculated, the number 1 is placed at each extreme end, and the process can go on indefinitely. (See dashed triangle, Table 5-3, for this relationship.)

From the foregoing it is obvious that if a histogram were plotted to show the number of combinations, these would be symmetrical about a median. However, a histogram representing the probabilities of occurrence of certain defects would not be symmetrical unless p and q were equal or n were very large. A glance at the solution of Example 5-2 will demonstrate this fact. As can be seen, the number of combinations are 1, 4, 6, 4, 1, which are symmetrical about the median 6, i.e., there are a 4 and 1 preceding and following 6. However, the probabilities of occurrence as seen in Fig. 5-1 are not symmetrical but markedly skewed. The reason for this should be obvious by examination of the calculations of Example 5-2.

PRACTICE PROBLEMS

5-1. A large lot of resistors is 5 per cent defective. Calculate the probability of occurrence of 0, 1, 2, 3, 4, and 5 defectives for samples of 5. Plot the resulting histogram.

5-2. Expand the binomial $(q + p)^8$. What is the probability of occurrence of 4 defectives if q equals 0.90?

5-3. A lot contains 100 balls. Ninety of the balls are white and 10 are black. Determine (by means of the binomial expansion) the number of combinations which will include only 1 black ball in samples of 4. Calculate the probability of this occurrence.

5-4. Plot the coefficients determined by Pascal's triangle for $n = 8$ in histogram form. Is this symmetrical?

5-5. Plot the probabilities of occurrence for the binomial expansion of Prob. 5-2 when $q = 0.98$; when $q = 0.50$. Are these symmetrical?

5-6. The 100-hr reliability of a batch of resistors is 0.98 per cent. What is the probability of 1 failure in a sample of 10? (*Note:* 100-hr reliability is discussed in Chap. 3.)

CHAPTER 6

THE POISSON DISTRIBUTION AND THE EXPONENTIAL FAILURE LAW

6-1. Introduction. In the case of the binomial distribution we took a sample of a definite size and observed the number of defectives in the sample, or, in general terms, the number of occurrences of an event. We also knew the number of times the event did not occur. Thus, in the case of a population consisting of black and white balls, a sample containing a certain number of black balls will also define the number of white because the sum of the two must equal the sample size. In the case of the Poisson distribution we are concerned with isolated events in a continuum of time, area, volume, or similar unit. Thus, the number of defects in a radio set is an example of the occurrence of the number of events, i.e., occurrence of defects, in a continuum of volume; the number of flaws in a given length of cloth is an example of the occurrence of isolated events in a continuum of area. Likewise, the number of failures per unit of time is representative of the occurrence of isolated events in a continuum of time. This latter case is what we shall be concerned with, since it is the basis for our reliability work. However, we must first learn something about the form of the Poisson distribution.

6-2. Form of the Poisson Distribution. The Poisson distribution is similar to the binomial in that it consists of a number of terms, each of which respectively gives the probability of 0, 1, 2, 3 or more occurrences per unit of measure. As in the case of the binomial, the sum of these probabilities equals unity. However, before the Poisson can predict probabilities of occurrence for each event, one must first determine the average number of occurrences per unit of measure. For example, if inspection data indicate that the average number of defects per unit for radio sets of a particular type is 1, then it is possible to predict the probability of 0, 1, 2, etc., defects per set randomly selected as a sample. However, before demonstrating the method of doing this, we must clarify certain terms.

The symbol for the average in terms of number of defects or defectives is a, which is computed by taking the product of the sample size n and the process average $\bar{p}$ when the latter is expressed as a decimal fraction. Thus, if $\bar{p}$ is indicated in terms of per cent defective or defects per 100 units, it must be converted to a decimal fraction before being multiplied by n in order to get the average value. In the first case, where $\bar{p}$ indicates per cent defective, the product of n and $\bar{p}$ will give a value of a in terms of average number of defectives. In the second case, where $\bar{p}$ indicates defects per 100 units, the product of n and $\bar{p}$ will give a value of a in terms of defects per unit. From this discussion, it can be seen that a defective is a unit which has one or more defects.

The following relationships will demonstrate the method of calculating $\bar{p}$ in terms of per cent defective or in terms of defects per 100 units.

In terms of per cent defective:

$$\bar{p} = \frac{\text{number of defectives}}{\text{total number of units inspected}} \times 100$$

In terms of defects per 100 units:

$$\bar{p} = \frac{\text{number of defects}}{\text{total number of units inspected}} \times 100$$

Thus, if we inspected 20 units and found 3 defects in one of them and 1 defect in another,

$$\bar{p} = 2/20 \times 100 = 10 \text{ per cent defective}$$

or

$$\bar{p} = 4/20 \times 100 = 20 \text{ defects per 100 units}$$

From the prior discussion, it is obvious that $\bar{p}$ is calculated from the available data. However, there are times when the value of the process average is prescribed by specification or experience. This is the case when an acceptable quality level (AQL) is established. In such instances, to distinguish it from a calculated $\bar{p}$, the process average is labeled p', which means that it is a standard, prescribed, or expected value.

The average value of the number of defects in a sample of size n is a and, to repeat, is calculated by taking the product of the sample size and process average. If p' is used instead of $\bar{p}$, then the product np' is called the expected average value or expectation. If we substitute the expectation a in the Poisson distribution it takes the form

$$1 = e^{-a} + ae^{-a} + \frac{a^2e^{-a}}{2!} + \frac{a^3e^{-a}}{3!} + \cdots + \frac{a^ce^{-a}}{c!} \tag{6-1}$$

where e^{-a} = probability of 0 defects per unit

ae^{-a} = probability of 1 defect per unit

$\frac{a^2e^{-a}}{2!}$ = probability of 2 defects per unit

$\frac{a^3e^{-a}}{3!}$ = probability of 3 defects per unit

and $\frac{a^ce^{-a}}{c!}$ = probability of c defects per unit

As is seen from Eq. (6-1), the sum of the probabilities is equal to unity. In the above expression, e represents the base of the Napierian or natural logarithms and is equal to 2.7183.

Now that we know something about the Poisson distribution, we are in a good position to solve our problem of a series of radio sets having an average of 1 defect per unit.

Example 6-1. The results of the inspection of 10 radio sets are shown in Table 6-1. Calculate the average number of defects per set and the probability of occurrence of 0, 1, 2, 3, and 4 defects in sets inspected at random.

TABLE 6-1. TABULATION OF INSPECTION DATA

Radio set no.	*No. of defects*
1	1
2	0
3	2
4	3
5	1
6	2
7	1
8	0
9	0
10	0
Total defects =	10

Solution. Average number of defects per unit $= 10/10 = 1 = a = n\bar{p}$. The calculations for the probabilities of occurrence are shown in Table 6-2.

The table demonstrates that the probability of a radio set's having 0 or 1 defect per unit is 37 per cent; of having 2 defects per unit is 18.5 per cent; of having 3 defects per unit is 6 per cent; and of having 4 defects per unit is 1.5 per cent. Moreover, the sum of all these probabilities is 100 per cent, which satisfies the stated condition that the sum of the probabilities should equal unity. These probabilities are accurate to only the first decimal place. Also they were not carried beyond the probability of 4 defects, since the occurrence of 5 or more defects is insignificantly small.

TABLE 6-2. TABULATIONS OF CALCULATIONS FOR EXAMPLE 6-1

No. of defects per unit	Calculations	Probability of occurrence, %
0	$e^{-a} = e^{-1} = \frac{1}{e} = \frac{1}{2.718} = 0.37$	37.00
1	$ae^{-a} = 1e^{-1} = \frac{1}{e} = \frac{1}{2.718} = 0.37$	37.00
2	$\frac{a^2e^{-a}}{2!} = \frac{1e^{-1}}{2!} = \frac{1}{2e} = \frac{1}{2(2.718)} = 0.185$	18.50
3	$\frac{a^3e^{-a}}{3!} = \frac{1e^{-1}}{3!} = \frac{1}{6e} = \frac{1}{6(2.718)} = 0.060$	6.00
4	$\frac{a^4e^{-a}}{4!} = \frac{1e^{-1}}{4!} = \frac{1}{24e} = \frac{1}{24(2.718)} = 0.015$	1.50
	Sum of probabilities =	100.00

6-3. The Poisson Probability Curves. The calculations shown for Example 6-1 are relatively simple, since in this problem $a = 1$. However, suppose $a = 0.5$; then these calculations would have been more difficult and a knowledge of logarithms would have been necessary in order to solve the problems, or else reference to tables of exponentials would have been required. The practical quality control man is often irked by these mathematical gymnastics since they are time-consuming, and time is an important factor in industry. Therefore, the use of the Poisson curves shown in Fig. 6-1 is a logical timesaving simplification. These curves provide at a glance the same information which can be obtained by solving Eq. (6-1) and therefore are of inestimable value in designing sampling plans. The abscissa (X axis) represents the average number of defects or failures in a sample of n units and p' fraction defective or a sample of T hours and a failure rate of r. The average number of defects is the product np', which we call a, and the average or expected number of failures is rT, which we refer to as d. The former product is used in quality control, while the latter is used in reliability evaluations. In each case we refer to this product as the expectation. Thus, for any value of the expectation the probability of c or f defects or failures respectively can be determined by reading the value of the ordinate labeled *probability of occurrence.*

It is important to realize that these are cumulative curves. For example, for an expectation of 2 the probability of 3 or less defects or failures is about 85 per cent, and the probability of 2 or less is approximately 67 per cent. Therefore, the difference between the two would be the probability of exactly 3 defects or failures, namely, about 18 per cent. Thus,

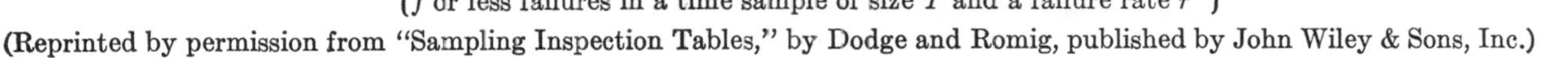

Fig. 6-1. Cumulative probability curves for Poisson's exponential binomial limit. May be used to determine the probability of

$$\begin{Bmatrix} c \text{ or less defects in a sample of size } n \text{ and a fraction defective } p' \\ f \text{ or less failures in a time sample of size } T \text{ and a failure rate } r \end{Bmatrix}$$

(Reprinted by permission from "Sampling Inspection Tables," by Dodge and Romig, published by John Wiley & Sons, Inc.)

we can see that these curves have a dual role; they can be used for designing both attributes as well as time sampling plans, as we shall see in Chaps. 10 and 11.

In Chaps. 8, 10, and 11 we shall become acquainted with the acceptance and rejection numbers A and R, respectively, which represent the limiting numbers of defects or failures which we are willing to accept in a sampling plan for a stated probability of occurrence. In summary, then, we reiterate that the abscissa represents the average number of defects, while the ordinate (Y axis) represents the probability of occurrence of 0, 1, 2, etc., or less defects, as indicated by the curves $c = 0$, $c = 1$, $c = 2$, etc.

The use of these Poisson curves can be effectively demonstrated by tracing the steps we would follow to solve Example 6-1. Reminding ourselves that $a = n\bar{p} = 1$, we proceed as follows:

Step 1. Locate $a = 1$ on the abscissa. Move up vertically to intersect the curve $c = 0$. Move horizontally to the left to intersect the ordinate at 0.37. The result checks with the calculated value of Example 6-1, which shows the probability of 0 defects to be 37 per cent.

Step 2. Proceed as in step 1, except that the point of intersection should now be with the curve $c = 1$. This corresponds to a probability of 74 per cent and represents the probability of a radio set's having 1 or less defects, or in other words, 0 or 1 defect. This indicates that these curves are cumulative. However, it was determined that the probability of 0 defects was exactly 37 per cent, and therefore the difference $74 - 37 = 37$ per cent represents the probability of a set's having exactly 1 defect per unit.

Step 3. In a similar manner the probabilities of 2, 3, and 4 defects per unit can be found. A check of results from the solution of Example 6-1 will demonstrate that they agree with those found by the use of the Poisson curves.

6-4. Application of the Poisson Distribution to Reliability Problems. From the previous discussion we have seen how the Poisson distribution can be used to predict the probability of occurrence of defects in a continuum of volume or area. This is a very valuable tool in the field of quality control and forms the basis for acceptance sampling plans. In a similar manner, since the probability of occurrence of failures in a continuum of time forms the basis for reliability prediction, it is only fitting that we consider the applicability of the Poisson to reliability analysis.

When we considered the quality control aspect, we used the symbol p' to represent defects per unit. In reliability work we shall talk about failures per unit of time, to which we shall assign the symbol r. This term is usually referred to as the failure rate or failure hazard.

6-5. Distinction between Failure Rate or Hazard and Ratio Failure Rate. Present-day reliability literature is not consistent with regard to the definition or use of the terms failure rate, hazard, and ratio failure rate. In this section, we shall distinguish between these terms in order to eliminate any source of confusion in subsequent discussions.

Failure Rate or Hazard r. The failure rate is expressed in terms of failures per unit of time, such as failures per hour or failures per 100 or 1,000 hr. It is computed as a simple ratio of the number of failures f, during a specified test interval t,* to the total or aggregate survival test time of the articles undergoing test during the test interval. The equation for this relationship is

$$r = \frac{f}{T} \tag{6-2}$$

where r = failure rate, fph
f = total number of failures for the test interval
T = total test hours

Equation (6-2) may also be written as $r = f/st$, since the product of the number of survivors s and the duration of the survival time of each during the test interval t in hours equals the total number of test hours T. Thus, if we began a reliability test at a particular time and recorded the actual time of survival for each unit being tested during the test interval, we would have the total or aggregate test time T. Another method of determining T is to calculate it as the product of the average number of survivors $\bar{s}$ and t. This would result in a new version of Eq. (6-2):

$$r = \frac{f}{\bar{s}t} \tag{6-3}$$

To illustrate how to compute $\bar{s}$, consider the following: Suppose there were 600 survivors at the beginning of a test interval of 1 hr and 50 failures occurred during this time. This means that there would be only 550 survivors left at the end of the interval. The average number of survivors $\bar{s}$ is

$$\bar{s} = \frac{600 + 550}{2} = 575$$

and the failure rate as calculated from this 1-hr interval is

$$r = \frac{f}{\bar{s}t} = \frac{50}{575(1)} = 0.087 \text{ fph}$$

When the design is mature, this failure rate is fairly constant during the operating or service life period (see Table 6-3). For this reason, the

* In this application t is used to indicate a small time interval. In Chap. 9 the same symbol is used to describe a maintenance time constraint. T is the sum of all test intervals t. When more than one unit is under test, this sum should be multiplied by n to obtain the total test time T.

failure rate is often referred to as the constant failure rate. It is also called the hazard. This latter name is used because there is always a probability or hazard that a predicted number of failures will occur during a particular time interval. In this text we shall consider the terms failure rate and hazard as synonymous; therefore we shall use the expression failure rate to mean either. Figure 6-2 shows a typical failure rate or hazard curve. From the figure we can see that the failure rate is constant during the operating or service period but changes rapidly during the wear-out period.

It should be noted at this time that in order to be able to demonstrate a constant failure rate, it is necessary to test a large number of parts in order to reduce the effects of sampling variations. If a small sample is used, the best that can be expected is the determination of an average failure rate within a degree of confidence and the assumption that r must be close to this average value as defined by the confidence limits.

Returning to our discussion of $\bar{s}$, we can see that there is a definite advantage in its use, since it eliminates the prior necessity of requiring or maintaining an accurate record of the exact time of failure for each unit undergoing test. Therefore, the fact that some units fail immediately after the start of a test interval while others fail at other times during the test interval is compensated for by the use of the average number of survivors $\bar{s}$.

Moreover, it is important to bear the distinction between t and T clearly in mind in order to avoid confusion when we study sampling and time samples in Chaps. 10 and 11. Actually, t represents a time or test interval during which one or more units may be undergoing test. If only one unit at a time is being tested, it is obvious that t and T are synonymous. However, when more than one unit is being tested during a test interval of t hours, the total test time T must be the sum of all the individual test times for each unit. Hence, if we are concerned with sampling or time samples, we are interested in the total test time, or, in other words, the sample T in hours, because the number of failures in any sample is a function of the sample size. However, if our concern is solely in determining the probability of failure of a particular part or group of identical parts, we use the time interval t in our calculations for each individual part. It is obvious, then, that for the group, $T = nt$, i.e., the product of the number of parts n and the time interval t. The reason for this is explained in the following theorem.

Theorem 6-1. *The probability of survival for a group of identical parts expressed as a percentage is equal to the probability of survival of each individual part composing the group.*

From this theorem it is apparent that if we are able to determine the probability of survival of an individual part, we shall then be able to predict the number of survivors during the same time in a sample of n parts.

Ratio Failure Rate λ. The ratio failure rate is the ratio of the number of failures which occur during a unit interval of time and the original number of items at the start of the reliability test. The symbol we shall use to represent it is lambda (λ). We shall also use subscripts when required for clarification. Thus, λ_p represents the ratio part failure rate, and λ_e would be the ratio equipment failure rate. The main advantage of λ is that it predicts the probability of failure during a particular time interval. (See solution to Example 6-2.)

The equation for ratio failure rate is

$$\lambda = \frac{f}{n} \tag{6-4}$$

where λ = ratio failure rate
f = total failures during a given interval of time
n = number of items originally placed on test

As an example of the use of Eq. (6-4), if 1,000 parts were originally placed on test and 46 failures were recorded during the interval between the seventh and eighth hour, then $\lambda_p = 46/1{,}000 = 0.046$ failures/hr per part for the stated interval.

The ratio part failure rate is useful in that it gives the fraction of original parts which are expected to survive during a particular time interval. It may also be expressed in per cent if desired. Its value lies mainly in the fact that it is also a probability-of-survival statistic. Thus, it is possible to predict the probability of survival of any specific part for any particular time interval, since it is equal to the probability of survival of the parts population.

6-6. Exponential Failure Law. The exponential failure law is used to predict the probability of survival of a part as a function of time. It is derived from the Poisson distribution in a manner similar to that described in Sec. 6-2, except that now we are concerned with failures instead of defects.

The failure rate r is, as we have discussed before, the number of failures per unit of time. Hence, rT must be the number of failures which will probably occur in time T. If the reliability test is of a destructive nature, then only one failure per unit is possible; if the test is not destructive, the unit may be repaired and returned to test, and therefore more than one failure is possible. Another alternative is to replace a failed unit with another similar one if it is desired to gather more data.

If the expected number of failures rT is substituted in the Poisson, then it is possible to calculate the probability of 0, 1, 2, 3, etc., failures in time T in a manner similar to that used in Sec. 6-2. If we make $d = rT$, the

Poisson may be written as

$$1 = e^{-d} + de^{-d} + \frac{d^2e^{-d}}{2!} + \frac{d^3e^{-d}}{3!} + \cdots + \frac{d^fe^{-d}}{f!} \tag{6-5}$$

where e^{-d} = probability of 0 failures in time t
de^{-d} = probability of 1 failure in time t
$\frac{d^2e^{-d}}{2!}$ = probability of 2 failures in time t
$\frac{d^3e^{-d}}{3!}$ = probability of 3 failures in time t
$\frac{d^fe^{-d}}{f!}$ = probability of f failures in time t

If in the last or general term of the Poisson we make $f = 0$, the expression reduces to the first term of Eq. (6-5). This is the probability of zero failures, which also represents a condition of survival, since the unit has not yet failed in test. Therefore, this first term is generally known as the exponential failure law. It is also called the probability of survival P_s and is written

$$P_s = e^{-rT} \qquad \text{or} \qquad P_s = e^{-d} \tag{6-6}$$

where P_s = probability of survival (equivalent to reliability)
T = total time in hours
e = base of Napierian or natural logarithms
r = failure rate in failures per hour
d = expected number of failures in time T

From the preceding discussion, it should be obvious that the Poisson curves can be used to predict the probability of occurrence of failures for any time T. If a large batch of parts are undergoing reliability testing, as we shall see is the case in Example 6-2, it is possible to predict the number of parts which are expected to fail in any time interval, since these probabilities are expressed by the separate terms of Eq. (6-5). Moreover, from Theorem 6-1 we see that the probabilities which are applicable to a batch of identical parts also apply to the probability of survival for any one part. This is found by substituting in Eq. (6-6). The relationship between the probability of survival P_s and the probability of failure P_f is $P_s = 1 - P_f$.

From the foregoing discussion, it should be obvious that reliability is also a probability statistic. Therefore, we have been using capital letter P to denote reliability in order to emphasize this fact. By using various subscripts for P we can indicate the applicability of P as required. Thus, P_s is usually used to indicate the probability of survival or probability of success. This is equivalent to no failures, or, in other words, it is the reliability of a device; and, as we explained before, it is also called the exponential failure law, which is expressed by Eq. (6-6). This equation tells us that if we know r, we can predict the reliability of an equipment for any time T, provided that r remains constant with time.

Tests in the field, on a variety of equipments, have indicated that if the design is mature, r will be essentially a constant throughout the prescribed operating period of the equipment. Moreover, it has been found that during the early life or infant stage of the equipment, early failures will occur more frequently than during the operating period. This results in a higher initial failure rate. Likewise, during the later part of the life of the equipment, i.e., the wear-out period, the frequency of failure is very high, and again the failure rate rises rapidly. Figure 6-2, a typical failure rate curve, shows the variation of failure rate with time.

6-7. Mean Time between Failures *m*. During the operating period when the failure rate is constant, the *mean time between failures* is the reciprocal of the *constant failure rate,* or the ratio of the total operating time to the total number of failures.

$$m = \frac{\bar{s}t}{f} = \frac{1}{r} = \frac{T}{f} \tag{6-7}$$

It is also defined as the average time of satisfactory operation of a population of equipments. On the other hand, the mean time *to* failure is defined as the measured operating time of a single device divided by the total number of failures during this period. This is accomplished by repairing the device after each failure and continuing the test.

Actually, the value of m as calculated from various samples will vary from sample to sample. However, the variation will be least when the sample size is large, and therefore when this occurs, it is simple to demonstrate that a constant failure rate exists or that its reciprocal m is the mean time between failures. For this reason, in Example 6-2, we selected 1,000 parts to facilitate the illustration of a constant failure rate. If we had not selected such a large sample, we could have calculated the failure rate of each of a series of small samples and then determined the grand average failure rate, which would essentially be equal to the constant failure rate. On this basis, we conclude that if the design is mature, the average failure rate is equal to the constant failure rate. This is the value that should be used in Eq. (6-6), and since the mean time between failures is the reciprocal of r, we can rewrite Eq. (6-6) in a more convenient form.

$$P_s = e^{-T/m} \tag{6-8}$$

A typical example will be used to illustrate the associated concepts more thoroughly. Actually, the example is purely hypothetical, convenient figures having been selected in order to illustrate several points with a minimum of arithmetic computation.

TABLE 6-3. RELIABILITY FAILURE DATA*

(1) Time interval, t	(2) No. of failures, f	(3) Cumulative failures, F	(4) No. of survivors, s	(5) Ratio part failure rate, λ	(6) Reliability, %, P	(7) Failure rate, r
0		0	1,000		100	
	130			0.130		0.139
1		130	870		87.0	
	83			0.083		0.101
2		213	787		78.7	
	75			0.075		0.100
3		288	712		71.2	
	68			0.068		0.100
4		356	644		64.4	
	62			0.062		0.101
5		418	582		58.2	
	56			0.056		0.101
6		474	526		52.6	
	51			0.051		0.101
7		525	475		47.5	
	46			0.046		0.101
8		571	429		42.9	
	41			0.041		0.100
9		612	388		38.8	
	37			0.037		0.100
10		649	351		35.1	
	34			0.034		0.101
11		683	317		31.7	
	31			0.031		0.103
12		714	286		28.6	
	28			0.028		0.103
13		742	258		25.8	
	64			0.064		0.283
14		806	194		19.4	
	76			0.076		0.486
15		882	118		11.8	
	62			0.062		0.714
16		944	56		5.6	
	40			0.040		1.110
17		984	16		1.6	
	12			0.012		1.200
18		996	4		0.4	
	4			0.004		2.000
19		1,000	0		0	
				1.000†		

* Figures appearing between any two values of t indicate that they are applicable to the interval so defined. All other figures are applicable to the beginning or end of an interval as indicated by their relative positions.

† Area under curve in Fig. 6-3.

Example 6-2. The reliability test of 1,000 parts resulted in the data shown in columns 1, 2, and 3 of Table 6-3.

Calculate the ratio part failure rate, the reliability at the end of each hour, the failure rate, and mean time between failures.

Plot the ratio part failure rate curve, the failure rate curve, and the reliability curve.

Solution. (*a*) Columns 1, 2, and 3 are the original data.

(*b*) Column 4 (Number of survivors) is obtained by subtracting the cumulative number of failures F from 1,000.

(*c*) Column 5 (Ratio part failure rate) is the ratio of the number of failures during a 1-hr interval (column 2) and 1,000, the original quantity of parts.

(*d*) Column 6 (Reliability in per cent) is calculated from the data as the ratio of the number of survivors for a given time (column 4) and 1,000 expressed in per cent.

(*e*) Column 7 (Failure rate) is the ratio of the number of failures during a 1-hr interval (column 2) and the average number of survivors for that period. A typical calculation for the time interval between 2 and 3 hr is

$$\frac{75}{(787 + 712)/2} = \frac{75}{749.5} = 0.100$$

(*f*) The mean time between failures m is the reciprocal of r during the operating period. Therefore

$$m = \frac{1}{r} = \frac{1}{0.10} = 10 \text{ hr}$$

(*g*) The curves for r, λ, and P are shown as Figs. 6-2 to 6-4, respectively.

6-8. Calculation of Reliability Curves. Figures 6-2 to 6-4 illustrate typical curves prepared from the data of Table 6-3.

Figure 6-3 is a plot of the ratio part failure rate. This curve actually represents the probability of failure of any part for any time t, since the probability of failure for a specific part is related to the over-all probability as ascertained by testing a large number of similar parts. The integral of the failure rate curve from the lower to the higher limit of any interval represents the probability of failure of a part during this time. This is called the unreliability of the part. This integral of a curve between any two limits is simply the area under the curve between these same two limits. This area may be calculated by resorting to the integral calculus or by using simple geometric facts. For example, suppose it was desired to find the area under the curve of Fig. 6-3 or 6-4 for the interval between 4 and 5 hr. We would find the area of a rectangle which is equal to the area under the curve. This is a simple technique, since all one has to do is divide the upper portion of the curve (between the two limits) into two parts so that the area excluded equals the area included. When this is done, it is obvious that the rectangle so created represents the

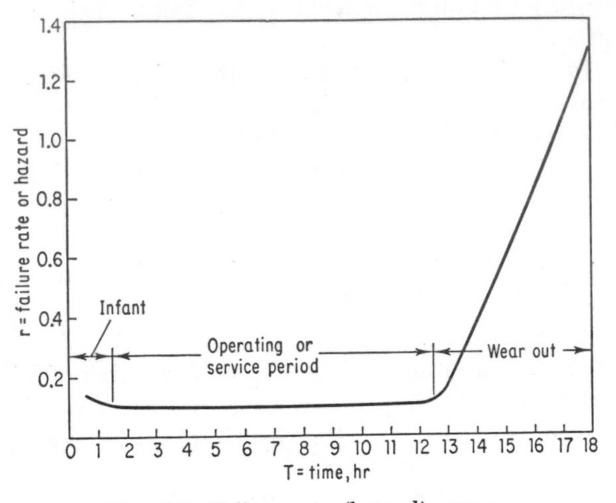

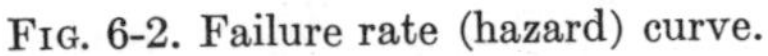

Fig. 6-2. Failure rate (hazard) curve.

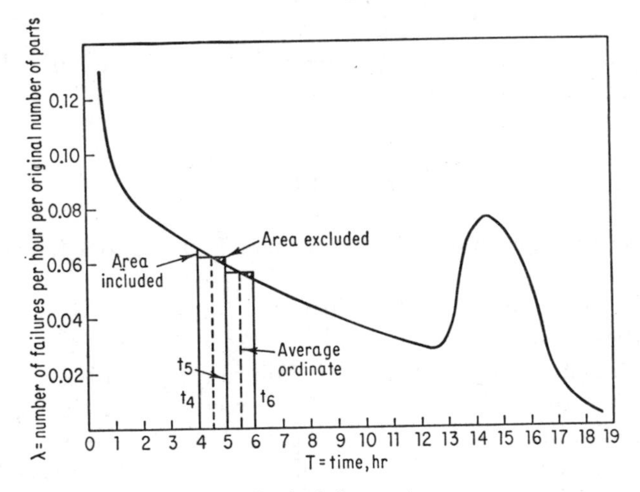

Fig. 6-3. Ratio failure rate curve.

desired area under the curve between t_4 and t_5. Now the formula for the area of a rectangle is the product of the base and the altitude. In this case, the base is unity, and the altitude is the value of the average ordinate. Therefore, since the base is unity, the average ordinate represents the area between t_4 and t_5. The same is true for any other interval.

Thus, if one wishes to determine the unreliability or probability of failure between $t = 0$ and $t = 5$, it is only necessary to sum all the average ordinates. A glance at Table 6-3, column 5, will show these average ordinates to be 0.130, 0.083, 0.075, 0.068, and 0.062, and their sum is 0.418. This means that the probability of failure between $t = 0$ and $t = 5$ is 41.8 per cent, which represents the unreliability. If the reliability for the same period is desired, it is only necessary to subtract the unreliability from unity. Thus, $1.000 - 0.418 = 0.582$, and the reliability is 58.2 per cent. The result obtained may be verified by glancing at column 6 of Table 6-3 for $t = 5$. It may also be observed in Fig. 6-4 as the reliability corresponding to $t = 5$.

It is interesting to note that the sum of all the data in Table 6-3, column 5, is unity, which indicates that the area under the curve repre-

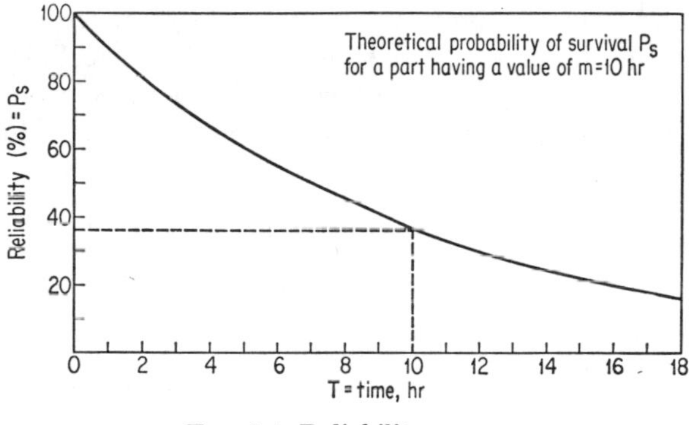

FIG. 6-4. Reliability curve.

sents the sum of the probabilities of failure from $t = 0$ to $t = 19$. From this we can conclude that the probability of failure is a certainty after 19 hr of test, or, in other words, there is no probability of survival after this time.

At this time a word of caution is in order regarding the interpretation of Figs. 6-2 to 6-4 as they relate to each other. For example, the data of columns 5 and 7 of Table 6-3 indicate that they represent the mid-point of an interval. These data therefore, should not be used with data in other columns unless they also are at the mid-point or average of the interval. In other words, data at the beginning of an interval should be compared with other data also at the beginning of an interval; data at the end of an interval, compared with corresponding data also at the end of an interval. It is for this very reason that the average value of the survivors was used in calculating the failure rate in Sec. 6-5 by the use of Eq. (6-3). Thus, if the data shown in Table 6-3 are not compatible because they are not at the proper points in the test interval, they may be rendered compatible either by calculation or by reading the appropriate values from the applicable graphs.

Figure 6-2 is a curve which demonstrates the failure rate with respect to operating time. It also demonstrates that the failure rate is constant during the operating or service life of the equipment and therefore is independent of time. During the early life of equipment, or what is commonly called the infant period, the failure rate is high. It is high also at the tail end of the curve, where wear-out or old age takes place. In this particular example, the magnitude of the early failure rate is not much greater than the constant failure rate. However, during the wear-out period it takes a sudden and pronounced climb upward. The shape of the curve depends upon the characteristics of the equipment or parts under test. In some cases the early failure rate is more pronounced than in others and vice versa.

In contrast with the failure rate curve of Fig. 6-2, we have the ratio failure rate curve of Fig. 6-3. It shows that the ratio failure rate is a function of time which appears to follow the exponential law but suddenly deviates and takes on the shape of the normal curve. This is indeed significant because, as was mentioned in Sec. 6-5, the ratio failure rate curve represents the probability of failure with respect to an interval of time. Therefore, the behavior of the ratio failure rate curve indicates that this probability of failure is a function of changing influences with respect to time.

If we examine Fig. 6-2, the failure rate curve, we shall notice that a definite relationship exists between this curve and Fig. 6-3. Somewhere between 12 and 13 hr the failure rate suddenly begins to increase in a rapid and practically linear manner. In a way, it reminds one of the behavior of steel when it is stretched beyond the elastic limit and rapidly loses its tensile strength. In a comparable sense this is actually what happens. The number of failures per unit of time increases so rapidly that we can conclude that we are at the wear-out stage because so many more failures are evident. We no longer have a constant failure rate, which is a characteristic of the operating period. Moreover, the failure rate increases so rapidly that it is practically impossible to maintain a high degree of reliability without exhaustive maintenance.

An obvious and interesting conclusion which can be drawn from Fig. 6-3 is that the average time of wear-out occurs at 14.5 hr. We can also conclude that this wear-out must be due to a constant-cause chance system, which is a characteristic of the normal distribution. Therefore, when a ratio failure rate curve changes from the exponential to the normal law, we can assume that we are at the wear-out point. We can also predict the frequency of occurrence of the time of wear-out in accordance with the principles explained in Chap. 4, where we discussed the normal curve.

The reason for a change from the exponential to a normal curve can be found in Table 6-3. Column 2 shows that the failures per interval get progressively less until the interval between 12 and 13 hr is reached, when, suddenly, the number of failures and the failure rate shoot sharply upward. This effect is also seen in Fig. 6-2, where the failure rate is seen to rise sharply. However, the ratio part failure rate (Fig. 6-3) cannot follow this sharp rise because, as the failure rate increases, the rate of depletion of survivors per interval also increases until a peak is reached between 14 and 15 hr. After this time, because of the increasing failure rate associated with a proportionately smaller number of survivors, the ratio part failure rate must decrease. This results in the shape of the normal curve we described earlier.

One of the advantages of this phenomenon is that it gives us a clear realization of what is happening, thereby providing us with the necessary tools to chart our future action. For example, in most instances the parts which comprise an electronic computer usually operate in the wear-out phase because of the tremendously long exposure in service. However, a knowledge of this fact makes possible the design of certain programmed preventive maintenance and marginal checking techniques which make the control of reliability practical even during the wear-out phase.

6-9. Comparison of Theoretical and Actual Reliability Curves. Figure 6-4 shows the reliability curve for Example 6-2. We also determined

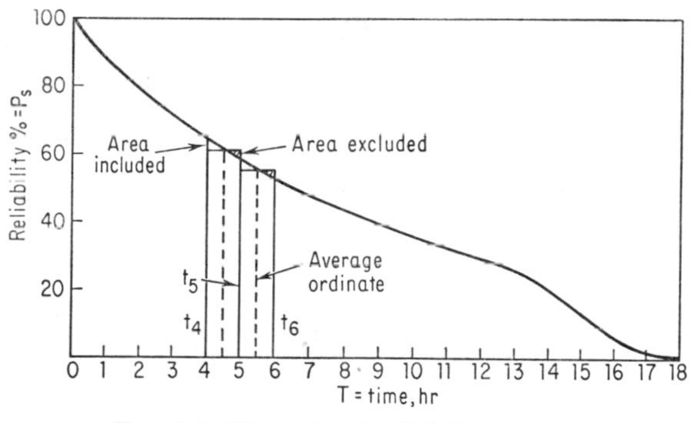

FIG. 6-5. Theoretical reliability curve.

from the solution of the example that the mean time between failures is 10 hr. Moreover, from Sec. 6-8, by the use of Eq. (6-8), we should be able to plot the theoretical reliability curve (exponential failure law) and compare it with the curve of Fig. 6-4.

Figure 6-5 is the result of such a plot, the data for which are tabulated in Table 6-4. A comparison between Figs. 6-4 and 6-5 shows excellent correlation within the operating period but a very poor correlation during the wear-out period. The slight variation between the two curves during the operating period is due to the effect of the early failures.

TABLE 6-4. DATA FOR THEORETICAL RELIABILITY CURVE WHEN $m = 10$ AND $r = 0.10$

Time, t	$e^{-T/m}$	P, %
0	$e^{-0/10}$	100
1	$e^{-1/10}$	90.5
2	$e^{-2/10}$	81.9
3	$e^{-3/10}$	74.1
4	$e^{-4/10}$	67.0
5	$e^{-5/10}$	60.7
6	$e^{-6/10}$	54.9
7	$e^{-7/10}$	49.7
8	$e^{-8/10}$	44.9
9	$e^{-9/10}$	40.7
10	$e^{-10/10}$	36.8*
11	$e^{-11/10}$	33.3
12	$e^{-12/10}$	30.1
13	$e^{-13/10}$	27.3
14	$e^{-14/10}$	24.7
15	$e^{-15/10}$	22.3
16	$e^{-16/10}$	20.2
17	$e^{-17/10}$	18.3
18	$e^{-18/10}$	16.5

* P_s = 37 per cent when $T = m$.

In most instances, when the design is mature, the effect of these early failures is negligible. Therefore, if the mean life is known, it is possible to predict the probability of survival with respect to time very accurately during the operating or service period of the equipment. Moreover, Fig. 6-5 indicates that if the wear-out point had not been reached, the probability of survival of some of the equipment would have been enhanced. It is interesting to note also that when $T = m$, i.e., when the operating time equals the mean life, the probability of survival is 37 per cent.

6-10. Combined Failure Rates. We have seen during our discussion on series reliability that the over-all reliability of an equipment is calculated as the product of the reliabilities of the respective parts, i.e.,

$$P = P_1 P_2 P_3 \cdots P_n$$

Substituting in accordance with Eq. (6-6), we have

$$P = e^{-r_1 T} e^{-r_2 T} e^{-r_3 T} \cdots e^{-r_n T}$$

$$P = e^{-(r_1 + r_2 + r_3 + \cdots + r_n)T} \tag{6-9}$$

Equation (6-9) indicates that if an equipment is composed of several different components, each of different failure rates, these can be added directly to obtain the over-all or equivalent failure rate. It should be apparent also that if there are many components, each of reliabilities r_1, r_2, r_3, etc., the expected failure contribution of each component to the equipment is n_1r_1, n_2r_2, n_3r_3, etc., and that then Eq. (6-9) becomes

$$P = e^{-(n_1r_1+n_2r_2+n_3r_3+\cdots)T} \tag{6-10}$$

This relationship is very useful to the design engineer, as we shall see later.

6-11. Combined Ages. If two or more populations are each of the Poisson variety, the combined distribution will not necessarily be a Poisson unless the mixture is of a random nature. This means that each element must have an equal chance of combining with every other element. In this case random samples of these random combinations will be of the Poisson variety.

In electronic equipment we can consider the combinations of various parts as being random, and therefore we can predict the probability of the number of failures with time by means of the Poisson. As we have seen, the first term of the Poisson gives us the exponential failure law, and therefore the entire equipment follows the exponential distribution.

It can also be shown that even in those cases where the individual parts do not follow the Poisson but are of mixed ages, random combinations will produce a Poisson. The reason is that the Poisson predicts the probability of failure in a continuum of time when the causative agent is randomness. This is a very important consideration in military equipment because of the variety and mixed ages of spare parts in various depots. Replacement of a defective part with a spare should therefore not affect the reliability of an equipment.

PRACTICE PROBLEMS

6-1. The failure rate for a television receiver is 0.02 failures/hr. Calculate the mean time between failures. What is the probability of its failing in 4 hr?

6-2. During the test of 1,000 resistors it was found that $m = 200{,}000$ hr. Find r. Determine the probability of 0, 1, 2, 3, and 4 failures when $T = 100{,}000$ hr.

6-3. Complete the following table and draw the exponential failure curve for $m = 20$ hr. What is P_s when $T = 20$?

T	$e^{-T/m}$	P_s, %
0	$e^{-0/20}$	100
2	$e^{-2/20}$	90.5
4	$e^{-4/20}$	81.9
. . . . .		

6-4. During a reliability test of 800 parts, the number of failures in the interval between 3 and 4 hr was 46. The number of survivors at the beginning of the interval was 475 and at the end of the interval 429. What is the failure rate? What is the ratio part failure rate?

6-5. From Fig. 6-3 find the reliability from $t = 0$ to $t = 10$. What is the unreliability from $t = 4$ to $t = 8$? Does this check by using column 5 of Table 6-3?

6-6. An equipment consists of 3 components connected functionally in series. Each component has 100 parts with respective failure rates for each part expressed in terms of per cent failures per 1,000 hr of 10, 20, and 40. Find the probability of survival P_s after 10 hr. (Convert to failures per hour before proceeding.)

6-7. In Prob. 6-6 suppose the first component had 150 parts and the third component had 200 parts, and the failure rates remain the same. Would the probability of survival improve? What would be its new value? What is the mean time between failures?

CHAPTER 7

RELIABILITY DATA

7-1. Purpose of Reliability Data. The primary advantage of accumulating reliability data is that these can be used to predict or calculate the reliability of a device when it is operated under the conditions which these data represent. It is therefore important that these data be as factual as possible in order that a high degree of confidence may be reposed in any derived conclusions. Data accuracy is particularly important when it is used for predicting reliability, because the reliability prediction technique, at best, gives us a broad estimate of the expected reliability. Therefore, it follows that more dependable data result in a higher degree of confidence in the reliability estimate.

Another reason for accumulating reliability data is to effect product improvement. This is usually accomplished by summarizing the results of our data analysis into a series of conclusions. These are then furnished to interested personnel or departments, such as the design engineer or the quality control department of the factory responsible for the research, development, and manufacture of the device undergoing evaluation, who follow through to effect any desired corrective action.

There is no fixed point in time for accumulating data. This is a determination which must be made by the individual desiring the information. For example, reliability tests may be conducted in the laboratory on prototypes or development models; or the testing might be done on randomly selected samples of preproduction or production units in the factory. In each instance, the purpose is to obtain data which can be used to determine statistically the expected reliability.

In the laboratory, reliability testing and evaluation provides the design engineer with insight into those factors which adversely affect reliability and which must be corrected. In the factory, data from the testing of random production samples provide a continuous measure of the reliability of the outgoing product. These also serve as a means of determining the assignable causes of reliability variation.

The ultimate measure of reliability is in the field, where the equipment is subjected to the actual conditions of use and operation for which it was designed. It is in this area that reliability data are most useful. Task group No. 4 of AGREE (The Advisory Group for Reliability of Electronic Equipment) has prepared a list of those factors which influence reliability data most. These were listed in a report issued by AGREE in September, 1956, which summarized them under the headings of *Use* and *Inherent* reliability. The product of these two reliabilities was labeled *Operational Reliability.*

The factors affecting use reliability were listed as capability of operating and maintenance personnel, operating and maintenance procedures, operational suitability (human engineering), maintainability, auxiliary and supporting equipment, installation environment, deterioration in storage, and the effects of shipping and handling.

The factors affecting inherent reliability were listed as selection and application of circuits, parts and tubes application, component operational parameters, mechanical structures, manufacturing techniques and workmanship.

7-2. Some Representative Data Sources. Reliability data come from many sources. The most important of these are government- and industry-sponsored associations and organizations which gather and disseminate reliability data to all interested parties. The most outstanding sources of reliability data are listed and discussed in the following paragraphs.

AGET. In 1953 the Advisory Group on Electron Tubes was formed under the auspices of the Assistant Secretary of Defense for Research and Development. The function of this group is to gather and publish data regarding the proper use and application of vacuum tubes. It also coordinates efforts between tube-applications engineers of various manufacturers and arranges symposia to correlate and disseminate information. Another contribution of AGET is to study the behavior of tubes in equipment when they are used under actual field conditions and to record and disseminate the information acquired during these studies. Information may be obtained from Secretary, Advisory Group on Electron Tubes, 346 Broadway, New York 13, New York.

ARINC (Aeronautical Radio, Incorporated). This group was organized in 1929 by the air transport industry of the United States as a joint nonprofit undertaking to provide the industry with aeronautical communications services. After World War II, ARINC investigated tube reliability for the commercial airlines. In April, 1957, the Navy Department, because of the effectiveness of the commercial airlines program, awarded ARINC a contract to study tube reliability at eight Army, Navy, and Air Force bases. The procedure required ARINC field representatives to remove failed tubes from their sockets and send them to ARINC or other laboratories for analysis. However, before shipping the tubes to a laboratory, the causes of failure were verified and recorded for each tube. The results of ARINC findings have been published, and the data are available. The address is Aeronautical Radio, Incorporated, 1700 K Street N.W., Washington 6, D.C.

Signal Corps–Cornell University. In July, 1951, the Signal Corps awarded Cornell University a contract to determine the reasons for tube rejections. Tubes were furnished for analysis by industry and by Aeronautical Radio, Incorporated (ARINC). The industry program, known as the *line-reject* program, permitted equipment manufacturers to participate in it whenever a pattern of tube failures occurred during production-line testing of equipment at a manufacturer's plant. Any contractor could qualify for Cornell services under this program, regardless of which government agency awarded the prime contract. This resulted in the development of a team of electronic manufacturers who worked closely with Cornell in the tube evaluation program. Each manufacturer would ship line-reject tubes to Cornell with all required data to permit a factual evaluation. The information included tube application, number of hours in use, associated circuitry, and apparent causes for rejection. These data facilitated the completion of the required analysis. Cornell University found that approximately 30 per cent of the tubes received by them had no defects when they were tested in accordance with the requirements of applicable military standards. These results were corroborated by some manufacturers who conducted tests in their own quality control laboratories. As a result, it was concluded that many equipment designs were marginal, and because of this tubes of the same type were not interchangeable. This resulted in tube selection, which is an undesirable procedure. The results also indicated that in some instances factory test personnel were incompetent and not capable as "trouble shooters." Consequently, they chose to correct trouble by randomly selecting tubes and trying each one until they obtained one which would work in the circuit. This was a good temporary way of correcting the trouble but did not solve the problem on a permanent basis, because it did not provide data for analytical purposes which could be used for future permanent corrective action.

The tubes furnished to Cornell by ARINC were called *controlled* and *uncontrolled* tubes. The controlled tubes had their parameters and length of time in service recorded prior to test. The uncontrolled tubes were those which had been used for an unspecified length of time. This method of tube collection enabled Cornell University to study the effects of usage and particular environments on the life and performance of each tube type.

The information obtained by Cornell was recorded on punched cards, which facilitated the tabulation and distribution of information to all participants in the program or, upon request, to other interested groups. Further information on this subject may be obtained from the School of Electrical Engineering, Cornell University, Ithaca, New York.

EIA (Electronic Industries Association). This organization, formerly known as RETMA, Radio-Electronic-Television Manufacturers Association, established a committee on electronic applications in March, 1953. The purpose of this committee was to formulate plans which would ensure cooperation between government agencies and industry in advancing the development of reliability. An additional objective was the planning of reliability training programs to educate technical personnel, including design engineers in both government and industry. They also established methods and procedures for gathering reliability data and for analyzing, tabulating, and publishing the results.

TABLE 7-1. RELIABILITY INDICES AT THE 90% CONFIDENCE LEVEL

Component part	Failures ÷ component hours	Adjusted index
Relays	21/20,700	1/740
Crystal diodes	18/57,500	1/2,340
Subminiature tubes	30/86,500	1/2,340
Miniature tubes	6/36,500	1/3,420
Potentiometers	10/76,500	1/4,950
Connectors and plugs	9/100,000	1/6,850
Transformers	3/47,000	1/7,050
Capacitors	13/545,000	1/28,700
Inductors and coils	0/90,500	1/39,400
Resistors	15/990,000	1/46,000
Solder connections	88/6,600,000	1/66,000

The results of the committee's investigations are published in bulletins entitled Electronics Applications Reliability Review. These bulletins are very comprehensive and contain much valuable information. Tables 7-1 and 7-2, originally Tables II and III from Vol. 4, No. 1, of Electronics Applications Reliability Review, are an example of the data available in these bulletins and represent the EIA method of tabulating information for the purpose of computing failure rates. These failure rates, shown as fractions, are expressed as an index for various components. The indices are listed in the last column of Tables 7-1 and 7-2. If we convert the index to a decimal fraction it will give us the failure rate in failures per hour. The index is adjusted to give a good estimate of the failure rate for various types of components from the same family. The address of the Electronic Industries Association is 1721 DeSales Street N.W., Washington, D.C.

AGREE (Advisory Group on Reliability of Electronic Equipment). In August, 1952, the Defense Department established the Advisory Group

TABLE 7-2. HOURLY RELIABILITY INDICES FOR MISSILE ELECTRONIC PARTS

Component part	Failures ÷ population	Failure rate per hour	Hourly index, p'
Relays	21/1,090	1/990	1/990
Delay lines	3/168	1/1,070	1/1,070
Rotating equipment:			
Motors	1/158	1/3,000	
Inverters	2/100	1/950	1/1,080
Dynamotors	1/58	1/1,100	
Rate gyros	4/142	1/685	
Subminiature tubes:			
5639	3/254	1/1,610	
5643	0/17	0/325	
5702	1/147	1/2,800	
5718	8/771	1/1,820	
5719	2/236	1/2,240	
5783	0/149	0/2,830	1/2,900
5784	0/135	0/2,570	
5840	2/780	1/7,400	
5896	0/37	0/700	
5902	1/135	1/2,570	
6021	10/1,685	1/3,200	
6112	3/104	1/660	
Crystal diodes:			
Silicon	14/878	1/1,200	
Selenium	3/400	1/2,530	1/3,180
Germanium	1/1,735	1/33,000	
Microswitches	2/400	1/3,800	1/3,800
Miniature tubes:			
5670	0/21	0/400	
5726	1/695	1/13,200	
5727	0/16	0/305	
5751	1/424	1/8,000	1/6,100
5814	2/484	1/4,600	
6005	2/264	1/2,500	
VC 1258	0/16	0/305	
Potentiometers:			
Linear plastic	2/152	1/1,440	
Wire wound	3/175	1/3,000	1/7,650
Composition	5/3,395	1/12,900	
Connectors and plugs	9/5,234	1/11,000	1/11,000
Transformers	3/2,476	1/15,700	1/15,700
Inductances	0/1,213	0/23,000	1/23,000*
Capacitors:			
Paper	10/15,677	1/30,000	
Ceramic	0/6,428	0/122,000	
Mica and glass	1/5,841	1/110,000	1/42,000
Tantalytic	2/678	1/6,500	
Tube sockets	1/2,944	1/56,000	1/56,000
Resistors:			
Composition	5/33,519	1/127,000	
Deposited carbon	7/16,911	1/46,000	1/66,000
Wire wound	3/1,787	1/11,300	
R-f Coils	0/3,542	0/67,500	1/67,500*
Solder joints and wires	88/346,700	1/75,000	1/75,000

* The occurrence of one failure has been taken arbitrarily until more data become available.

on Reliability of Electronic Equipment. This group was delegated the responsibility of studying the causes for unreliability and making recommendations to government and industry on the various ways of improving reliability of equipment. Several committees responsible for the study of some phase of the over-all program were organized. These subcommittees were composed of members from industry and government, and each issued reports of its findings, which have been published. The publications have been of much value in advancing reliability improvement. If additional information is desired, it may be obtained from the Research and Development Board, Office of the Assistant Secretary of Defense, Washington, D.C.

Industry and Government. The Department of Defense conducts planned programs to acquaint the military services and industrial suppliers with reliability requirements. In turn, each of the services has participated in reliability symposia and published bulletins describing methods of achieving better reliability. They have also prepared a series of reliability specifications which are included in government contracts to assure that reliable products are delivered to the government.

The electronics industry has also demonstrated an appreciation of reliability by participating in national symposia sponsored by professional societies such as the American Society for Quality Control. Moreover, many national manufacturers and designers of electronic equipment have established reliability departments to conduct research and develop methods and procedures for improving the reliability of their products. An outstanding example of such a company is the International Telephone and Telegraph Corporation (ITT), which is actively engaged in improving the reliability of its products. Federal Electric Corporation (FEC), as the ITT Service Division, has established the Equipment and Systems Evaluation Division, which evaluates the reliability of ITT products. This is accomplished by gathering reliability data in the field and performing statistical evaluations of operational reliability. These studies have resulted in production changes and modification programs which improved the reliability of existing equipments. The data obtained are also useful for guiding design personnel in developing new products, because these data are representative of the reliability which can be expected if comparable equipment is subjected to similar operational requirements.

As a result of its vast experience with electronic equipment in the field, Federal Electric has tabulated many useful and meaningful statistics. Typical of such data is the tabulation, on page 83, of failure rates which were used in calculating the predicted reliability of a complex radar set.

Vitro Corporation, Silver Spring, Maryland; Rand Corporation, Santa Monica, California; Bell Telephone Laboratories, New York City; and the Radio Corporation of America, Camden, New Jersey, are actively engaged in advancing the science of reliability and have been awarded contracts to study the various facets of reliability and to issue reports based on their findings. As an example, U.S. Air Force contract AF 30(602)-1623, awarded to RCA, resulted in an interim report dated October 15, 1957, entitled Philosophy and Guidelines for Reliability Prediction of Ground Electronics Equipments. This publication takes cognizance of the performance of parts under differing environmental conditions and recommends the range for their most reliable use.

Component part	*Failure rate per 1,000 hr*
Capacitors	0.1517
Resistors	0.2210
Connectors	0.3083
Coils	0.3893
Crystal diodes	0.0750
Motors	0.1410
Relays	0.4623
Transformers	0.2837
Tubes	3.3000

Another publication, prepared under the auspices of the Rome Air Development Center, Griffiss Air Force Base, New York, and published by the McGraw-Hill Book Company, is "Reliability Factors for Ground Electronic Equipment." This book, a pioneering effort in this field, covers such topics as basic reliability concepts, systems aspects, human engineering, mathematical approach, mechanical and environmental factors, and other related subjects.

Publications printed under a government contract usually can be acquired from the Armed Service Technical Information Agency with regional offices as follows:

ASTIA Washington Regional Office, Arlington Hall Station, Arlington 12, Virginia

ASTIA New York Regional Office, 346 Broadway, Room 804, New York 13, New York

ASTIA Dayton Regional Office, Building 275, Area A, Wright-Patterson Air Force Base, Ohio

ASTIA San Francisco Regional Office, Building 1, Wing 2, Oakland Army Terminal, Oakland 14, California

ASTIA Los Angeles Regional Office, Building 1, Room 112, 125 South Grand Avenue, Pasadena, California

From the foregoing discussion, it is apparent there is much literature available to the engineer for reference which should prove of inestimable value to him in improving the reliability aspects of his designs.

7-3. Inherent Reliability Data. The inherent reliability of an equipment is predictable within certain confidence levels when this prediction is based on data obtained from previous experience. This implies that data used for reliability prediction should be collected from supervised sources. Therefore, the methods, procedures, and environmental conditions applicable to these data should be clearly specified in order to assure objective information. In this manner, the design engineer is assured that the data furnished him are accurate and substantial and can be used to evaluate the effectiveness of a projected design. Moreover, the design engineer has a greater confidence in the selection and application of parts if he knows the prevalent conditions under which the data were recorded. Another necessary requirement to assure bona fide data is to assign competent personnel, familiar with proper test methods and having the ability to diagnose failures in accordance with preestablished standards. These individuals should also be capable of differentiating between various types of failures and determining the true cause of trouble. Moreover, the personnel gathering the data should not confuse failures of a primary nature with failures of a secondary nature. The ability to discriminate between causes of failure is a most important attribute of operating personnel. In this manner, good factual information is obtained.

7-4. Nature of Failures. Failures are usually classified as catastrophic (chance), degradation, or wear-out, and they can be further classified as independent or secondary.

Catastrophic (Chance) Failures. These are failures which cause a normally operating system to suddenly become completely inoperative, for example, a blown fuse or a random "open" occurring in a wire resistor after several hundred hours of operation. This type of failure is usually caused by chance and therefore cannot be predicted in advance for a particular time and a specific part. However, if we know the failure rate, we can statistically predict the probability of occurrence of one or more catastrophic failures for any time period.

Degradation (Creeping) Failures. These are the type of failures which occur gradually because of the change of some parameter with time. For example, a decrease in the value of the transconductance of a radio tube will cause a drop in power output. A change in the value of the resistance of a resistor might cause excessive hum or frequency shift in the same receiver. This type of failure may usually be detected in advance by proper techniques. Marginal testing is such a method of detection. By its means it is possible to plan an effective preventive maintenance program to weed out degradation failures.

Physical Wear-out Failures. These are failures which can be predicted on the basis of a known wear-out characteristic and which can therefore be prevented by means of appropriate preventive maintenance. For example, if excessive wear of the brushes of an electric motor is observed, they should be changed during the preventive maintenance routine in order to forestall the occurrence of failure. Another example is excessive wear of an automobile tire. Experience dictates to the average driver that if he observes excessive wear, the tire should be replaced if future blowouts are to be avoided.

Independent Failures. An independent failure may be of the catastrophic, degradation, or wear-out type. It is called independent because it does not occur as a result of the effects generated by other failures. For example, the failure of the door latch of an automobile is not related in any manner to the failure of its radio receiver. Each of these failures is therefore not mutually exclusive.

Secondary Failures. A secondary failure occurs as a result of some primary failure. To illustrate, suppose that a tire on an automobile traveling at high speed blew out and the wheel rim bent as a result; the bent rim would be considered a secondary failure. Or if a resistor in an electronic circuit shorted, causing an excessive drain on a tube, the tube failure would be classified as secondary because it was the result of the primary resistor failure.

7-5. Use-reliability Data. As we have seen in Sec. 7-1, use reliability is associated with those parameters which are not an inherent part of the equipment but whose effects are very important in the over-all assessment of the reliability of a system. The factors that constitute use reliability are difficult to evaluate because they involve intangibles which cannot be expressed quantitatively. As examples, the effects of the capability of operating personnel or of maintenance procedures are difficult to assess. It is necessary, therefore, that the data assessing the contribution of use reliability to operational reliability be interpreted in the light of experience with similar equipment under like conditions of use. This requires competent and experienced personnel if meaningful conclusions are to be expected.

Usually military technicians and their counterparts from industry, such as field technicians and field engineers, have little interest in accurate procedures for gathering data because they consider their primary mission to be to maintain the equipment. Consequently, a program of indoctrination of field personnel is necessary for obtaining meaningful data. This indoctrination program should outline the importance of data as well as describe applicable failure criteria and the methods of recording and reporting failures. Moreover, the entire program should be supervised by responsible representatives of the agency responsible for data analysis and evaluation. Without such a controlled data collection system, any program of reliability assessment will become ineffective, and inaccurate results must be the end product.

Another method of evaluating use reliability is to attempt to simulate field conditions in a laboratory. However, it must be recognized that laboratory data are limited in application, because even the best simulated environmental tests do not consider the various conditions involved, such as the competence, intelligence, and training of maintenance personnel and the adequacy or availability of instruction books.

7-6. Factory Data. Assuming that the methods described in the previous sections result in the accumulation of data which are useful for reliability calculations, it now becomes important to monitor the effectiveness of the factory production setup in order to maintain a reliable product through production. Therefore, methods of gathering reliability data on current production should be specified and implemented throughout the production cycle. It should be pointed out that the same pitfalls that exist in acquiring accurate and complete failure data for evaluating use reliability exist in acquiring factory data, and proper controls to guard against these pitfalls should be developed. This means that test technicians and engineers in the factory should be indoctrinated on the value and importance of accurate data recording and evaluation.

In general, one of the best ways of monitoring reliability is to establish a system of control charting. This is done through the medium of *time sample* control charts. The general methods are similar to those used for plotting control charts in quality control work. The entire subject is explained in Chaps. 10 and 11. However, it is again emphasized that in order for these charts to be of any value, it is imperative that the data be gathered by competent personnel capable not only of effectively making analysis of failures but also of interpreting the chart results.

7-7. Commercial Data. Some commercial organizations are now actively engaged in providing reliable components. Catalogues have been distributed which show parts and components furnished in the form of subassemblies which allegedly have a specified failure rate under particular conditions of usage. These conditions of usage encompass such things as vibration, shock, temperature, humidity, and associated parameters. Such data are useful as a starting point because of the assistance provided to engineers in simplifying their task of designing reliable equipment. This is due to the fact that a considerable part of the work of predicting the inherent reliability has been accomplished in advance.

7-8. Methods of Gathering Reliability Data—Field Data. There are two basic methods of gathering data: (1) at random from various installations or (2) on the basis of controlled programs using data from units functioning under operational conditions in accordance with a fixed routine.

The method used is normally a function of the requirements of a particular contract with the customer. In some studies contracts may require that the data be gathered by using both methods. Experience has shown that it is usually more advantageous to gather data by sampling techniques, because they are more rapid and less costly to the customer. However, sampling experiments cannot always be arranged, and difficulty has been encountered even when the government is the customer. Therefore, other methods, such as simulating field conditions in a laboratory and calculating the resultant reliability, or determining the failure rates of parts or components under simulated conditions of operation and calculating the system reliability, must be relied upon. Simulated experiments, if performed properly, can provide an estimate of operational reliability, but it must be borne in mind that the methods used for simulating actual field conditions are at best a guess. There is no known substitute for the actual thing. Nevertheless, in some instances, the results obtained by simulation methods have often been proven to be more realistic than those based on actual field data gathered through uncontrolled programs. The reason for this situation is that field personnel usually consider the technical aspects of their jobs their primary mission and are not too sympathetic with requests for filling out forms to provide data for reliability calculations. Consequently, these forms often contain incomplete or conflicting information. Field failure data obtained through a controlled program would, however, be superior to laboratory data because the former reflect actual operational conditions. The factors to be considered in choosing the type of program are (1) the degree of assurance required that the data obtained reflect an accurate picture of equipment reliability, (2) the amount of data required, (3) the period of time over which the data must be accumulated, and (4) the relative costs of various programs. When time permits, a controlled field program or a field and laboratory combination program should be used.

7-9. Methods of Recording Reliability Data. The methods of recording reliability data should be engineered to fit the study. The major consideration is that the forms utilized should be designed to provide all the required information.

The Defense Department has standardized on one form which is designed to cover the general situation at many installations. The DD form 787-1, illustrated in Fig. 7-1, is the electronic-failure report form used in the government's Product Improvement Program (PIP). It is used to report the failure of only one part or vacuum tube. Another typical form used for product improvement is shown in Fig. 7-2.

The government furnishes completed DD form 787-1 to all interested military and naval installations and upon request will also supply the equipment manufacturer with the same information. This information is usually supplied in the form of punched-card decks.

Another type of report form is illustrated in Fig. 7-3. This form provides space to record practically all the pertinent details regarding failures

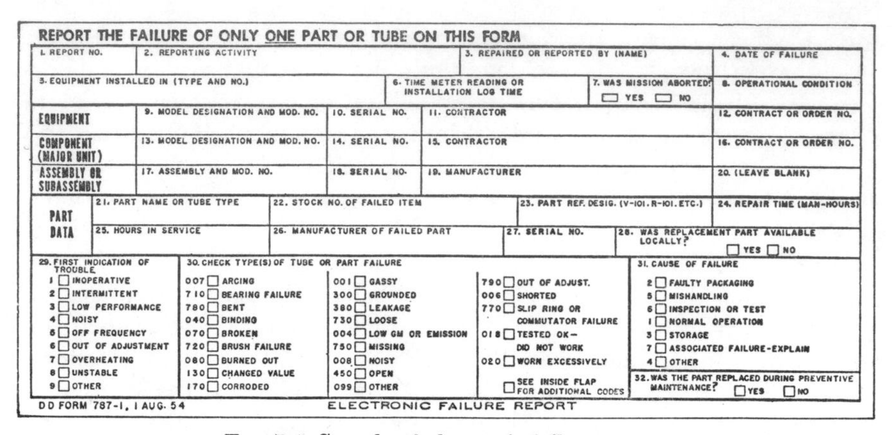

REPORT THE FAILURE OF ONLY ONE PART OR TUBE ON THIS FORM

1. REPORT NO. | 2. REPORTING ACTIVITY | 3. REPAIRED OR REPORTED BY (NAME) | 4. DATE OF FAILURE

5. EQUIPMENT INSTALLED IN (TYPE AND NO.) | 6. TIME METER READING OR INSTALLATION LOG TIME | 7. WAS MISSION ABORTED? ☐ YES ☐ NO | 8. OPERATIONAL CONDITION

EQUIPMENT | 9. MODEL DESIGNATION AND MOD. NO. | 10. SERIAL NO. | 11. CONTRACTOR | 12. CONTRACT OR ORDER NO.

COMPONENT (MAJOR UNIT) | 13. MODEL DESIGNATION AND MOD. NO. | 14. SERIAL NO. | 15. CONTRACTOR | 16. CONTRACT OR ORDER NO.

ASSEMBLY OR SUBASSEMBLY | 17. ASSEMBLY AND MOD. NO. | 18. SERIAL NO. | 19. MANUFACTURER | 20. (LEAVE BLANK)

PART DATA | 21. PART NAME OR TUBE TYPE | 22. STOCK NO. OF FAILED ITEM | 23. PART REF. DESIG. (V-101, R-101, ETC.) | 24. REPAIR TIME (MAN-HOURS)

25. HOURS IN SERVICE | 26. MANUFACTURER OF FAILED PART | 27. SERIAL NO. | 28. WAS REPLACEMENT PART AVAILABLE LOCALLY? ☐ YES ☐ NO

29. FIRST INDICATION OF TROUBLE
1 ☐ INOPERATIVE
2 ☐ INTERMITTENT
3 ☐ LOW PERFORMANCE
4 ☐ NOISY
5 ☐ OFF FREQUENCY
6 ☐ OUT OF ADJUSTMENT
7 ☐ OVERHEATING
8 ☐ UNSTABLE
9 ☐ OTHER

30. CHECK TYPE(S) OF TUBE OR PART FAILURE
007 ☐ ARCING
710 ☐ BEARING FAILURE
780 ☐ BENT
040 ☐ BINDING
070 ☐ BROKEN
720 ☐ BRUSH FAILURE
080 ☐ BURNED OUT
130 ☐ CHANGED VALUE
170 ☐ CORRODED
001 ☐ GASSY
300 ☐ GROUNDED
380 ☐ LEAKAGE
730 ☐ LOOSE
004 ☐ LOW GM OR EMISSION
750 ☐ MISSING
008 ☐ NOISY
450 ☐ OPEN
099 ☐ OTHER
790 ☐ OUT OF ADJUST.
006 ☐ SHORTED
770 ☐ SLIP RING OR COMMUTATOR FAILURE
018 ☐ TESTED OK—DID NOT WORK
020 ☐ WORN EXCESSIVELY
☐ SEE INSIDE FLAP FOR ADDITIONAL CODES

31. CAUSE OF FAILURE
2 ☐ FAULTY PACKAGING
5 ☐ MISHANDLING
6 ☐ INSPECTION OR TEST
1 ☐ NORMAL OPERATION
3 ☐ STORAGE
7 ☐ ASSOCIATED FAILURE-EXPLAIN
4 ☐ OTHER

32. WAS THE PART REPLACED DURING PREVENTIVE MAINTENANCE? ☐ YES ☐ NO

DD FORM 787-1, 1 AUG. 54 ELECTRONIC FAILURE REPORT

Fig. 7-1. Sample of electronic-failure report.

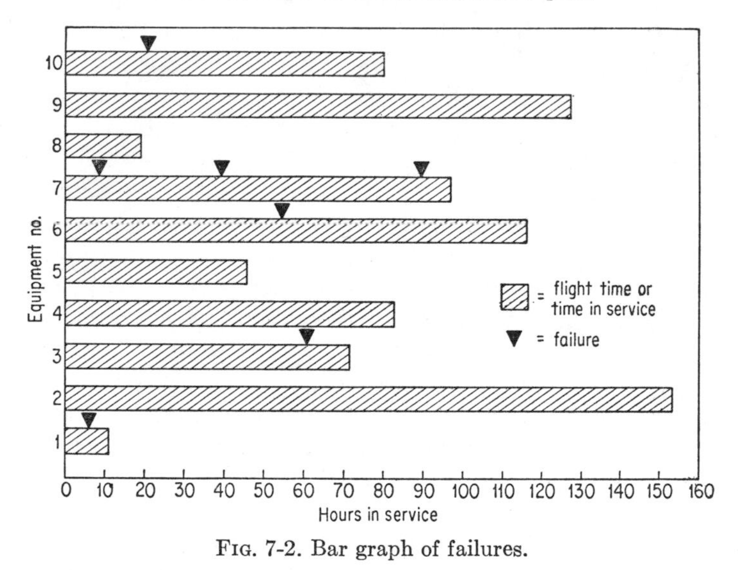

Fig. 7-2. Bar graph of failures.

or preventive maintenance activities. It was developed by Federal Electric Corporation for use on the Dewline project, for which FEC is the maintenance and operation contractor for the Air Force. In addition to providing the necessary data for supplying the United States Air Force with punched-card part failure data, this form also supplies data that can be used for reliability purposes, logistics, operations, engineering investigations, and management control. While this form was developed for a specific project, it illustrates a type which could be adaptable to other similar programs.

DEWLINE C & E REPORT USE NOTATION: ABCDEFGHIJKLMNOPQRSTUVWXYZ 1234567890 DL 161 (REVISED 1-60)

1. | 2. STA. CODE | 3. REPORT NO. Nº 8508 | 4. REPORT TYPE (CIRCLE ONE) MAINTENANCE: PREV. Y, CORRECT N; OTHER: PWR. J, PROP. P, INT. L; CONTROL: FACILITY K, MOD. M | 5. SYSTEM | 6. IDENT. CODE | 7. SERVICE OUT-TROUBLE/PM/MOD-START MO. DAY YR. GMT | BY

8. INITIAL TROUBLE INDICATION (BE SPECIFIC)

9. SERVICE RESTORED MO. DAY GMT | BY | 10. TROUBLE CLEARED-PM COMPL. MO. DAY GMT | BY | 11. ASSIGN/CIRCUIT CODE CODE | 12. ESR NO. | 13. ACTION CODE | 14. RELEASE NO. | 15. PM ROUTINE/MOD. NO.

16. STEP BY STEP ACTION TO RESTORE SERVICE AND CORRECT TROUBLE-IMMEDIATE ACTION CODE A B C D E F G H I J K L M N O (CIRCLE) IF MODIFICATION ENTER REMARKS

17. DECIMAL REPAIR MAN-HOURS | 18. USE CODE
19. LOWEST SUB-ASSEMBLY | 20. S/A MFR
21. SUB-ASSEMBLY SERIAL NO. | 22. NO. PARTS REPLACED
23. DECIMAL DELAY HOURS AWAITING (1) PARTS HRS TENTHS (2) ASSISTANCE HRS TENTHS | 24. METER TIME
25. CHECKED BY—SIGNATURE | DATE

26. REFERENCE DESIG.	27. PART NAME OR TUBE TYPE	28. FED/MFR. STOCK NO./FIG. INDEX	29. MFR	30. AVAIL.	31. TYPE FAILURE	32. CAUSE	33. ROUT. NO.

Fig. 7-3. Dewline maintenance report.

7-10. Desired Characteristics of Field Reports. Field data have limited value for reliability purposes unless they furnish accurate information about the existent situation in sufficient detail to provide a basis for analysis and corrective action.

In general, good field reports provide the information necessary to compute the number of failures occurring within a specific period of equipment operating time. The report should also include, and identify as such, preventive maintenance outages and part replacements. It should contain a clear description of the cause of failure in addition to the type of failure, such as catastrophic, degradation, or wear-out, and classification of the parts failures as independent or secondary. The failed part or component should also be identified by indicating subassembly and circuit designation and the effect of the failure on the subsystem or system. The time to failure and, if possible, the total operating time before failure should also be specified. Completely prepared field reports will summarize the conditions under which the failures occur. In other words, was the ambient temperature high because of lack of ventilation? Was the equipment subjected to extremes of salt air or excessive moisture?

Or were the environmental conditions extremely rigorous, and in what manner?

7-11. Effects and Usages of Reliability Data. The value of accurate reliability data should not be underestimated, since analysis of these data can provide much useful information. These data can tell us how close the equipment comes to meeting its reliability requirements with respect to maintainability, availability, and mean time between failures. The information can also highlight the causes of failure and relegate them to their proper area; i.e., are they due to design, usage, or other factors? From these data it is also possible to determine areas needing improvement, effects of subsystems and interconnecting links on the system reliability, as well as the effects of modifications on reliability. Another important use of a failure analysis is the feedback of practical experience in summary form. This is useful to design engineers working on comparable projects. A realistic evaluation of the preliminary or applicable specification is another benefit of analyzing failure data. By this means it is possible to assess the original concepts and requirements in terms of practical experience. Moreover, a technical analysis of field data provides a realistic means of comparing results with those obtained by simulated tests in the factory. This comparison is useful for assessing the effectiveness of the simulated test program. Thus, if it is found that good correlation exists between simulated and field testing, subsequent production may be evaluated through simulation tests with a corresponding high degree of confidence in the results.

Reliability data are also useful in providing information about logistics, maintenance, and operations. These data can provide a good estimate of spare parts requirements. In many instances in the past, "good practice" was the basis for estimating spare parts requirements. The result was either a shortage of critical parts or much waste of material and useless expenditures of funds for unnecessary spares. With respect to maintenance, reliability data make it possible to estimate the degradation and wear-out characteristics of parts and components. From this information, not only can effective preventive maintenance routines to control frequent trouble areas be developed, but also an estimate can be obtained of the number of maintenance man-hours required to assure a desired level of reliability. These data can influence operations by providing a good estimate of the probability of readiness or availability. This information is useful in calculating the number of equipments necessary at any site to ensure the success of a mission.

7-12. Shortcomings of Data Based on Present-day Field Reporting Systems. Most present-day routine reporting systems are deficient in some respects. The systems seem to lack completeness of detail, particularly when tabulating systems are used to process the data. The reason is that tabulated data do not provide sufficiently detailed information which describes the causes of failure. Nor do they generally include recommendations for corrective action. Their principal advantage is that they highlight repetitive trouble areas and thus pinpoint those factors which require further study, investigation, or analysis.

Experience has shown that the deficiencies of a tabulated field-failure reporting system can be overcome by assigning a resident engineer to the project. The duties of such an engineer are to conduct on-the-spot preliminary analyses, report operational performance data in narrative form, and accumulate supplemental reliability data not provided by the routine reporting system. Most routine data systems thus appear to be valuable for collecting data which are usable more as a reliability control medium, while the method of using specialists assigned to a project to supplement data gained through existing reporting systems is more effective as a reliability design improvement method. A system which provides routine data as well as detailed analysis of failures and recommendations for improvements has been found to be the most practicable and effective. As an illustration, we shall outline a successful field reporting system used to evaluate the operational reliability of the AN/ARN-21 radio set. The AN/ARN-21 is an airborne receiver-transmitter commonly labeled as a "navigational aid" designed to provide distance and azimuth indications to aircraft from identifiable ground or shipboard beacons. Two methods were used for gathering data. The first collected data under actual flight conditions, while the second obtained information from simulated tests. This was one of the first such programs undertaken jointly by a private contractor and the government. The ITT Federal Division, formerly known as Federal Telephone and Radio Company, as prime contractor, was the leader in a leader-follower type of contract. The two followers were other contractors obliged to take directions from the leader.

The Navy authorized the leader to produce 200 AN/ARN-21 sets and each of the follower-contractors to produce 100 sets. These sets were known as the November 1954 series. It was realized, at that time, that these sets were being built essentially in accordance with initial design and that probably much reliability improvement could be made if failure data were accumulated, failure trends analyzed, and corrective action initiated to eliminate deficiencies. Therefore, the Navy authorized the reliability evaluation study. All contractors understood that at a later date all sets of the November 1954 series would be returned from the field to their respective factories to be reworked in accordance with the latest design improvements generated as a result of the data obtained from the reliability evaluation study.

The large initial production was permitted in order to ensure a sufficiently large sample to obtain data not only for reliability evaluation but for any other studies which might be considered necessary.

The first objective was to demonstrate electrical and mechanical interchangeability of the radio sets of all three manufacturers. The leader assured interchangeability of sets by developing an interchangeability specification which was submitted to the Navy for approval. The interchangeability tests were conducted in accordance with this specification and ultimately approved. Once electrical and mechanical interchangeability had been established, the study considered the next problem of evaluating operational and use reliability. To assist in this phase of the program, the Navy made available the services of an experimental squadron in the United States and another in Alaska. The purpose of the squadrons was to fly AN/ARN-21 units under actual tactical conditions to accumulate reliability data under realistic conditions. To complement this effort, a staff of engineers and statisticians from each of the contractors was organized to work with naval personnel at each of the field locations. Operational procedures and data accumulation systems were established for the purpose of systematically gathering reliability data and for issuing periodic reports. The operational procedures assured that failures of the AN/ARN-21 equipment were attributable solely to the airborne receiver-transmitter and not caused by any of the auxiliary equipment. Each failure, when verified, was recorded. The equipment was then repaired and returned to operation.

The plan of accumulating actual flight data was sound, since the AN/ARN-21 was operated under actual field conditions, but the need for this type of equipment was so great that the delivery schedules which were established did not allow the time necessary for further accumulation of data. But field data had been coming in for some time, and practically all the problem areas were becoming evident. Simulated flight testing procedures were initiated at this time in order to permit more rapid analysis and prompt corrective action. It should be pointed out, however, that although the completion of a field study ordinarily takes more time than a simulated test, most of the corrective measures resulting in higher equipment reliability are determined in the early stages of the study. The continuation of the test leads to subtle changes that will not increase the reliability of the equipment by as large a value as does correction of the more apparent weaknesses of the equipment uncovered in the early stages of the test.

The simulated test was organized at a naval development laboratory. The length of the test period was empirically planned to exceed 10 times the estimated mean time between failures. Each failure was verified and recorded. The equipment was repaired after each failure and the reliability test continued. The Navy and contractors staffed this laboratory facility with sufficient equipment, material, and personnel for a complete simulated test program. The airborne receiver-transmitters (AN/ARN-21) were operated at a simulated distance of approximately

100 miles from an AN/URN-3 beacon, which is the ground equipment. The following separate environmental groups were used in the simulated tests:

1. Hot-humid environment (80 per cent relative humidity, 50° centigrade)
2. Hot (50° centigrade)
3. Cold (−40° centigrade)
4. Ambient

Fifteen positions were equipped for group 1 operation, twenty-three for group 2, and six each for groups 3 and 4.

While it was not known whether the simulated program would approximate the various flight environmental conditions encountered in actual use, it was felt that any shortcomings would be uncovered as the data were accumulated and compared with field data.

A joint team of government and industry personnel was used in the simulated program to collect the data in a manner similar to that used in gathering actual flight data. Emphasis was placed on evaluating malfunctions properly to determine whether these actually constituted a failure, because only bona fide failures were recorded. The data from both the simulated and flight tests were compared in order to determine the changes necessary to improve the reliability and to ascertain the degree of correlation between actual and simulated reliability tests.

At various times equipments in different stages of development were introduced into the tests to determine the effects of improvements on the degree of reliability. The series were as follows:

November, 1954. Original Pilot Equipment
March, 1955. Advanced Pilot Equipment
July, 1955. Production Equipment
February, 1956. Latest Production Equipment

The observation unit used throughout the tests was the time between equipment failures. The reliability of the equipment was expressed as the probability that it would survive by performing satisfactorily for a given period of time after being put in service. The theoretical function for this relationship is

$$P_s = e^{-T/m}$$

In this equation, we reiterate, P_s is the probability of survival and m is the mean time between failures. The latter is calculated as the ratio of the total time of a reliability test T to the total number of failures observed during the test. The method of calculating the confidence levels for the mean time between failures is explained in Chap. 8.

The following tabulation shows the number of equipments per group and the total number of hours of operational data accumulated for the various series for both simulated and actual flight tests.

SIMULATION TESTS		
November 1954 series	3 sets	3,000 hr of operation
March 1955 series	3 sets	2,500 hr of operation
July 1955 series	3 sets	2,000 hr of operation
February 1956 series	50 sets	23,000 hr of operation
ACTUAL FLIGHT TESTS		
November 1954 series	26 sets	2,200 hr of operation
March 1955 series	65 sets	5,600 hr of operation

Each time a set failed, the cause of the failure was recorded, and the set was repaired and returned to test. The time required to make the

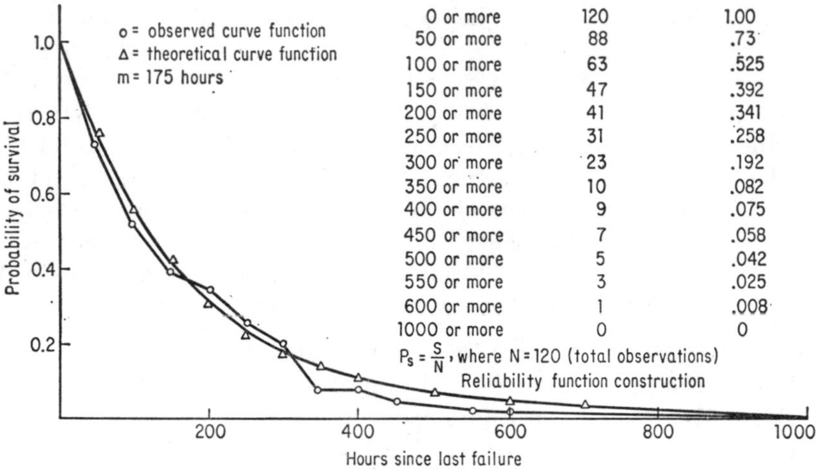

Hours between malfunctions	Number of observations	P_s observed reliability function
0 or more	120	1.00
50 or more	88	.73
100 or more	63	.525
150 or more	47	.392
200 or more	41	.341
250 or more	31	.258
300 or more	23	.192
350 or more	10	.082
400 or more	9	.075
450 or more	7	.058
500 or more	5	.042
550 or more	3	.025
600 or more	1	.008
1000 or more	0	0

$P_s = \frac{S}{N}$, where N = 120 (total observations)

Reliability function construction

o = observed curve function
Δ = theoretical curve function
m = 175 hours

Probability of survival

Hours since last failure

FIG. 7-4. Reliability functions for simulated data versus theoretical curve for February 1956 series.

repair was also recorded. The latter data were used to calculate up-time ratio and spare parts requirements. The analysis of causes of failures and the subsequent corrective action resulted in substantial improvement in reliability. For example, the mean time between failures at the inception of the test for the November 1954 series was approximately 50 hr, and after corrective action was taken, the mean time between failures for the 1956 series was calculated to be 175 hr.

The data for the February 1956 series is shown in Fig. 7-4. Although only 50 sets were involved, the total number of observations was 120, because failures were repaired and the sets were returned to test. During the operational period, this was equivalent to testing 120 sets for a total or aggregate time of 23,000 hr.

To illustrate the method of computations, the data show that 47 sets were still functioning after 150 hr of operation, and since the total number beginning the test can be considered as 120, then P_s is calculated as the ratio of 47 and 120, which is 0.392. The rest of the points for the observed reliability function were determined in a similar manner. By comparison, the theoretical curve based on $m = 175$ hr is also shown in Fig. 7-4, and, as can be seen, the agreement is close. Although there are more elaborate methods for plotting this curve, this was considered sufficiently accurate for this purpose at that time.

As we mentioned previously, throughout the tests the results obtained by simulated flight tests and actual flight data were compared in order to determine whether any correlation existed between the two. As can be seen from Fig. 7-5, the correlation was found to be exceedingly good.

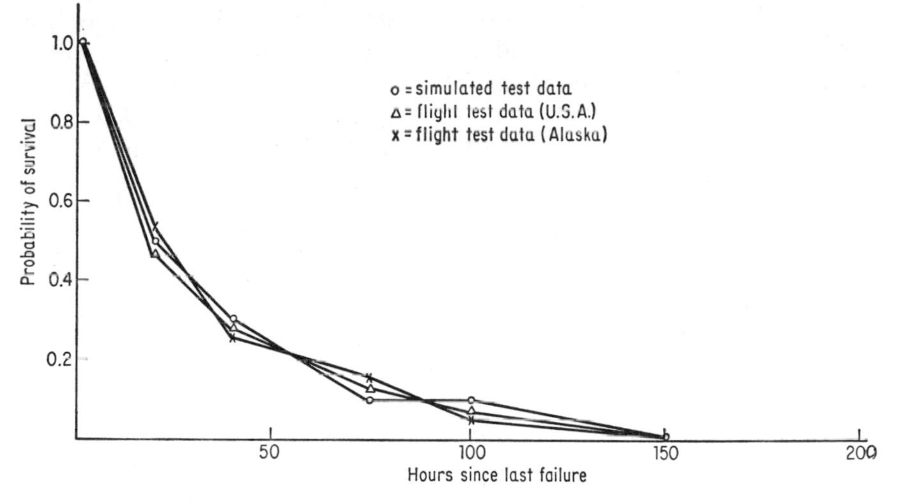

FIG. 7-5. Reliability functions for simulated versus flight test data for March 1955 series.

CHAPTER 8

ANALYSIS OF RELIABILITY DATA

8-1. The Purpose of Reliability Data. The principal reason for gathering reliability data is to use it for making various determinations. Among these are such items as failure rate of parts and components, operational readiness, mission success, reliability, maintenance ratio, serviceability, and maintainability. These terms are defined in App. 1. From these factors it is possible to evaluate the effectiveness of the design and, in cases of inadequacy, recommend improvements. In general, almost any desired information can be gleaned from these data if they are obtained in a planned and organized manner and carefully recorded and collated. However, the methods of analysis must be clearly understood in order to interpret properly the results obtained. This requires some knowledge of basic statistical and analytical concepts and confidence levels. A confidence level is associated with an appreciation of odds, that is, the number of times out of a stated number of trials that an event is expected to occur. In other words, a 90 per cent confidence level means that a particular occurrence is expected to occur 90 times out of 100 trials; or that a specific experiment, if repeated 100 times, will yield a predicted goal 90 times out of 100.

In measuring the mean time between failures of a device, we are also concerned with confidence levels. In this case, if the analysis of data from a reliability experiment of the device indicates that it has a certain mean life, it is necessary that we know the degree of validity of this prediction with respect to certain limiting values. In other words, we must know the degree of confidence associated with the estimate of mean life based on our calculations. The terms *mean time between failures* and *mean life* are often used synonymously and will be so considered in our discussions. We shall use the symbol m to represent the specified or standard value of mean time between failures, and $\bar{m}$ as the estimate of mean life based on observed data.

We shall call our estimate of the mean time between failures the mean life and consequently we shall use the symbol $\bar{m}$ to represent it. This is also referred to as the estimator of the mean. The minimum and maximum values of $\bar{m}$ will be referred to as the confidence limits of the mean; and the probability that a value of $\bar{m}$ will be between these two limits, the *confidence level.*

8-2. Mean Life. The mean life, i.e., estimator of the mean, is given by Eq. (6-7), which, when expressed verbally, is

$$\bar{m} = \frac{\text{total reliability test time}}{\text{total number of failures}}$$

In mathematical language, when the probability distribution of time between failures is exponential, the mean life is

$$\bar{m} = \frac{\sum_{i=1}^{f} t_i + \sum_{f=1}^{s} t_j}{f} \tag{8-1}$$

where f = total number of observed failures
$i = 1, 2, 3, \ldots, f$
t_i = time to failure of the ith observation
s = number of observations terminated without failures (survivors)
$j = 1, 2, 3, \ldots, s$
t_j = time to termination (without failure) of the jth observation

In Eq. (8-1) the numerator is simply the total observed life, while the denominator is the total number of failures. The numerator consists of two components: the uncensored observations, represented by

$$\sum_{i=1}^{f} t_i$$

and the censored observations, represented by

$$\sum_{j=1}^{s} t_j$$

An uncensored observation is one terminating in a failure; i.e., the article under test had definitely failed at the time the test was discontinued or when it was removed from test. A censored observation is one which has been terminated for extraneous reasons other than failure. This is usually the result of a demand for the articles' being tested for other purposes or applications. On this basis, Eq. (8-1) may also be written

$$\bar{m} = \frac{\text{sum of uncensored observations} + \text{sum of censored observations}}{\text{total failures}}$$

8-3. The Chi-square (χ^2) Distribution. The χ^2 distribution is a very effective statistical tool which has a variety of applications, among which is its usefulness in determining the confidence limits of the estimator of the mean. This is what we will be specifically interested in at this time, and as the discussion progresses we will attempt to explain what is meant by confidence limits. For the present, it appears apropos to discuss the nature of χ^2 itself and its relationship to the variance.

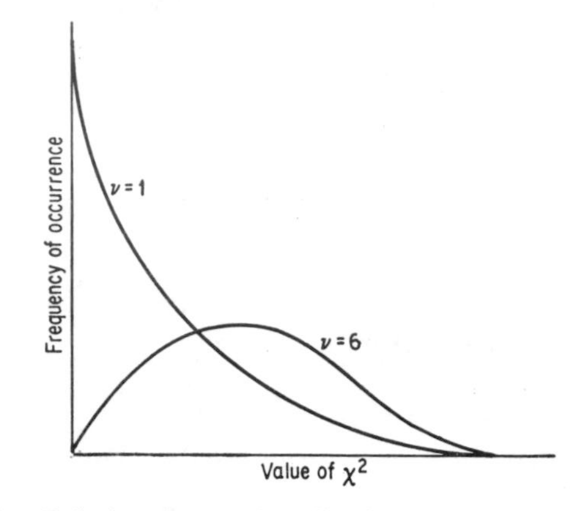

FIG. 8-1. Relative shapes of χ^2 distribution for $\nu = 1$, $\nu = 6$.

Chi square may be defined as the sum of squares of ν independent normal deviates with zero mean and unit standard deviation. Mathematically it can be expressed as

$$\chi^2 = \sum_{1}^{\nu} \left(\frac{x - \bar{x}}{\sigma} \right)^2 \tag{8-1a}$$

The value of χ^2, accordingly, is a function of the sample size n. Thus, for any desired confidence level, the larger the value of n, the greater will be the value of χ^2. Tables of χ^2 show this relationship but make reference to degrees of freedom ν instead of sample size n. In this application the degrees of freedom are calculated as equal to $n - 1$. However, this is not always the case, as we shall see in Sec. 8-5. For the present discussion, it is beyond our scope to discuss degrees of freedom in detail, but if the reader is interested, he may refer to any good text on statistics for additional information.

Figure 8-1 shows the relative shapes of the χ^2 distribution when $\nu = 1$ and $\nu = 6$. It should be noted that for $\nu = 1$ the curve is very unsymmetrical, but for $\nu = 6$ it approaches symmetry. These curves were not drawn accurately to scale but are intended to show only relative shapes. When ν is equal to or greater than 30, the χ^2 curve becomes normal, and therefore we can calculate the value of χ^2 directly by using our knowledge of the areas under the normal curve. The methods of doing this will be discussed in Sec. 8-8.

However, when ν is less than 30, it is necessary to determine the value of χ^2 from a table. Table 8-1 is an abstract of Table 2, App. 7, and shows cumulative values of χ^2 for $\nu = 5$ and various levels of significance α.[1]

TABLE 8-1. ABSTRACT OF TABLE 2, APP. 7

	Probability (level of significance α)													
ν	0.99	0.98	0.95	0.90	0.80	0.70	0.50	0.30	0.20	0.10	0.05	0.02	0.01	0.001
5	0.554	0.752	1.145	1.610	2.343	3.000	4.351	6.064	7.289	9.236	11.070	13.389	15.086	20.517

It is important to note that this table is cumulative. This means that there is a probability α that a value will exceed that shown in the table. For example, when $\alpha = 0.05$ and $\nu = 5$, the corresponding value of 11.070 will probably be exceeded only 5 per cent of the time. This probability is the level of significance previously referred to and is labeled as such in Fig. 8-2. The confidence level is the remaining area under the curve,

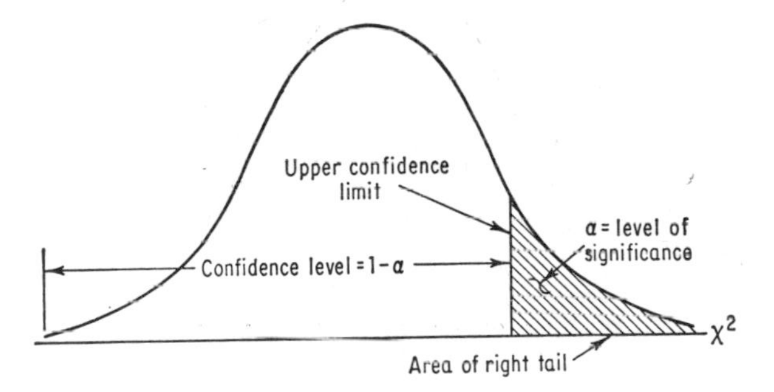

FIG. 8-2. One-tail confidence level (right tail).

that is, $1 - \alpha$. This is logical, since the total area under the curve for a probability distribution must be unity, and if the area in the right tail is α, then the remaining area must be $1 - \alpha$. Figure 8-2 shows the proper relationships. It also demonstrates what is meant by the upper confidence limit, and it is labeled accordingly. This case is known as the one-tail confidence level or interval.

In our discussions, whenever we speak of α, we will be referring to the area of the curve in the right tail, or to the right of the upper confidence limit.

[1] α as used in this application should not be confused with the producer's risk, which is also symbolized by α. This subject is discussed in Chap. 10.

Similarly, Fig. 8-3 shows another example of a one-tail confidence level, except that now we are considering the left tail. In this case, the area in the left tail, or the area to the left of the lower confidence limit, is the level of significance for the left tail and is labeled β. Generally speaking, therefore, we use the term level of significance to indicate the area in the tails and the term confidence level to indicate the area outside of the tails. Thus, to find the confidence level for β, we look up $1 - \beta$ in Table 2.

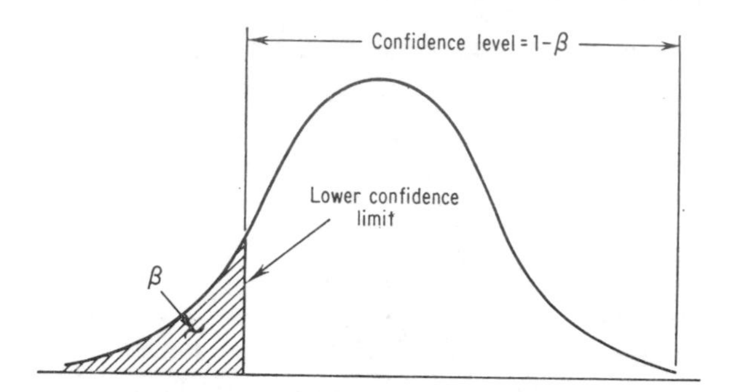

FIG. 8-3. One-tail confidence level (left tail).

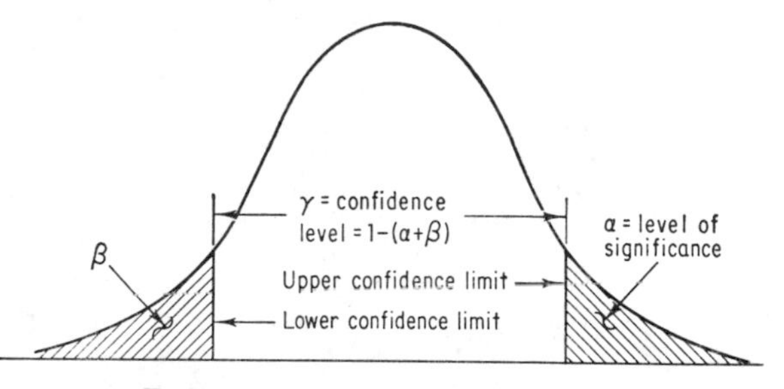

FIG. 8-4. Two-tail confidence levels.

To get two-tail confidence limits, we can combine our knowledge of the one-tail cases, and the result is as illustrated in Fig. 8-4. Thus, it is obvious that the confidence level gamma (γ) is proportional to that area of the total curve lying between the lower and upper confidence limits.

The reason for going through this detailed explanation is to clarify the nomenclature we will use in this text for confidence estimates, since it varies with different authors. For example, in most instances, α is considered the total area outside the confidence limits, and for the two-tail case it is customary to allocate half of α to each tail, and therefore β is never used. This is based upon the assumption that the areas of the two tails are equal, which is not always the case. Therefore, the author has chosen to use α and β to fix the exact areas of each tail, thereby also fixing the confidence level γ.

To summarize we can state that $\gamma = 1 - (\alpha + \beta)$ is the two-tail confidence level, or interval; α is the right-tail area, or area to the right of the upper confidence limit, and is also called the level of significance; and β is the left-tail area, or area to the left of the lower confidence limit. Moreover, it is obvious that when α or β is considered as zero, the two-tail case reduces to the one-tail case accordingly. This makes the use of α and β quite general and applicable in all cases. We can also see that the level of significance may be defined as the probability of a value of χ^2 lying outside the limiting values (i.e., upper or lower confidence limit) represented by the area in the tails. The portion of this area above the high limit is α, and that below the low limit is β.

We will now give some examples of the use of symbolism and also of the use of Table 2. The symbol $\chi^2_{\nu,\alpha}$ means the value of χ^2 with ν degrees of freedom and a level of significance equal to α. Thus, to find the value of $\chi^2_{5,(0.02)}$, we refer to Table 2 or the abstract shown in Table 8-1 and find a value of χ^2 equal to 13.389. This means that 98 per cent of the values of χ^2 when $\nu = 5$ will be equal to or less than this value, or, conversely, that it will be exceeded only 2 per cent of the time. Likewise, ${}_{\beta}\chi^2_{\nu}$ means that value of χ^2 which will probably be less than the lower confidence limit β per cent of the time or which will be exceeded by the lower confidence limit $1 - \beta$ per cent of the time. Therefore, to find ${}_{0.02}\chi^2_5$, we look up the value of $\chi^2_{5,(0.98)}$ and find it to be 0.752. This means that it is expected that 0.752 will be exceeded 98 per cent of the time. Note the technique that the author uses here. When the level of significance α and degrees of freedom ν represent the right tail, they both appear as subscripts to the right, and we have $\chi^2_{\nu,\alpha}$. Likewise, when we are concerned with the left tail, we shift the subscript for the level of significance to the left and write ${}_{\beta}\chi^2_{\nu}$. Also, to shift from the left side to the right, we subtract the value of β from 1, or ${}_{\beta}\chi^2_{\nu} = \chi^2_{\nu,(1-\beta)}$, and look this latter value up in Table 2 for the cumulative χ^2 distribution which gives the lower confidence limit.

Example 8-1. Find the two-tail confidence limits for $\gamma = 90$ per cent; $\alpha = 0.05$; $\nu = 5$.

Solution. Since $\gamma = 1 - (\alpha + \beta)$, then

$$\beta = 1 - (\alpha + \gamma) = 1 - 0.95 = 0.05$$

From Table 2, $\chi^2_{5,(0.05)} = 11.070 = \text{UCL}$ (upper confidence limit) and ${}_{0.05}\chi^2_5 = \chi^2_{5,(0.95)} = 1.145 = \text{LCL}$ (lower confidence limit).

8-4. Use of χ^2 to Establish Confidence Limits for the Variance. Chi square can be used to establish confidence limits of the variance of a population, on the basis of data from a sample of size n. The basic relationship is

$$_{\beta}\hat{\sigma}^2 = \frac{n\sigma^2}{\chi^2_{\nu,\alpha}} \tag{8-2}$$

where $\hat{\sigma}^2$ = estimate of the population variance
σ^2 = variance of the sample
n = number of items in the sample
ν = $(n - 1)$, or degrees of freedom
α = right-tail significance level of χ^2
β = left-tail significance level of $\hat{\sigma}^2$

It must be borne in mind that the significance levels represent the areas in the tails. In the case of $_{\beta}\hat{\sigma}^2$ no degrees of freedom are shown, since this is applicable only to χ^2. Therefore, β in this case represents the left tail of the distribution of the population variances. This means that since it is the left tail, it must define the lower confidence limit. Accordingly, the upper confidence limit would be

$$\hat{\sigma}^2_{\alpha} = \frac{n\sigma^2}{_{\beta}\chi^2_{\nu}} \tag{8-3}$$

In order to illustrate the use of Eqs. (8-2) and (8-3), we shall do an example.

Example 8-2. The calculated variance from the data of a sample of 10 items is $\sigma^2 = 40$. What is the estimate of the population variance which will be exceeded 95 per cent of the time? 5 per cent of the time?

Solution. (*a*) Record the data: $n = 10$; $\nu = 9$; $\sigma^2 = 40$. From Table 2, $\chi^2_{9,(0.05)} = 16.9$. Substituting in Eq. (8-2), we get

$$_{0.05}\hat{\sigma} = \frac{10(40)}{16.9} = 23.6 = \text{lower confidence limit}$$

Since 23.6 is the lower confidence limit, it is probable that it will be exceeded 95 per cent of the time and that it will not be exceeded 5 per cent of the time, which is the value of β, the left-tail area.

(*b*) To get the upper confidence limit, we substitute in Eq. (8-3) from Table 2, $_{0.05}\chi^2_9 = 3.33$ [by looking up value of $\chi^2_{9,(1-0.05)}$ or $\chi^2_{9,(0.95)}$] and $\hat{\sigma}^2_{0.05} = 10(40)/3.33 = 120$ = upper confidence limit.

Thus, it is apparent that since the sum of α and β is 10 per cent, then γ = 90 per cent, which is the confidence level. This assures us that 90 per cent of the time the population variance will be between 23.6 and 120.

The solution of Example 8-2 should point to the advantage of the nomenclature methods which we are using. The important thing to remember is that we can indicate the tail areas we are referring to by placing the subscript representing the level of significance either to the right for the right tail or to the left for the left-tail area. It should also be noted that since the population variance estimate is inversely proportional to the value of χ^2, then when χ^2 is large, $\hat{\sigma}^2$ is small and its level of significance is 1 minus the significance level of χ^2 and vice versa.

Equations (8-2) and (8-3) may also be used to determine what confidence can be placed in a claim that a variance lies between two limits. The next example will be used to demonstrate the application of Eqs. (8-2) and (8-3) in determining this degree of confidence.

Example 8-3. Using the facts of Example 8-2, what confidence can we place in the assumption that the true estimate of the population variance lies between 30 and 150?

Solution. In this case we must solve for $\chi^2_{\nu,\alpha}$ and ${}_\beta\chi^2_\nu$ and look up their values in Table 2. Thus

$$\chi^2_{9,\alpha} = \frac{n\sigma^2}{{}_\beta\hat{\sigma}^2} = \frac{10 \times 40}{30} = 13.3$$

Referring to Table 2 for $\nu = 9$ and $\chi^2 = 13.3$, we find by interpolation that $\alpha = 16$ per cent. Similarly

$${}_\beta\chi^2_9 = \frac{n\sigma^2}{\hat{\sigma}^2_\alpha} = \frac{10 \times 40}{150} = 2.66$$

and from Table 2 for $\nu = 9$ and $\chi^2 = 2.66$, we get $\beta = 2.5$ per cent.

$$\gamma = 1 - (\alpha + \beta) = 1 - (0.16 + 0.025) = 81.5 \text{ per cent}$$

and therefore the confidence level γ is 81.5 per cent, which means that this is the degree of confidence we can repose in the fact that the population variance is between 30 and 150.

8-5. Relationship of χ^2 and Poisson Distributions. A definite relationship exists between the χ^2 and Poisson distributions when R is a whole number. This relationship is expressed as

$$\chi^2_{2R,\alpha} = 2rT \tag{8-4}$$

This equation states that the value of χ^2 corresponding to $2R$ degrees of freedom and a level of significance of α is equal to 2 times the expected number of failures rT. If we make the lower-case letter d equal the expected number of failures, we can rewrite this equation as follows:

$$\chi^2_{2R,\alpha} = 2d \tag{8-5}$$

In order to understand this equation, we must remember that the significance level α, for any statistic, is the probability of exceeding some limiting value. For example, when made applicable to χ^2 as a subscript, it represents the probability of a value of χ^2 for $2R$ degrees of freedom exceeding the value $2d$.

The quality control man will remember that the rejection number R is used extensively in acceptance sampling. It in turn is associated with a significance level commonly referred to as α. In the field of quality control, where attributes sampling is of major importance, we recognize α as what is commonly referred to as the producer's risk (see Chap. 10). It is called this because it represents the probability of rejection of a random sample because it has R or more defects. Likewise, in reliability, reference is made to the producer's reliability risk R_α, because if a random sample of T hours has R or more failures it is rejected. However, if the sample has $R - 1$ or less defects or failures, $R - 1$ is called A, the acceptance number, and in this case the sample is accepted. Thus, we see that a definite boundary or limit for failures is established. If we equal or exceed the R number of failures dictated by the boundary, we reject the sample as nonconforming, and the probability of doing this is R_α. If we have A or less failures, we accept the sample, and the probability of this occurring is $1 - R_\alpha$. Let us see how this works out. Suppose rT equals 2.3; from Fig. 6-1 we see that if the probability of 3 or less failures is 0.80, then the probability of 4 or more failures must be 0.20. This latter probability is R_α, because it represents the area in the tail and fits our past definition. However, this value of R_α is applicable to the curves for the distribution of failures and not to χ^2. The level of significance or the α applicable to χ^2 would be $\alpha = 1 - R_\alpha$. This means that when the significance level for R is small (i.e., R_α is small), the comparable α for χ^2 in Eq. (8-5) must be large. This relationship appears to be logical, because for a given value of d, if R_α is small, then R must be large and the corresponding value of χ^2 should be small; but the value of χ^2 is equal to $2d$, and therefore the probability of this value being exceeded is $1 - R_\alpha$.

In order to make certain that the relationships of Eq. (8-5) are clear, we shall do an example.

Example 8-4. Find the value of $\chi^2_{20,(0.80)}$.

Solution. Solving for R, since

$$2R = 20 \text{ is the degrees of freedom}$$

then $$R = 10 \qquad \text{and} \qquad A = R - 1 = 9$$

But since the level of significance for R is the unity complement of the level of significance of χ^2, this must equal $1 - 0.80 = 0.20 = R_\alpha$. Therefore, this corresponds to a probability of 10 or more failures of 20 per cent or a probability of 9 or less failures of 80 per cent.

From Fig. 6-1, if we follow the line $f = 9$ until it intersects the horizontal line corresponding to a probability of occurrence of 80 per cent (which is the same as the probability of occurrence of 10 or more failures of 20 per cent), we find the value of rT to be 7.3. Since $\chi^2 = 2rT$, the value of $\chi^2_{20,(0.80)}$ therefore is

$$\chi^2_{20,(0.80)} = 2(7.3) = 14.60$$

In order to clarify the relationship between rejection numbers and degrees of freedom for χ^2, let us consider the situation when $R_\alpha = 0.50$. An examination of Fig. 6-1 will show that for this significance level the value of the average number of failures $d = rT$ is equal to the rejection number R. This means that if we selected random samples from a population of d expected failures, 50 per cent of the time we would get R or more failures and 50 per cent of the time we would have A or less failures. Thus, if we knew the average number of failures as the result of a reliability experiment, we would also know R, since $R = d$. From this we could determine the degrees of freedom needed in Eq. (8-5), since $n = 2R$, and determine the confidence limits for the estimator of the mean. We shall cover this subject in Sec. 8-6. It must be remembered, however, that this relationship between d and R is based on the use of integers for R. For example, when $d = 0.5$ we cannot talk in terms of an acceptance or rejection number of 0.5 of a failure unless we do so in a purely abstract sense, as the author does in App. 2 when he discusses his concept of the working acceptance number A_w. However, for whole numbers the relationship between d and R is logical when it is considered that d represents the average or expectation of the Poisson, and therefore half of the total failures occurring in various samples of equal size should be A or less in number, while the other half should be R or more, even though the shape of the Poisson might not be symmetrical.

8-6. Two-tailed Confidence Limits for the Estimator of the Mean. An estimator, as the name implies, is simply an estimate. We ask ourselves about how much confidence we can place in the result. This is no different from evaluating factors in everyday life by considering their source. If the source is reliable, we have a high degree of confidence, and conversely if the source is unreliable, we have little or no confidence. Therefore, the technique which will now be explained is intended to demonstrate the appropriate confidence techniques involved.

Equations (8-2) and (8-3) can be used as a basis for developing the equations for determining the two-tailed limits for the estimator of the mean. In this instance, we will follow the common practice of splitting α equally between the two tails instead of using α and β as we did in Eqs. (8-2) and (8-3). The reason for doing so at this time is that these equations are quite commonplace and the author wishes to avoid confusion for the reader.

As we have seen, α is that portion of the area of the curve which is outside the limiting value. Thus, if we have a lower and an upper value, we can apportion α between the two tails. If we make this apportionment equal, we will thus divide α into two equal parts. Thus, half the area outside the limiting values will be in the right tail, while half will be in the left tail. In other words, we will have $\alpha/2$ in each tail. Figure 8-5 illustrates this case. It also shows that when this is done, the confidence level, which is the area between the limits, must be $1 - 2(\alpha/2) = 1 - \alpha$.

From the foregoing discussion and from a consideration of the facts as described in Sec. 8-5 concerning the relationship between χ^2 and d, we are

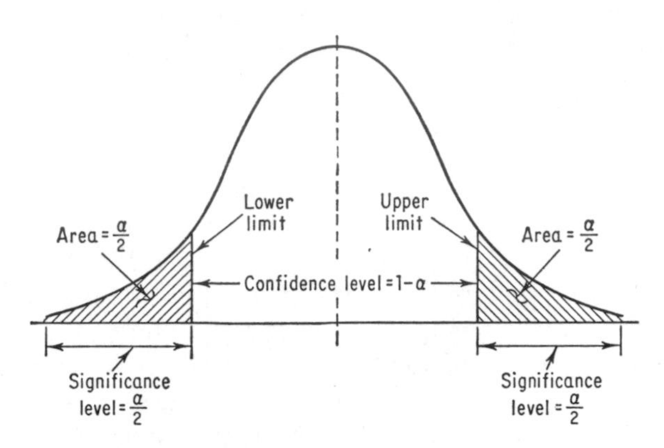

FIG. 8-5. Two-tail confidence level with $\alpha/2$ of the area in each tail.

in a good position to write the final formulas for the two-tailed confidence limits of the estimator of the mean. These formulas are

$$L = \frac{2R\bar{m}}{\chi^2_{2R,(\alpha/2)}} \tag{8-6}$$

$$U = \frac{2R\bar{m}}{\chi^2_{2R(1-\alpha/2)}} \tag{8-7}$$

where L = lower confidence limit
U = upper confidence limit
R = rejection number
α = significance level
$\bar{m}$ = estimator of the mean

An analysis of the above equations reveals that they are practically identical to Eqs. (8-2) and (8-3). This equality is obvious if we realize that in the latter equations we made $n = 2R$ and $\sigma^2 = \bar{m}$ and distributed α between the two tails as $\alpha/2$. The use of these equations will be illustrated by Example 8-5.

Example 8-5. (Data taken from RETMA, Guide for Reporting Reliability.)

The data of the results of an experiment which involved censored observations are presented in Table 8-2. What are the 95 per cent confidence limits for the mean time between failures?

TABLE 8-2. RELIABILITY DATA

Hours to failure	*Hours to withdrawal*
1	2
3	15
10	21
12	29
16	41
20	52
23	65
34	66
36	83
40	112
63	
78	
78	
83	
94	
102	
111	
117	
118	
125	

Solution. The total observed hours (sum of both columns of Table 8-2) = 1,650. Number of failures = 20. Rejection number $R = 20$; $A = R - 1 = 19$; $n = 2R = 40$. $\bar{m} = 1{,}650/20 = 82.5$. From Sec. 8-5 and Fig. 6-1 we can solve for $\chi^2_{2R,(\alpha/2)}$ by looking up the value of rT corresponding to R for a level of significance of $1 - \alpha/2$ which represents a probability of 0.975 of getting 20 or more failures, or a probability of 0.025 of getting 19 or less failures. Referring to Fig. 6-1, we find $d = 29.65$; hence

$$\chi^2_{2R,(\alpha/2)} = 2d = \chi^2_{40,(0.025)} = 2(29.65) = 59.3$$

Similarly

$$\chi^2_{2R,(1-\alpha/2)} = 2d = \chi^2_{40,(0.975)} = 2(12.2) = 24.4$$

Substituting in Eqs. (8-6) and (8-7),

$$L = \frac{2R\bar{m}}{\chi^2_{2R,(\alpha/2)}} = \frac{40(82.5)}{59.3} = 55.6$$

$$U = \frac{2R\bar{m}}{\chi^2_{2R,(1-\alpha/2)}} = \frac{40(82.5)}{24.4} = 135.2$$

8-7. Approximate Methods for Determining Confidence Limits for the Estimator of the Mean. We shall discuss two approximate methods for determining the confidence limits of the estimator of the mean. The first depends upon the use of the Poisson curves shown in Fig. 6-1, while the second utilizes Eq. (1) of App. 2. We shall first consider the former method. As was noted in Example 8-5, the average number of failures d was 20 for a total test time T of 1,650 hr. Assuming that the expectation remains constant, we can also assume that if we randomly selected several samples of 1,650 hr, the number of failures would vary between an upper and lower limit about the mean of 20 failures. The extent of these limits would also be a function of the desired confidence level. Therefore, if we were to divide the total test time by the number of failures at each end of the confidence interval (i.e., the confidence limits), we would get the confidence limits for the mean. Our result would be still more accurate if we decided to use a fictional or working value for the number of failures we observe on the Poisson curve; i.e., if the working value lies between two values of f which are integers, we could visually interpolate from Fig. 6-1 and thus obtain a decimal estimate. Strictly speaking, it is improper to do this, because the number of failures is obviously a whole number. However, by doing so, we very closely approximate the results obtained by using χ^2. We will call our fictional or working value of the acceptance number A_w and the rejection number $R_w = A_w + 1$. [See Eq. (1), App. 2, and Chap. 10 for further application.] This value of A_w may be determined from Fig. 6-1 by visual interpolation or from the method for determining A_w used in App. 2; however, the results obtained by either method are always reasonably close.

Example 8-6. Solve Example 8-5 by the use of the Poisson curves of Fig. 6-1 in lieu of using χ^2.

Solution

$$rT = 20 \qquad \frac{\alpha}{2} = 0.025 \qquad 1 - \frac{\alpha}{2} = 0.975 \qquad T = 1{,}650 \text{ hr}$$

Since $rT = 20$, the rejection number must be 20, and the acceptance number 19. Therefore, we examine Fig. 6-1 for a probability of occurrence of 19 or less failures of 0.975 and 0.025, respectively, in order to get the confidence limits. This is synonymous with saying that the probability of occurrence of 20 or more failures is 0.025 and 0.975, respectively. Therefore, from Fig. 6-1,

$$\text{for } R_\alpha = 0.025 \qquad rT = 12$$
$$\text{for } R_\alpha = 0.975 \qquad rT = 30$$

The limits of the mean can, therefore, be computed by dividing the test time, 1,650 hr, by each of the above values. Thus

$$L \text{ (lower limit)} = \frac{1{,}650}{30} = 55 \text{ hr}$$
$$U \text{ (upper limit)} = \frac{1{,}650}{12} = 137.5 \text{ hr}$$

which result compares favorably with the results obtained by χ^2 in the solution of Example 8-5. If Fig. 6-1 could have been read more accurately by visual methods, the results for both methods would compare exactly.

Example 8-7. Solve Example 8-5 by using Eq. (1) of App. 2.
Solution. Since $\beta = \alpha/2$,

$$Z_\alpha = +1.96$$
$$Z_\beta = -1.96$$

When $d = 20$, the lower limit of d is

$$A_w = d - Z_\beta \sqrt{d} = 20 - 1.96\sqrt{20} = 11.24$$

and
$$U = \frac{1{,}650}{11.24} = 146 = \text{upper limit of the mean}$$

To solve for the lower limit of the mean,

$$A_w = d + Z_\alpha \sqrt{d} = 20 + 1.96\sqrt{20} = 28.76$$
$$L = \frac{1{,}650}{28.76} = 57.5 = \text{lower limit of the mean}$$

It should be noted that the results obtained by the two approximate methods are very close to the exact solution using χ^2. The results from the Poisson curves will always be better than those obtained by Eq. (1) of App. 2, because the equation is only reasonably accurate for higher values of d. In any event, it should not be used when d is smaller than 9 unless a lower confidence estimate is tolerable. In this case, where d equals 20, it can be seen that the results obtained are quite good.

To reiterate, the Poisson curve method would give results as good as χ^2 if it were possible to interpolate to decimal fractions visually. However, since this is impossible, we call it an approximate method of extremely good accuracy.

8-8. Determination of χ^2 for Degrees of Freedom Greater Than 30. It should be noted that Table 2, App. 7, does not go beyond $\nu = 30$. When the degrees of freedom exceed 30, we must therefore resort to other methods for finding χ^2.

It was shown by Fisher that when ν is greater than 30, the $\sqrt{2\chi^2}$ forms a normal distribution with a mean value of $\sqrt{2\nu - 1}$. This is a very handy tool, since now it is possible to find the mean value of χ^2, simply and easily, for any value of ν. For example, if we use the value of $\nu = 30$, the mean value of χ^2 is found as follows:

$$\sqrt{2\chi^2} = \sqrt{2\nu - 1}$$

Substituting, $\sqrt{2\chi^2} = \sqrt{2(30) - 1} = \sqrt{59} = 7.67$

Solving for χ^2, $\chi^2 = \dfrac{7.67 \times 7.67}{2} = 29.4$

For $\nu = 30$ and $\alpha = 0.50$ Table 2 gives, by interpolation, a value of 29.3, which compares very favorably with the calculated value of 29.4.

For the mean value it is obvious that α must be equal to 0.5, since the probability of a value of χ^2 being greater than the mean value is 50 per cent. Thus, it should be possible to determine the value of χ^2 for any probability. The general equation that is applicable is

$$\sqrt{2\chi^2} = \sqrt{2\nu - 1} + Z \tag{8-8}$$

In this equation Z is defined as in App. 2. Equation (8-8) thus becomes a very handy tool. For example, if it is desired to find the value of χ^2 that would be exceeded only 5 per cent of the time when $\nu = 30$, we would proceed as follows: From Table 3, $Z = 1.64$. Substituting in Eq. (8-8), we get

$$\begin{aligned} \sqrt{2\chi^2} &= \sqrt{2(30) - 1} + 1.64 \\ &= 7.67 + 1.64 = 9.31 \\ \chi^2 &= \frac{9.31 \times 9.31}{2} = 43.5 \end{aligned}$$

The actual value as found in Table 3 for the 5 per cent level is 43.7. Therefore, it is apparent that Eq. (8-8) is very effective. The accuracy is still more exact as ν increases beyond 30.

8-9. Approximate Determination of Degrees of Freedom and χ^2 for Any Significance Level of rT. In Sec. 8-5 we explained the method of determining χ^2 for any value of np' or rT and also how to determine the degrees of freedom associated with this value of χ^2. However, in this case we had access to the Poisson curves. Now we shall show how this can be approximately done, without resort to the Poisson curves, by using Eqs. (1) to (3) of App. 2. This technique is particularly important, as we shall see in Chaps. 10 and 11 when we deal with reliability sampling and time samples.

Appendix 2 gives a number of equations. We shall repeat Eqs. (1) to (3) here and label them (8-9) to (8-11), respectively, for easy reference.

$$A_w = rT + Z_\alpha \sqrt{rT} \tag{8-9}$$

$$A = A_w - 0.5 \tag{8-10}$$

$$rT = \frac{2A_w + Z_\alpha^2 - \sqrt{4A_w Z_\alpha^2 + Z_\alpha^4}}{2} \tag{8-11}$$

Example 8-8. Determine the value of χ^2, degrees of freedom, and significance level for the data of Example 8-5 by using the above equations. ($rT = 20$; $\alpha = 0.025$; $1 - \alpha = 0.975$.)

Solution. Refer to App. 2. $Z_\alpha = 1.96$; $Z_\beta = -1.96$.

To obtain degrees of freedom, we must find the value of R. This is done by finding A_w when $\alpha = 0.50$ or $Z_\alpha = 0$ by substituting in Eq. (8-9). Thus

$$A_w = 20 + 0\sqrt{20} = 20$$

$$A = A_w - 0.5 = 19.5 \qquad \therefore A = 19 \text{ to nearest integer}$$

$$R = A + 1 = 19 + 1 = 20$$

$$\nu = 2R = 2(20) = 40 = \text{degrees of freedom}$$

To find $\chi_{40,(0.025)}$, we substitute in Eq. (8-11). In this equation we will use the integer value, i.e., the value $A = 19$ instead of $A_w = 19.5$, because we are making a comparison with Example 8-5 and a comparison of results is desirable. Actually if we used A_w associated with 41 degrees of freedom, which would have resulted had we calculated $2R = 2(19.5 + 1)$, we would have gotten a perfectly acceptable result; but since we are comparing, for integer values we shall use $A = 19$, $R = 20$, $\nu = 40$.

Substituting in (8-12) ($Z_\beta = -1.96$),

$$rT = \frac{2(19) + (1.96)^2 - \sqrt{4(19)(1.96)^2 + (1.96)^4}}{2}$$

$$= \frac{41.93 - 17.4}{2} = \frac{24.53}{2}$$

But

$$\chi^2_{40,(0.975)} = 2rT = 2\left(\frac{24.53}{2}\right) = 24.53$$

which value compares with 24.4 used in Example 8-5.

Similarly, to get $\chi^2_{40,(0.025)}$, we proceed as follows:

$$\chi^2_{40,(0.025)} = 2rT \qquad (Z_\alpha = 1.96)$$

$$rT = \frac{2(19) + (1.96)^2 + \sqrt{4(19)(1.96)^2 + (1.96)^4}}{2}$$

$$= \frac{41.93 + 17.4}{2} = \frac{59.33}{2}$$

Therefore

$$\chi^2_{40,(0.025)} = 2\left(\frac{59.33}{2}\right) = 59.33$$

which equals the value used in the solution of Example 8-5.

These values of χ^2 could also have been obtained by using Eq. (8-8). It is left as an exercise for the reader to solve for χ^2 by the use of this equation and to compare with the above results.

PRACTICE PROBLEMS

8-1. One hundred tubes were tested for 100 hr each under ambient conditions. Ten of these tubes failed after the following hours: 1, 6, 10, 20, 50, 70, 80, 85, 90, 99. What is the mean time between failures? What is the 100-hr reliability? How many censored observations were there? How many uncensored?

8-2. The tubes of Prob. 8-1 were continued on test until all tubes failed. The total accumulated test time was calculated to be 48,286 hr. What is the $\bar{m}$? What are the two-tailed confidence limits when $\alpha = 5$ per cent?

8-3. Find the value of χ^2 when ν equals 50 for a one-tailed confidence level of 90 per cent; a two-tailed confidence level of 90 per cent. Use Eq. (8-9) as a basis.

8-4. A reliability test of 50 units indicates that $\bar{m}$ is 165 hr. The total number of failures at the termination of the test was 25. What confidence can we place in the assumption that the true estimate of m lies between 117 and 275 hr?

8-5. Find the value of $\chi^2_{2R,(0.20)}$ and the degrees of freedom for a value of d equal to 16.

8-6. By the use of Eq. (8-11) determine the value of rT when $A = 5$ and $Z_\alpha = 1.96$. From this result determine the value of χ^2 and its one-tailed confidence level and degrees of freedom.

8-7. Demonstrate by means of Eq. (8-9) that when $rT = 9$, for the special situation when $\alpha = 0.50$, $R = 9$. (Refer to Example 8-8.)

8-8. Use the approximate method demonstrated in Example 8-7 when $\alpha = 0.05$; $1 - \alpha = 0.95$; $\bar{m} = 1{,}200$; $f = 12$ to find the lower and upper confidence limits of $\bar{m}$.

CHAPTER 9

MAINTAINABILITY AND AVAILABILITY

9-1. Definition of Maintainability. Unlike reliability, whose definition is universally accepted, maintainability has still to approach that goal. There are a number of definitions of maintainability presently in use, but there does not seem to be any universal agreement on which one is correct. The Department of Defense defines maintainability as *a quality of the combined features and characteristics of equipment design which permits or enhances the accomplishment of maintenance by personnel of average skill under natural and environmental conditions under which it will operate.*

As can be seen, this definition is certainly broad in concept and therefore of little use to the design engineer. The engineer is not interested in generalities. He desires a definition which is simple and which uses quantitative terms so that he can calculate, measure, and assure an adequate design from the maintainability standpoint. Accordingly, the author would like to suggest a definition which he believes will satisfy the engineer's requirements. Moreover, as the text develops, this definition will be expressed in mathematical terms. The relationship between maintainability and reliability will also be demonstrated.

Maintainability M is the probability that a device will be restored to operational effectiveness within a given period of time when the maintenance action is performed in accordance with prescribed procedures.

It is important to note that this definition refers to maintenance action and not to repair. This is an important distinction, because a device does not always fail to accomplish its mission because of failure requiring a repair action. For example, the power output of a radio transmitter may drop because of drift in the master oscillator. In this case all that is necessary is a slight turn of a dial to tune the circuit, and the transmitter is back to operating efficiency. On the other hand, a screech or hunting of the rotor of a motor may be a sign that lubrication is required, the lack of which would ultimately result in failure. Therefore, the required maintenance action would be to oil the rotor. It is obvious from these examples that the term maintenance action is a general expression which is used to describe any type of maintenance activity, whether it involves a preventive or a repair action. This is a very useful term, as we shall see later, because it can be expressed in a quantitative sense and used in formulas which show the relationship between maintainability and reliability.

As in the case of reliability, maintainability is a probability statistic. The basic difference between the two is that in the case of maintainability we are interested in the probability of restoring a device which has failed or is functioning abnormally to its full operating effectiveness within a period of time, whereas reliability is concerned with the probability of survival of an operating unit with respect to time. We have also learned that the probability of survival is greatest when the mean time between failures is large or the failure rate is small. Conversely, the maintainability is greatest when the mean time of maintenance actions ϕ is small and the maintenance action rate μ is large.

9-2. The Nature and Importance of Maintainability. As of this writing, to the author's knowledge, no simple quantitative definition of maintainability has been agreed upon despite the fact that the term has been widely used. Perhaps the reason for neglecting maintainability is that most of the past effort has been devoted to reliability. However, it is now realized that maintainability is equally important because in many instances the cost of maintenance is prohibitively high. Moreover, even if the cost were tolerable, delays due to equipment malfunction or failure are not desirable, particularly to military personnel, who must plan for an immediate availability of weapons in case of a national emergency.

In general, maintainability may be considered as consisting of two types: (1) preventive and (2) corrective. The purpose of preventive maintenance is, as its name implies, to forestall the occurrence of a failure or malfunction by means of preventive methods such as tuning or adjusting, lubrication, inspection and corrective action, and cleaning. Preventive maintenance may also involve the replacement of marginal parts or components, even though they might still be functioning, because they are suspect as a result of routine checks.

On the other hand, corrective maintenance is undertaken only when it is necessary because of a malfunction or failure. Generally, the corrective action consists in replacing parts and components which have failed or making general repairs such as splicing a broken lead or replacing a worn part.

Both types of maintenance activity result in down time. In the case of scheduled or preventive maintenance, the down time is called *scheduled interruption of service,* whereas in the case of corrective or unscheduled maintenance, it is called *unscheduled or emergency interruption of service.* In any event, regardless of the cause, if we keep a log of down time, we can easily compute the mean time of maintenance action ϕ and the maintenance-action rate μ. The former is calculated as the ratio of the total maintenance-action time to the number of maintenance actions.

$$\phi = \frac{\text{total maintenance-action time in hours}}{\text{number of maintenance actions}} \tag{9-1}$$

The maintenance-action rate is calculated as the reciprocal of the mean time of maintenance action and is expressed in terms of maintenance actions per hour, viz.:

$$\mu = \frac{1}{\phi} \tag{9-2}$$

9-3. Factors Affecting Maintainability. The several factors which affect maintainability may be grouped under the two major headings of design and installation. Typical of those which are related to design are reliability, complexity, interchangeability and replaceability, compatibility, visibility, and configuration. Each of these factors in turn is very complex and is deserving of an individual discussion. This is done to a limited degree in the discussion of reliability design in Chap. 13, to which the reader is referred if he desires more information at this time.

The installation factors generally relate to the human being who is charged with the equipment, as well as with the associated environment. In summary form they may be listed as the experience, training, skill, and supervision of maintenance personnel, as well as the techniques used for maintenance and logistic support. Other factors include environment, publications such as the equipment overhaul and modification schedule, and available test and calibration techniques.

From an analysis of the above factors, it should be possible to make some maintainability predictions. However, since many of the factors listed are intangibles which have not yet been measured in specific numerical terms, it is often impossible to assess their importance to maintainability with any degree of confidence. For example, it would be difficult to objectively and quantitatively evaluate the effects of the experience and training of maintenance personnel on maintainability unless it were done by trial and error, or by means of data analysis. For this reason, a mature design should incorporate elements which make maintainability less dependent on such intangibles. In the next section, we will discuss some of the most outstanding of these elements.

9-4. Methods of Achieving Optimum Maintainability. In general, the number of required maintenance hours can be reduced by using modern techniques to locate or anticipate failures rapidly, speed up corrective or emergency maintenance, and provide for the deferment of immediate repair by the use of interchangeable and replaceable units.

Fault-location and isolation devices are utilized in complex equipment to speed up maintenance. They are generally of two types: (1) built-in test equipment and (2) marginal checking. The value of built-in test equipment lies in the fact that it can be used either for scheduled preventive maintenance or for rapid location of the cause of trouble. On the other hand, marginal checking is used to anticipate failures before they are expected to occur. Both of these methods therefore have as their objective the reduction of the number of maintenance hours, which is a definite advantage because the more hours spent in maintenance, the higher the cost of maintainability.

When using built-in test equipment, the designer should specify reliable test instruments, preferably of the go–no-go variety, since operating conditions usually require qualitative and not quantitative data and since operating personnel are usually interested only in equipment performance and not in data analysis. Therefore, if the parameter being checked is within usable limits, it is satisfactory. This is similar to a radio tube checker or a meter which simply displays on a "good or bad" basis which tube is good or bad without indicating any specific values of emission current on the meter face. Moreover, test equipment must be simple to operate, have a minimum of multipurpose features, and be conveniently located so that it can be readily used for a preventive maintenance schedule. Complicated or bulky test equipment defeats the very purpose for which it was designed. If operating personnel are required to follow an involved test equipment manual or manipulate a large number of multipurpose controls, they cannot do an effective job of maintenance. Under these conditions, it might have been better not to have provided the test equipment at all.

Marginal checking has been very successfully employed on large digital computers and various other devices. This is a special diagnostic program whose purpose is to isolate wear-out failures. Another type of program is called the *automatic recovery program.* This is customarily referred to as a "fix" program because it provides for operation around failure.

Another important attribute of maintainability is accessibility. Adequate accessibility means that parts and components should be readily accessible without the necessity for removing adjacent parts. We have all had the experience at one time or another of repairing one of the various types of modern automobiles and discovering that the accessibility of parts is poor. In most instances, a simple part like a spark plug requires a special tool, because without it the plug is inaccessible. Likewise, such items as the coil and generator are not always easy to reach. On older model cars the battery was located below the floor boards, which had to be completely removed before it became accessible. Present-day portable television sets are another example of devices with poor part accessibility. This is particularly true in the case of vacuum tubes, which require a major job of disassembly of adjacent parts before they can be replaced.

In military equipment, good accessibility is a necessity in order to reduce maintenance time. In designing for accessibility, one should consider all those sequences of operations necessary to effect replacements. The size of the part should not be the only governing factor, since a small part such as a nut can hold up the maintenance job just as much as a large part. Visibility is also an important consideration. Without visibility a part which is physically accessible may not be readily replaced unless a flashlight or a built-in light source is available.

Another important consideration in speeding the maintainability task is described by the terms interchangeability and replaceability. Most military procurement contracts call for interchangeable and replaceable parts. A part may be interchangeable and yet not replaceable, as is particularly true in the case of electrical interchangeability. It is not at all uncommon for two or more electrical components purchased to the same specification to produce different results. Radio tubes are typical illustrations. Although the two parts might appear to be interchangeable when referred to the basic tube specification, they might not be replaceable in the equipment if their outputs when in use are different or if other obstructions prevent the replacement. Likewise, two mechanical assemblies fabricated to the same drawings might be interchangeable with respect to themselves but might not be replaceable because of a clash of tolerances of adjacent parts. Moreover, these adjacent parts might be so located that replacement of the spare part in question is either impracticable or impossible. The designer should give this subject much thought in order to avoid having many parts in stock which are neither interchangeable nor replaceable.

Another technique that has been successfully utilized to decrease maintenance man-hours is that of *unitization of equipment.* Unitized equipment is composed of a series of pluggable parts or components which are easily removed and replaced with working units in case of failure. In this manner, maintainability is enhanced because in the event of a failure, the repair of the defective pluggable unit can be deferred to a later period when more time, tools, and equipment are available. This type of maintenance is ideal because it assures the operation of the equipment without the delay of waiting for repairs. However, in order for this technique to be effective, the design must be mature so that when pluggable units are inserted, they will restore the equipment's function immediately without any special tuning or selection of other pluggable components. In other words, the component must be electrically and mechanically interchangeable or else the method is of little value. As a word of caution, care should be exercised in design to ensure that only the proper components can be plugged into a particular cavity. This is usually accomplished through a series of keying features to ensure against error. The chassis and clamps used to hold the pluggable component in place must also be of such type and construction that the component will clamp easily into place and be removable without the use of tools other than the human hand.

The type, caliber, and training of maintenance personnel is another important factor which affects maintainability. Untrained personnel not only will delay a maintenance action, but, because of their clumsiness or unfamiliarity with the equipment, may actually cause other problems. For this reason, it is recommended that a planned program be prescribed for either preventive or corrective maintenance before these activities are implemented. These programs should spell out the times, tools, manuals, data, and test equipment required, as well as the methods of recording these data and diagnosing trouble. The experience gained and information gleaned from the analysis of data are useful in improving future maintainability. Moreover, if adequate data are fed back to the design engineers responsible for the equipment, many future improvements can be effected. These data may be recorded, studied, and analyzed in a manner similar to that demonstrated in Chap. 8. Usually such analysis pinpoints specific areas of improvement. For example, the analysis might result in a human engineering study which might determine that an individual has been asked to conduct a task which is too difficult for the average human or has been required to exercise too much concentration or attention, which tends to increase the probability of error; or the study might show that an operator cannot adequately view instruments or distinguish their readings, or that the accessibility of controls is cumbersome.

From the preceding discussion, it should be obvious that maintainability consists of many factors, only a partial number of which have been mentioned here. Moreover, the techniques which are employed in assuring maintainability are not always the same. For example, in simple equipment it might be determined that it would be cheaper and more expeditious to replace a failed part or small subassembly rather than repair it. In this case, the very act of replacing the part instead of repairing it is a facet of the maintenance technique. This type of maintenance action requires an adequate supply of spare parts in order to assure its effectiveness.

This means that the failure incidence of parts must be reasonably predicted in order to assure the availability of spares when required. Thus, we can see that maintenance per se is not solely restricted to repair; this is the reason for using the term maintenance action in our definition of maintainability. To reiterate, maintenance action is not necessarily restricted to a repair of the failed part, but rather it is a method or means used to restore or assure the continued operation of the equipment.

Another example of maintenance action is found in large complex equipment where redundancy within the equipment or exterior to the equipment is utilized. Thus, in the case of large computers, by proper programming, one section of the computer may be used to check the operation of another of its sections while the load is switched to still another section of the computer until the initial trouble is isolated and repaired. On the other hand, in the case of radio transmitters, it is a common practice to employ a standby or redundant transmitter in order that the reliability of the broadcast may be assured. In the event of failure, the alternate or redundant transmitter is switched on to carry out the function of the transmitter which failed. The type of maintenance action used in this case is called standby or switchover redundancy.

In general, any type of maintenance action is satisfactory provided that it is fast and economical and that it tends to minimize down time and thus maximize operational reliability (i.e., availability).

9-5. The Maintainability Equation. The maintainability equation is expressed as

$$M = (1 - e^{-\mu t}) \tag{9-3}$$

and since $\mu = 1/\phi$, this may also be written as

$$M = (1 - e^{-t/\phi}) \tag{9-4}$$

In either case, the second term of the parenthesis represents the probability of not performing a repair or maintenance action within the maximum allowable time interval t, and therefore $1 - e^{-\mu t}$ must equal the probability of completing the action in this time. The maximum allowable time t is called the *maintenance time constraint.* It should not be confused with the mission time T, which by comparison with t is very large. The value of t is usually arbitrarily prescribed as a requirement of a mission. In essence, it is the maximum allowable time after a failure occurs during which it is mandatory that a repair or maintenance action be completed. In other words, it is the permissible repair time interval. The number of these time intervals in a mission of duration T hours will therefore be equal to the number of failures or number of incidents requiring a maintenance action. Thus, if the average number of failures in a mission of T hours is rT, the number which cannot be repaired in time t

is $rTe^{-\mu t}$, and the number which can be repaired is $rT(1 - e^{-\mu t})$. This is the basic reason for our labeling the expression $1 - e^{-\mu t}$ maintainability, since it represents the proportion or probability of the number of failures which can be repaired in time t. [See App. 2, Sec. 6*a*.]

A cursory examination of the maintainability equations should convince the reader that they are intuitively correct, because the more maintenance actions which we can complete during a time interval t, the greater should be the maintainability. Equation (9-3) verifies this conclusion. Similarly, the smaller the mean time of a maintenance action, the greater the maintainability. This is shown to be the case by Eq. (9-4).

For military applications the duration of the maintenance time constraint is dictated by the requirements of the mission. These usually specify the proportion of failures in time T which must be repairable in time t. In other words, in order for the mission to be considered successful, a specified proportion of failures must not exceed t time to repair. These requirements may not always be consistent with cost considerations since better maintainability is more expensive. Maintainability costs increase very rapidly as t becomes smaller because more expensive spares and maintenance personnel are required. For example, one method of improving μ is to provide pluggable interchangeable units in lieu of parts replacement, because it is faster to plug in a subassembly than it is to remove a defective part and repair it. However, the cost of the pluggable unit exceeds the cost of an individual part.

When military considerations are not involved, in the interest of economy, it may be desirable to achieve the required maintainability by increasing the maintenance time constraint. This approach may also be utilized for those military operations which are not of a critical nature, such as training and exercise missions. The end effect when t is increased is a greater down or repair time; however, the proportion of failures which are repairable for a required maintainability equals those for a smaller t and a greater μ. Thus we see that maintainability is a function of the product of μ and t, and therefore when these products are equal for different devices, their maintainabilities are also equal.

The following example will illustrate an application of the maintainability equation.

Example 9-1. The data from a reliability test were collated and tabulated below. Calculate ϕ and μ. What is M expressed in per cent for 1 hr? 2 hr? 10 hr? What conclusions can be drawn from the calculated results?

Solution. The raw data were rearranged as shown. Column 1 shows the number of times that a maintenance action occurred, with a duration as indicated in column 2. Column 3 is the product of columns 1 and 2.

The totals for each column are shown. The mean time of maintenance action ϕ is the ratio of the sum of column 3 to the sum of column 1.[1]

$$\phi = \frac{646}{106} = 6.09 \text{ hr}$$

$$\mu = \frac{1}{\phi} = \frac{1}{6.09} = 0.162 \text{ maintenance actions per hour}$$

Substituting in Eq. (9-3), we get

$$\begin{aligned} M\ (1\text{ hr}) &= 1 - e^{-0.162} = 15 \text{ per cent} \\ M\ (2\text{ hr}) &= 1 - e^{-2(0.162)} = 28 \text{ per cent} \\ M\ (10\text{ hr}) &= 1 - e^{-10(0.162)} = 80 \text{ per cent} \end{aligned}$$

The conclusion we can draw is that for a given value of μ, the maintainability increases exponentially with time. Again this appears to be logical, because the greater the time one has available to perform a maintenance action, the greater should be the probability of successfully performing the maintenance activity. Thus, given sufficient time, any device can be restored to its operational effectiveness. However, in actual practice we cannot always afford to give the maintenance man

TABLE 9-1. DATA OF EXAMPLE 9-1*

(1) Frequency of occurrence	(2) Duration of each maintenance action in hours	(3) Product of columns 1 and 2
2	1	2
4	2	8
7	3	21
13	4	52
16	5	80
16	6	96
24	7	168
10	8	80
6	9	54
4	10	40
3	11	33
1	12	12
106 Total number of occurrences	78	646 Total maintenance hours

* These data were rounded off to the nearest hour merely to simplify calculations.

[1] If confidence limits are desired, they can be determined in accordance with the procedure described in Sec. 8-6.

long periods of time to perform his work because this may adversely affect the mission requirements. Therefore, particularly in those cases where the inherent reliability is low, it is most important for us to find the ways and means of improving the maintainability.

Since the maintainability is constant for particular values of t and μ as shown in Eqs. (9-3) and (9-4), it provides us with a good index of the probability of restoring a device to operational effectiveness within a given time constraint. This concept of the time constraint is most important in military tactical operations, because in most instances there is a finite time t during which a specified proportion of malfunctioning or inoperative equipment must be restored to service if the mission is to be accomplished successfully. Thus, as mentioned previously, if we desire to keep M constant, the product of μt must also be a constant. This means that as t becomes smaller, as might be necessary for the requirements of the mission, μ must in turn become larger if the relationship of constant μt is to be maintained. If we make μt or t/ϕ equal a constant b, then we can rewrite Eqs. (9-3) and (9-4) as

$$M = 1 - e^{-b} \tag{9-5}$$

There is a definite advantage to expressing Eq. (9-5) in this form, as we shall see when we study reliability and availability prediction in Chap. 12.

9-6. The Maintainability Increment M_Δ. In the past discussion, we indicated that maintainability is expressed as the probability that when maintenance is performed, a device is restored to operational effectiveness within a time interval t. It is important to realize that the time interval we are referring to is the time measured from the point at which a maintenance action begins to the time specified for its completion. This is the maintenance time constraint t. The total time T is measured from the beginning of the reliability test. The relationship between an interval t and T is very clearly shown in App. 2, Sec. 6, and therefore we shall not amplify this point any further here.

The maintainability increment defines the proportion of failures in time T which will be restored to operational effectiveness in an interval of time t, as a result of maintenance activity. Therefore, as shown in the derivation in App. 2, Sec. 6, the product of the probability of performing one or more maintenance actions in time t and the probability of one or more failures in time T is called the maintainability increment M_Δ. Unlike maintainability, which is solely a function of μ and t, the maintainability increment is a function of μ, r, t, and T. Actually it represents the proportion of unsatisfactory equipment expected to be restored to service within the maintenance time permitted by the maintenance time constraint t. The equation for M_Δ is

$$M_\Delta = 1 - e^{-rT} - e^{-\mu t}(1 - e^{-rT}) \tag{9-6}$$

The calculation of M_Δ for a given set of parameters provides a measure of the per cent of total failures in time T which can be repaired in time t. The next example will demonstrate the method of calculating M_Δ.

Example 9-2. Data from a reliability test of a prototype were evaluated and the following was determined: $r = 0.001$ failures/hr; $\mu = 1$ maintenance action/hr. Calculate M_Δ when $T = 100$ hr and $t = 1$ hr.

Solution

$$\begin{aligned} M_\Delta &= 1 - e^{-rT} - e^{-\mu t}(1 - e^{-rT}) \\ &= 1 - e^{-0.001(100)} - e^{-1}[1 - e^{-0.001(100)}] \\ &= 1 - 0.904 - 0.368(1 - 0.904) = 0.061 \\ &= 6.1 \text{ per cent} \end{aligned}$$

Thus, we see that if we were allowed a 1-hr maintenance time constraint, $M_\Delta = 6.1$ per cent. We shall see how this becomes meaningful when we discuss availability in the next section.

9-7. Equipment and Mission Availability. Equipment and mission availability are respectively the probability that a stated per cent of equipment or missions will provide adequate performance in time T with no down-time interval exceeding the maintenance time constraint. This definition makes no mention of how the availability is achieved, principally because this factor should be of little interest to the ultimate consumer. As an illustration, a telephone subscriber would judge the availability of telephone service on the basis of performance of the handset. If it operated well with little delay for most of the time, he would conclude that he had a high telephone availability. The fact that there might have been failures occurring within the telephone exchange itself would be of little interest, since the subscriber has no knowledge of this occurring. Thus, it is possible that in an automatic dial exchange of the Strowger type, some of the step-by-step switches might fail, but when this occurs, other switches take over to complete the call. This type of switching action is what is referred to as automatic switchover redundancy, a subject which we will discuss later in this chapter.

On the other hand, the subscriber's success in completing his calls, instead of being due to redundancy alone, might also be due to a very high equipment reliability which results in a reduced probability of failure. Thus, we see that availability actually consists of two components: maintainability and reliability. This means that poor reliability can be offset by correspondingly improved maintainability. The faster the maintenance-action rate, the better the resulting availability.

However, there are cases when such rapid maintenance action is not possible, and because of this there will be some down time. If this down time is high, the availability will be poor, whereas if it is low, the availability will be good. For illustrative purposes, let us consider the practical example of an automobile race. Basically, there are two methods for assuring that we approach the 100 per cent availability of a racing car. First, we can theoretically design an automobile of such high reliability that its availability throughout the race will be practically 100 per cent; or on the other hand, we can provide for very rapid maintenance which will also result in a high availability and therefore still make possible the winning of the race. Oftentimes, this latter alternative is the only possible course, since the initial cost of designing high reliability into the racing car might be prohibitive. As an illustration, if a racing car were to get a flat tire and the change to a new tire could be made in seconds, the end effect would be a high degree of availability of the car which should enhance the probability of its winning the race.

As was mentioned previously and from the foregoing illustration, it should be apparent that availability in general consists of two basic factors, i.e., reliability and maintainability. The equipment availability equation which is derived in App. 2, Sec. 6, namely, Eq. (18), is repeated here for ready reference.[1]

$$\Lambda_E = 1 - e^{-\mu t} + e^{-(\mu t + rT)} \tag{9-7}$$

The equation may also be written as

$$\Lambda_E = 1 - e^{-\mu t}(1 - e^{-rT}) \tag{9-7a}$$

and if we make $\mu t = b$ and $rT = d$, we can rewrite Eq. (9-7a) as

$$\Lambda_E = 1 - e^{-b}(1 - e^{-d}) \tag{9-7b}$$

In mathematical language, these equations indicate that the equipment availability at any time T is a function of r, μ, and the maintenance time constraint t.

The derivation of these equations in App. 2, Sec. 6c, involves some degree of mathematical sophistication. However, the equations can be intuitively deduced in other ways. For example, suppose that maintain-

[1] When there is no time constraint for effecting a maintenance action, the reliability of a system at time T is

$$\Lambda = 1 - \frac{r}{r + \mu}[1 - e^{-(r+\mu)T}]$$

and when $T \to \infty$ this becomes $1 - [r/(r + \mu)]$, which is the time availability expression. This equals the UTR (see Sec. 9-10). See R. E. Barlow and L. C. Hunter, Mathematical Models for System Reliability, *The Sylvania Technologist*, vol. 13, January, 1960.

ability were not considered; the probability of survival of a device in time T would be $P_s = e^{-rT}$. However, in Sec. 9-6 we explained that if maintainability were practiced the maintainability increment M_Δ, as shown in Eq. (9-6), would result. Therefore, the equipment availability should equal the sum of P_s and M_Δ. Thus

$$\begin{aligned}\Lambda &= P_s + M_\Delta \\ &= e^{-rT} + 1 - e^{-rT} - e^{-\mu t}(1 - e^{-rT}) \\ &= 1 - e^{-\mu t}(1 - e^{-rT})\end{aligned}$$

The latter equation agrees with availability equation (9-7*a*).

If we examine this availability equation, we should intuitively be able to verify that it is correct, because the term $e^{-\mu t}$ represents the probability of performing zero maintenance actions in time t, and the expression $(1 - e^{-rT})$ represents the probability of failure in time T. Therefore, the product of the two is the probability of not being able to restore the inoperative units to use, and if this probability is subtracted from unity, what remains is Λ_E, the equipment availability.

Although we have been discussing equipment availability up to this point, prior to discussing mission availability, we shall distinguish between the two in a more precise manner.

Equipment availability is the probability that a stated percentage of equipment will be available for use at time T due to the combined effect of the survivors and units restored to service through maintainability in a time equal to, or less than, the time constraint t.

From the above definition it is obvious that the percentage of units which we can repair in time t added to those surviving in time T gives us the equipment availability. If we know μ, r, t, and T, we can calculate the equipment availability by using Eq. (9-7). We can also verify the results by testing actual equipment. If we recorded the per cent of failures occurring in time T which could be repaired in t or less time and added them to the per cent of survivors for time T, the result should be the equipment availability.

A knowledge of equipment availability is very useful because it is a measure of the percentage of equipments which are expected to be in use at time T with no interruption to service exceeding the time interval t. We shall now discuss mission availability and see how it differs with equipment availability.

Mission availability is the probability that a stated percentage of missions of time duration T will not have any failure in any mission which cannot, through maintainability, be restored to service in a time equal to, or less than, the time constraint t.

This means that regardless of the number of failures which occur during the mission time T, we cannot consider the mission successful if any of these failures took more than t time to repair or correct. For example, suppose our mission were 24 hr long and that the average number of failures were 3; then for some missions it would be expected that we would have exactly 3 failures, while for others we would have more or less than 3. However, in any case in order for the mission to be considered successful, the failures for any mission, regardless of their number, would all have to be repaired in t or less time.

The equation for mission availability is

$$\Lambda_M = \exp(-rTe^{-\mu t}) \tag{9-8}$$

This equation is derived in Sec. 6 of App. 2.

It should be noted from Fig. 9-1 that, unlike the equipment availability equation which we discussed previously [Eq. (9-7)], Eq. (9-8) approaches

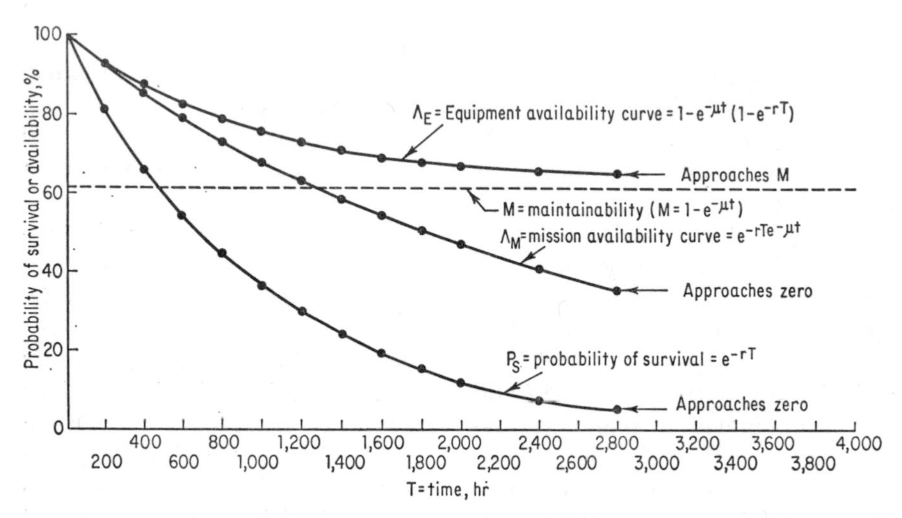

FIG. 9-1. Relationship between Λ and P_s. Availability and probability-of-survival curves. $\mu = 4$, $t = 0.25$ hr, $r = 0.001$ fph.

zero as $T \to \infty$. This should be obvious, since as T gets larger, we should expect more failures for the longer missions, and the probability of restoring all of them in t or less time obviously approaches zero.

The relationship of the availability and reliability characteristics for constant values of $r = 0.001$ fph, $\mu = 6$, and $t = 1/6$ hr is shown in Fig. 9-1.

Table 9-2 lists the data from which the curves of Fig. 9-1 were plotted. The equipment availability Λ_E was calculated by using Eq. (9-7*a*), while the mission availability Λ_M was computed by using Eq. (9-8).

A study of the curves of Fig. 9-1 leads to many interesting conclusions. In the first case, it is obvious that the equipment availability curve approaches the maintainability line as a limit. On the other hand, the mission availability curve approaches zero as a limit. This is also true of the probability-of-survival curve.

TABLE 9-2. DATA FOR PLOTTING Λ AND P_s CURVES SHOWN IN FIG. 9-1

T	$b = \mu t$	$d = rT$	$P_s = e^{-rT}$	$\Lambda_E = 1 - e^{-\mu t}(1 - e^{-rT})$	$\Lambda_M = \exp(-rTe^{-\mu t})$
0	1	0	1	1	1
200	1	0.2	81.9	93.3	92.6
400	1	0.4	67.0	87.9	86.0
600	1	0.6	54.9	83.4	80.3
800	1	0.8	44.9	79.7	74.4
1000	1	1.0	36.8	76.7	69.0
1200	1	1.2	30.1	74.4	64.2
1400	1	1.4	24.7	72.3	59.0
1600	1	1.6	20.1	70.7	55.0
1800	1	1.8	16.5	69.4	51.4
2000	1	2.0	13.5	68.2	47.9
2400	1	2.4	9.1	66.5	41.3
2800	1	2.8	6.1	65.4	35.7

Let us try to interpret these results. The Λ_E curve represents the percentage of equipment which should be available after T hours of operation. When T approaches infinity, the probability of survival approaches zero, as can be seen from the P_s curve. Therefore, any equipment which is available for use after an extended period of time is the direct result of maintainability or the ability to repair or maintain. Obviously, the proportion that can be repaired is $1 - e^{-\mu t}$. This percentage, therefore, becomes the equipment availability as T approaches infinity, which is the steady-state condition. This is tantamount to saying that M is the availability of equipment for large mission times if no preventive maintenance is practiced and all failures are of a catastrophic nature. When scheduled maintenance is practiced, the operating periods between such scheduled maintenance activity may be considered as the mission time. In this case Λ_E may be calculated by direct substitution in Eq. (9-7*a*).

The mission availability Λ_M curve defines that proportion of missions which it is expected will be successful for T hours without any failure which exceeds the maintenance time constraint t. It differs from the equipment availability curve because it is concerned with the percentage of successful missions, whereas in equipment availability one is concerned with the percentage of operating equipments which can be expected at time T, or the percentage of equipment which will be available. It is obvious that as T gets larger, Λ_M approaches zero, since the chance of failure increases with time and the probability of repairing all failures within time t decreases. Under these conditions, since the definition of a successful mission requires that all failures be repaired within t time, the mission availability must also decrease, and as T approaches infinity, Λ_M approaches zero.

In any event, whether we are discussing Λ_E or Λ_M, Fig. 9-1 demonstrates that the effect of maintainability is pronounced, since in each instance the resulting availability is far better than what might have been expected from the effects of reliability alone. This can be seen if one observes the relatively rapid manner in which the P_s curve approaches zero.

In general, when the probability of one or more failures is small, Λ_E and Λ_M are essentially equal. The reason is that if the probability of getting more than one failure is small, then each failure is also tantamount to a failure of the mission, since it is highly improbable that any mission will have more than one failure. Usually when $d = rT$ is less than 0.2, Λ_E and Λ_M are approximately equal. In many instances the agreement is fairly good up to a value of $d < 0.4$ (see Fig. 9-1).

From the foregoing discussion of equipment and mission availability, it is apparent that both of these measures of availability are actually measures of reliability, since they represent the probability of survival when failures which are repaired in t or less time are not considered ultimate failures. In a sense, then, Λ_E and Λ_M are really measures of reliability. This becomes obvious if we make $t = 0$, in which case Eq. (9-8) reduces to $\Lambda_M = e^{-rt}$ and Eq. (9-7a) also becomes $\Lambda_E = e^{-rt}$. We recognize both of these to be the standard probability-of-survival characteristic. Thus we see that the basic difference of Λ_E and Λ_M when compared with P_s is that the former considers the effects of a time constraint $t > 0$, whereas the latter does not recognize any time constraint, and therefore in effect $t = 0$, in which case repair is impossible.

In Sec. 9-10 we shall discuss a measure of availability which is customarily called up-time ratio (UTR). The UTR is the per cent of the operating time that equipment is expected to be operational. In a sense, it is a point estimate that represents the probability that equipment will be functioning for a per cent of the mission time T. For lack of a better term, we shall label UTR *time availability*.

9-8. Scheduled Maintenance. Although for mathematical comparisons we talk in terms of missions approaching infinity, these are not practical considerations. A mission must be finite in time in order to achieve a stated purpose. In most instances, when we consider continuously operating equipment, scheduled maintenance is practiced. In this case, the time interval between scheduled maintenance periods can be considered as being equivalent to a mission. For example, suppose that a computer is scheduled to operate continuously but receives 2 hr of scheduled maintenance every 24 hr; we can then consider the mission time as being composed of 22 hr of operating time and 2 hr of scheduled maintenance time.

The purpose of the scheduled maintenance is to cull and eliminate the wear-out failures. It has no effect on catastrophic failures, which are due purely to chance. Therefore, if our maintenance activity is effective, we can interpret the operating periods as separate and distinct missions because of the refurbishing of the equipment with respect to wear-out failures which occurs during scheduled maintenance. Therefore, if we are still within the period in which a constant failure rate exists, each of the separate missions will be described by the shape of the curves of Fig. 9-1.

In the event that no scheduled maintenance is practiced, the probability of obtaining a specific value of Λ_E will be as shown in Fig. 9-1 for the operating time T.

There are some devices which, by the very nature of their short mission times and operational requirements, cannot be maintained on a scheduled basis. A missile is such a device. Once it is fired, no maintainability by the human hand is possible. The only possible type of maintainability that can be utilized must be designed into the circuitry including such techniques as the use of redundant functional elements or automatic switchover. However, because of space and weight considerations these techniques are usually not employed except on a minimal basis.

On the other hand, as we mentioned previously, when we consider continuously operated equipment our problem is different. In this case, we cannot depend on reliability alone but must also count on maintainability. Theoretically, if we could restore a malfunctioning device to operational effectiveness instantaneously we would have 100 per cent availability, and any lesser degree of maintainability would represent a corresponding decrease in availability. The case of instantaneous restoration represents a high degree of redundancy which is equivalent to a high maintenance-action rate. It is also indicative of a high order of interchangeability and replaceability.

There are two types of unscheduled maintenance generally practiced. The first is called *primary maintenance* and the other is called *secondary maintenance*. Primary maintenance is the technique of performing a maintenance action in minimum time without isolating the specific cause of failure. This is usually done by plugging in a replaceable part or by switching to another component in case of failure. In this manner, any impairment to service is minimized.

Secondary maintenance follows primary maintenance and is practiced on those plug-in units which were removed during the primary maintenance period. Usually, it is a bench operation which is performed at some other location removed from the point of original failure. The maintenance-action rate for secondary maintenance is much less than for primary maintenance because the mean time of maintenance is greater for secondary maintenance.

As an example, it is impracticable to perform secondary maintenance on electronic gear of a high-speed jet fighter while in flight. Therefore, this is done when the plane returns to its base, at which time the defective component is removed and repaired on the bench. As another illustration, if we wanted to minimize the down time of a radio transmitter, we would resort to the technique of replacing malfunctioning pluggable units with working replacements. In this instance, if a failure occurs, we would perform a primary maintenance by immediately plugging an operable unit in place of the one which failed and a secondary maintenance when the defective original is repaired on the bench at a later time.

In electronic gear, it is recommended that the minimum maintainability be restricted to the component or pluggable unit level not at the part level. The principal reason for this maintenance concept is based on statistics which show that about 80 per cent of maintenance time is spent in locating the trouble and 20 per cent in actually making the repair. Therefore, it is not practical to resort to primary maintenance at the part level when a high availability is required because it takes too long a time to complete. On the other hand, if the devices we are using work at intermittent periods with long periods of shutdown or if excessive down time is not important, then we can resort to primary maintenance at the part level during the shutdown period.

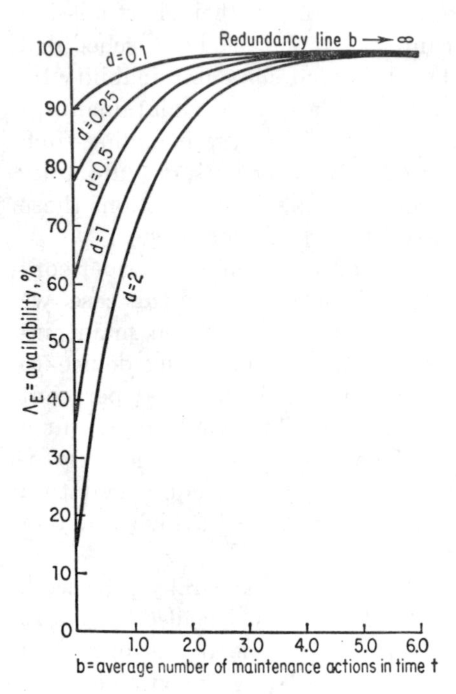

FIG. 9-2. Relationship between Λ_E, b, and d. Range from $\Lambda = 0$ to 100 per cent. Based on equation $\Lambda_E = 1 - e^{-\mu t}(1 - e^{-rT})$. Λ_E = equipment availability.

9-9. Availability as a Function of b and d. Equation (9-7b) shows the relationship between Λ_E, b, and d. Figures 9-2 to 9-4 also show this relationship graphically as a family of curves.

Figures 9-3 and 9-4 represent portions of Fig. 9-2. These figures provide a greater degree of clarity in the ranges of Λ_E from 99 to 100 per cent, since Fig. 9-2 is lacking in detail in this area.

This family of curves is useful for rapidly determining the availability Λ_E at any time T for specific values of μ, r, and t. Or, conversely, if the availability is known corresponding to a specific value of T and a maintenance time constraint of t, it is possible to determine rapidly a set of values of μ and r which will satisfy this value of availability. As can be seen from the curves, there are several possible sets of r and μ which will provide the desired availability. The most optimum values of r and μ should therefore be selected, with all trade-offs being taken into consideration.

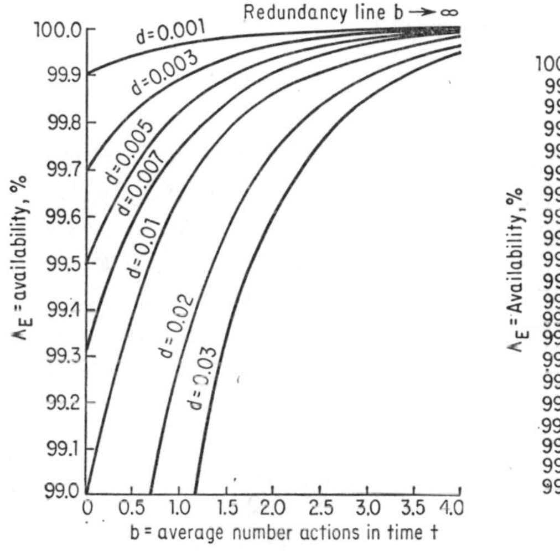

Fig. 9-3. Relationship between Λ_E, b, and d. Range from $\Lambda = 99.0$ to 100 per cent. Based on equation $\Lambda = 1 - e^{-\mu T}(1 - e^{-rT})$. Λ_E = equipment availability.

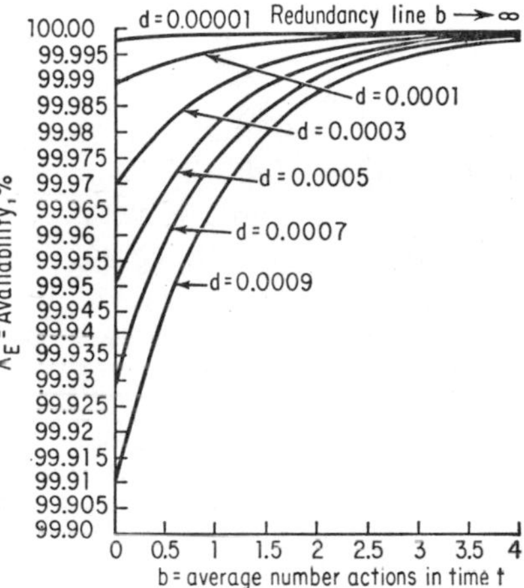

Fig. 9-4. Relationship between Λ_E, b, and d. Range from $\Lambda = 99.90$ to 100 per cent. Based on equation $\Lambda = 1 - e^{-\mu t}(1 - e^{-rT})$. Λ_E = equipment availability.

The principles involved can best be demonstrated by means of an illustrative example.

Example 9-3. Calculate the value of μ and r which is required to assure that a device will have an equipment availability of 99.95 per cent when $T = 100$ hr and $t = 0.25$ hr.

Solution. Refer to Fig. 9-4. Locate $\Lambda_E = 99.95$ per cent on the ordinate, and follow it across until it intersects the curve $d = 0.0007$. Read the value of b on the abscissa as $b = 0.33$.

Since $b = \mu t$, when $t = 0.25$ hr,

$$\mu = \frac{b}{t} = \frac{0.33}{0.25} = 1.32 \text{ maintenance actions per hr}$$

Also, since $d = rT$, when $T = 100$ hr,

$$r = \frac{d}{T} = \frac{0.0007}{100} = 0.000007 \text{ failures/hr}$$
$$= 0.007 \text{ failures/1,000 hr}$$

Another solution would be to use curve $d = 0.0009$. In this case $b = 0.575$, $\mu = 0.575/0.25 = 2.3$ maintenance actions per hr, and

$$r = \frac{d}{T} = \frac{0.0009}{100} = 0.000009 \text{ failures/hr}$$
$$= 0.009 \text{ failures/1,000 hr}$$

Alternative solutions are possible by the use of other curves for the various values of d. The correct solution is the most practical and feasible at minimum cost. In this problem, for example, if the device can be maintained by performing 2.3 maintenance actions per hr and if the cost of designing to the lower reliability of 0.009 failures/1,000 hr is less than is required to design for an r of 0.007 failures/1,000 hr, perhaps the second solution is more desirable, provided that the additional cost of spares does not offset the advantage gained in economy of design.

The solution to the preceding problem by means of the curves represents a short cut over solving it by means of Eq. (9-7b). In this regard, a nomograph has been developed to further simplify solutions of this type by presenting all the information on one form. This nomograph is shown in Fig. 9-5.[1]

Its use is in accordance with accepted practice for the use of nomographs. The left-hand scale indicates the value of d; the center scale Λ_E is availability; and the right scale represents the value of b. By connection of any two points on any two scales by means of a straightedge, the point on the third scale is found to be the point of intersection of the straightedge and the third scale. Thus, if we had used the nomograph to solve Example 9-3, it is obvious that again we could get a series of values of d and b which would satisfy the requirements of the problem. The use of this nomograph is left as an exercise for the reader, since it is obvious that the steps are similar to those used in Example 9-3 after the desired values of b and d are obtained from it.

An interesting sidelight when we use either the curves or the nomograph is to observe that if we set $\mu = 0$, the value of Λ_E which we will read for any desired value of d will be equal to the inherent reliability or proba-

[1] This nomograph has been developed for Λ_E; however, when $d < 0.4$, it can also be used for Λ_M with reasonably good accuracy.

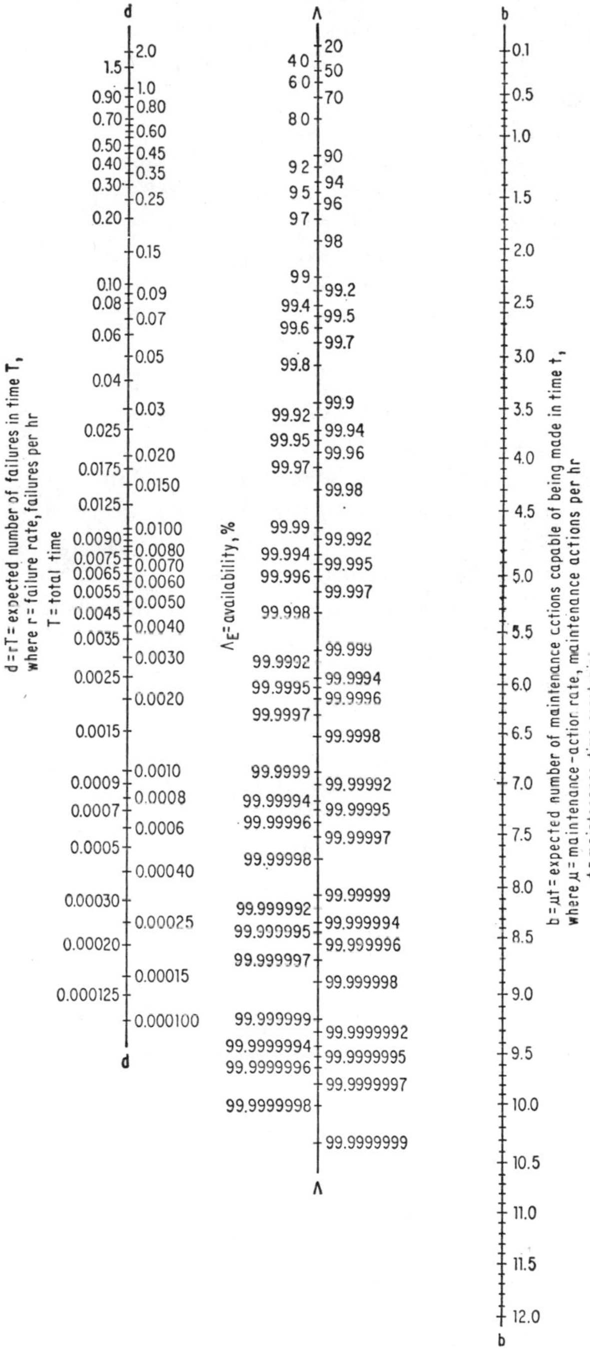

d

1.5, 0.90, 0.70, 0.50, 0.40, 0.30, 0.20, 0.10, 0.08, 0.06, 0.04, 0.025, 0.0175, 0.0125, 0.0090, 0.0075, 0.0065, 0.0055, 0.0045, 0.0035, 0.0025, 0.0015, 0.0009, 0.0007, 0.0005, 0.00030, 0.00020, 0.000125

2.0, 1.0, 0.80, 0.60, 0.45, 0.35, 0.25, 0.15, 0.09, 0.07, 0.05, 0.03, 0.020, 0.0150, 0.0100, 0.0080, 0.0070, 0.0060, 0.0050, 0.0040, 0.0030, 0.0020, 0.0010, 0.0008, 0.0006, 0.00040, 0.00025, 0.00015, 0.000100

d

$d = rT$ = expected number of failures in time T,
where r = failure rate, failures per hr
T = total time

Λ

40, 60, 80, 92, 95, 97, 99, 99.4, 99.6, 99.8, 99.92, 99.95, 99.97, 99.99, 99.994, 99.996, 99.998, 99.9992, 99.9995, 99.9997, 99.9999, 99.99994, 99.99996, 99.99998, 99.999992, 99.999995, 99.999997, 99.999999, 99.9999994, 99.9999996, 99.9999998

20, 50, 70, 90, 94, 96, 98, 99.2, 99.5, 99.7, 99.9, 99.94, 99.96, 99.98, 99.992, 99.995, 99.997, 99.999, 99.9994, 99.9996, 99.9998, 99.99992, 99.99995, 99.99997, 99.99999, 99.999994, 99.999996, 99.999998, 99.9999992, 99.9999995, 99.9999997, 99.9999999

Λ

Λ_E = availability, %

b

0.1, 0.5, 1.0, 1.5, 2.0, 2.5, 3.0, 3.5, 4.0, 4.5, 5.0, 5.5, 6.0, 6.5, 7.0, 7.5, 8.0, 8.5, 9.0, 9.5, 10.0, 10.5, 11.0, 11.5, 12.0

b

$b = \mu t$ = expected number of maintenance actions capable of being made in time t,
where μ = maintenance-action rate, maintenance actions per hr
t = maintenance time constraint

Fig. 9-5. Nomograph showing relationship between Λ_E, d, and b.

bility of survival in time T corresponding to $P_s = e^{-rT}$. This is a logical result, since when $\mu = 0$ there is no maintenance activity and the Λ_E equals the reliability.

9-10. Time Availability (Up-time Ratio). In the preceding sections we have been discussing mission and equipment reliability with respect to a time constraint. That is, we have been concerned with the probability of survival of a mission in time T with no failure requiring more than t hours to repair; or we were concerned with the percentage of failures in time T which could be repaired in t or less time. The advantage of mission availability is that it is a measure of the probability of the successful completion of a mission of a required time duration and time constraint. Likewise, equipment reliability is valuable to the military because it provides an indication of the percentage of weapons which will not fail for a period exceeding the time constraint during a given mission.

There is another measure of availability which is commonly applied to continuously operable maintained systems. This is usually called the up-time ratio, or time availability. The up-time ratio consists of a steady-state component and a transient component. The steady state is merely the ratio of the up, or operable, time to the sum of the up and down, or inoperable, time. If the mean time between failures is considered the up time and the mean time to repair ϕ the down time, the steady-state equation for up-time ratio can be expressed as

$$\text{UTR} = \frac{m}{m + \phi} \tag{9-9}$$

However, since $r = 1/m$ and $\phi = 1/\mu$, it can also be expressed as

$$\text{UTR} = \frac{\mu}{r + \mu} \tag{9-10}$$

Similarly, since the down-time ratio DTR is the ratio of the down time to the sum of the up and down time it may be written as

$$\text{DTR} = \frac{\phi}{m + \phi} \tag{9-11}$$

or

$$\text{DTR} = \frac{r}{r + \mu} \tag{9-12}$$

At the beginning of a mission it is obvious that the probability of an equipment's operating at time T is high. This is due to the contribution of the transient component of the UTR, because there is a smaller probability of failure when T is small than when it is large. It can be shown that the complete expression for the UTR is[1]

$$\text{UTR} = \frac{\mu}{r + \mu} + \frac{r}{T(r + \mu)^2} - \frac{r}{T(r + \mu)^2} \exp [-(r + \mu)T] \quad (9\text{-}13)$$

This represents a measure of the availability of the system, since it gives the probability that it is on at time T. Similarly, the expression for down-time ratio (DTR) is[1]

$$\text{DTR} = \frac{r}{r + \mu} - \frac{r}{T(r + \mu)^2} + \frac{r}{T(r + \mu)^2} \exp [-(r + \mu)T] \quad (9\text{-}14)$$

It should be noted that as $T \rightarrow \infty$ the transient state disappears and the general expression for UTR and DTR reduces to the steady-state condition of Eqs. (9-9) and (9-11).

The UTR and DTR are very useful measures of the availability of maintainable devices at time T. However, in the steady state it should be recognized that the ratios are not time-dependent and, therefore, do not represent the reliability of a system for any particular time interval. Rather, the UTR represents the percentage of time in a continuum of operating time that it is expected the equipment will be in an operable state. Similarly, the down-time ratio is the percentage of time that it is expected the equipment will not be functioning. It is obvious, therefore, that two similar equipments could have the same DTR and yet the down time for one interval could be much greater than the down time for another interval. Therefore, when it is desired to determine the probability of survival over a specific interval of time, it is necessary that the DTR be converted to a failure rate and used in exponential formulas as hitherto explained. We shall illustrate how this is done in the following paragraphs.

It can be shown by simple mathematics that the DTR divided by the mean time to repair is approximately equal to the equivalent failure rate. This can be simply derived as follows:

$$\text{Average number of failures} = \frac{\text{DTR}(T)}{\phi} = rT \quad (9\text{-}15)$$

$$\text{Equivalent failure rate} = \frac{\text{DTR}}{\phi} = r_1 \quad (9\text{-}16)$$

This can also be written in terms of μ as

$$r_1 = \frac{\text{DTR}}{1/\mu} = \mu(\text{DTR}) \quad (9\text{-}17)$$

[1] See R. E. Barlow and L. C. Hunter, Mathematical Models for System Reliability, *The Sylvania Technologist*, vol. 13, January, 1960.

Therefore, from the relationship shown, it is obvious that for a simplex system the product of the DTR and maintenance-action rate is equal to the equivalent failure rate. We use r with 1 as a subscript to indicate the equivalent failure rate for a simplex system. Subsequently we shall use the subscript 2 for a duplex system and subscript 3 for a triplex system, etc.[1] The equivalent failure rate for multiplex systems equals the value of the failure rate which a simplex system should have in order to achieve the same availability as the multiplex system. Therefore, in a sense, the equivalent system can be thought of as a device of greater inherent reliability whose availability equals that of the multiplex system. It is interesting to note, therefore, that if we use reliability in the broad sense, it is achievable either as the result of a high degree of inherent reliability or as a combination of a lower reliability and specified maintainability. The deciding factor with regard to whether it is better to design high inherent availability into a system or resort to maintainable systems as a function of the mission is cost.

9-11. Duplex System Time Availability. In Sec. 9-13 it will be explained that Epstein and Hosford calculated the equivalent failure rate of a duplex system by using certain mathematical techniques which were outlined in their paper entitled "Reliability of Some Two Unit Redundant Systems." The equivalent failure rate which they derived in this paper for a duplex system is

$$r_2 = \frac{2r^2}{3r + \mu} \tag{9-18}$$

It can be shown that through much simpler methods we can attain another equation which is approximately equal to the Epstein equation, which is

$$r_2 = \frac{2\mu r^2}{(r + \mu)^2} \tag{9-19}$$

This equation is obtained by calculating the steady-state down-time ratio of a duplex system and dividing it by the equivalent maintenance-action rate. It can be shown that the equivalent maintenance-action rate is equal to $n\mu$. Although this appears to be a simple relationship, it can be proved by rigorous mathematical methods. Thus, in the case of a duplex system, since n is equal to 2, the equivalent maintenance-action rate $\mu_2 = 2\mu$.

The duplex down-time ratio DTR_2 is the product of the two simplex down-time ratios.

$$\text{DTR}_2 = \left(\frac{r}{r + \mu}\right)^2 = \frac{r^2}{(r + \mu)^2} \tag{9-20}$$

[1] r_1 has also been used for ARL, and r_2 for LTFRD.

But from Eq. (9-17), the duplex failure rate is the product of the duplex DTR_2 and the equivalent duplex maintenance-action rate μ_2. Hence

$$r_2 = \frac{2\mu r^2}{(r + \mu)^2}$$

By this means, it is possible, therefore, to convert from time availability to either equipment or mission availability very simply by substituting the equivalent failure rate in the proper exponential equations. By way of illustration, if the mission availability of a duplex system over time T is desired, we can use the basic mission availability equation (9-8) and substitute the equivalent failure rate r_2. This gives

$$\Lambda_M = \exp\left[\frac{-2\mu r^2 T}{(r + \mu)^2} e^{-2\mu t}\right] \tag{9-21}$$

Similarly other exponential forms can be written by substituting the equivalent failure rate for the simplex failure rate and the equivalent maintenance-action rate for the simplex μ.

9-12. The Combinatorial. In maintainable systems it is often desired to determine the probability that a certain number of units will be operable at time T or that a certain minimum number of units will be operable. The easiest method of achieving this objective is to use some of the techniques that have been described in the previous sections in conjunction with the binomial expansion. For example, in a duplex system, if we arbitrarily make UTR equal Ψ and DTR equal D, the expansion would be as follows:

$$(\Psi + D)^2 = \Psi^2 + 2\Psi D + D^2 = 1 \tag{9-22}$$

but since

$$\Psi = \frac{\mu}{r + \mu} \qquad \text{and} \qquad D = \frac{r}{r + \mu} \tag{9-23}$$

the expression for any time T can be written as

$$\frac{\mu^2}{(r + \mu)^2} + \frac{2\mu r}{(r + \mu)^2} + \frac{r^2}{(r + \mu)^2} = 1 \tag{9-24}$$

The first term of Eq. (9-24) represents the probability of the two equipments being operable, or up, for that percentage of time. The second expression represents the probability that one will be operable and one inoperable (i.e., one up and one down) for that per cent of time; and the third term, the probability that both equipments are inoperable (i.e., both down) for that per cent of time. The sum of the probabilities must be equal to unity.

It should be noted that Eq. (9-24) is dimensionless with respect to time. That is, each of the fractions represents a percentage of time. However, if we are interested in probabilities of survival, we can write the corresponding equations by using the equivalent failure rate. For the duplex case this is shown in Eq. (9-19). A little thought should convince the reader that for the duplex case the probability P_2 of both units surviving for a time T is

$$P_2 = e^{-2rT} \tag{9-25}$$

The probability of one surviving and one failing in time T, P_1 is

$$P_1 = \exp\left[-\frac{2\mu r^2 T}{(r+\mu)^2}\right] - e^{-2rT} \tag{9-26}$$

and the probability of both failing P_0 is

$$P_0 = 1 - \exp\left[-\frac{2\mu r^2 T}{(r+\mu)^2}\right] \tag{9-27}$$

It is obvious, therefore, that by the use of the binomial expansion it is possible to convert from time availability to mission availability or equipment availability, as required. It should also be noted that in small mission times, that is, during the transient state, the time availability and probability of survival as given in prior equations are practically equivalent in value.

It should be pointed out that when the equivalent failure rate is used in the exponential form, we are actually attempting to find out the probability of survival, that is, the probability of not having any failure for an interval of time T. We are not concerned with the steady-state probability over a continuum of time which gives us the percentage of time that the equipment is in operation or the percentage of time that it is down.

As an interesting sidelight, it can be shown that for a continuum of time the number of failures is equal to

$$\frac{T}{m+\phi} \tag{9-28}$$

This can also be determined on the basis of common sense if we consider the sum of the mean time between failures m and the mean time to repair ϕ as constituting one cycle; and thus, the total time divided by the cycle of operation and repair should give us the total number of failures. The equation can also be written

$$\frac{T}{1/r + 1/\mu} = \frac{T\mu r}{\mu + r} \tag{9-29}$$

In effect, this is the equation for a simplex system with repair, the equivalent failure rate being $\mu r/(\mu + r)$. Again we see that the equivalent failure rate is the product of DTR and μ.

The equations developed in this section by means of the DTR are very practical and useful. This is particularly true of the combinatorial and the methods which can be used to convert time availability to mission and equipment availability. However, in the next section we shall introduce other equations for equivalent failure rates which were developed by other methods. In any event, the results obtained by the former and the latter are comparable.

9-13. Redundancy as a Technique of Maintainability. As is probably realized by this time, redundancy is the use of one or more additional elements functionally in parallel, so that if one or more should fail, the remaining redundant elements will carry on the function. In this fashion continuity of operation is preserved and ceases only when all elements become inoperative at the same time.

There are several specific methods of achieving better availability by the use of redundant elements. However, since each presents its own particular problems, we can discuss only general cases at this time and leave particular applications completely in the hands of the design engineer.

Equation (19) of Sec. 6*e*, App. 2 gives a practical derivation of the general redundancy equation for any number of redundant elements. Hence we will refer to it as the n-plex or multiplex redundancy equation. Its chief advantage is that it can be used to solve redundancy problems for any number of redundant elements, such as duplex (two elements), triplex (three elements), or quadruplex (four elements). This is done by substituting the proper value of n for the desired application. This equation is based on the assumption that all redundant elements are functioning at the start of a mission and that if a failure occurs, maintenance action is instituted immediately; and as long as a minimum of one redundant element is working, the mission is considered successful. Failure occurs when all elements are in a state of failure at the same time.

Equation (20) of App. 2, which represents the mission availability of a subsystem of n redundant elements, is repeated here and labeled (9-30).

$$\Lambda_M = \exp\left[\frac{-nr^nT}{n\left(\sum_{i=1}^{n} 1/i\right) r^{n-1} + \mu^{n-1}}\right] \tag{9-30}$$

This particular equation does not provide for a maintenance time constraint.

When a time constraint t is permitted, the general mission availability equation is as shown in Eq. (9-31) below. This equation is identical to Eq. (21) of App. 2 and indicates that if all elements of a redundant system have failed, the system is not considered a failure if at least one of the elements can be restored to operation within the time t.

$$\Lambda_M = \exp\left[\frac{-nr^nTe^{-n\mu t}}{n\left(\sum_{i=1}^{n} 1/i\right) r^{n-1} + \mu^{n-1}}\right] \tag{9-31}$$

The general equation for equipment availability, shown in App. 2 as Eq. (22), is

$$\Lambda_E = 1 - e^{-n\mu t}\left\{1 - \exp\left[\frac{-nr^nT}{n\left(\sum_{i=1}^{n} 1/i\right) r^{n-1} + \mu^{n-1}}\right]\right\} \tag{9-32}$$

Professor Benjamin Epstein, in a paper on two-unit redundant systems, uses a rigorous approach to derive the duplex redundancy case. His techniques involve the solution of differential difference equations by the use of the Laplace transforms. This type of derivation is beyond the scope of this text and will therefore not be explained at this time, since simpler approximate methods can be used such as are shown in App. 2 in Sec. 6*e*.

In Epstein's paper[1] it was pointed out that the equivalent failure rate for a duplex system consisting of two units, both of which are assumed to be operating at the beginning of the mission, is $2r^2/3r + \mu$. Therefore, since availability is expressed as an exponential, we can write the duplex mission availability equation as follows:

Duplex case
$$\Lambda_2 = \exp\left(\frac{-2r^2T}{3r + \mu}\right) \tag{9-33}$$

Similarly, the mission availability expressions for a triplex, quadruplex, and pentaplex system are

Triplex case
$$\Lambda_3 = \exp\left[\frac{-6r^3T}{11r^2 + 7r\mu + 2\mu^2}\right] \tag{9-34}$$

Quadruplex case
$$\Lambda_4 = \exp\left[\frac{-12r^4T}{25r^3 + 23r^2\mu + 13r\mu^2 + 3\mu^3}\right] \tag{9-35}$$

Pentaplex case
$$\Lambda_5 = \exp\left[\frac{-60r^5T}{137r^4 + 163r^3\mu + 137r^2\mu^2 + 60r\mu^3 + 12\mu^4}\right] \tag{9-36}$$

[1] B. Epstein and J. Hosford, "Reliability of Some Two Unit Redundant Systems," Sixth National Symposium on Reliability and Quality Control, January, 1960.

The above equations were derived by solving difference equations by means of Laplace transforms in a manner similar to that used by Epstein for the duplex case. It should be noted that, with the exception of the duplex case, each of these equations contains cross products of r and μ. However, these cross products do not appear in any expression which is derived from Eq. (9-30); but since their effect is negligible, they can be neglected. As an illustration the quadruplex expression can be determined as follows: Substituting in Eq. (9-30), for $n = 4$ we get

$$\Lambda_M = \exp\left[\frac{-4r^4T}{4(25/12)r^3 + \mu^3}\right] = \exp\left(\frac{-12r^4T}{25r^3 + 3\mu^3}\right)$$

Since $$\left(\sum_{1}^{4} \frac{1}{i}\right) = 1 + \frac{1}{2} + \frac{1}{3} + \frac{1}{4} = \frac{25}{12}$$

It should be noted that the approximate expression obtained above is similar to that of Eq. (9-35) minus the cross products $23r^2\mu$ and $13r\mu^2$. The deletion of the cross products is insignificant for all practical purposes, particularly when it is realized that reliability prediction techniques at best are mere approximations since the failure rate data which are used as a basis for computations are never exact.

At this point, it should be interesting to note that, as explained in App. 2, Sec. 6, availability equations are made up of two components. These are the reliability component P_c and the maintainability component M_c of the effective mean time between failures m_n, shown as follows:

$$P_c = \frac{1}{r}\left(1 + \frac{1}{2} + \frac{1}{3} + \cdots + \frac{1}{n}\right) = \frac{1}{r}\left(\sum_{i=1}^{n} \frac{1}{i}\right)$$

and $$M_c = \frac{\mu^{n-1}}{nr^n}$$

This is an important concept because it illustrates the fact that in most practical applications the major contribution to availability is made by the maintainability component and relatively little is due to the reliability component.

Let us illustrate for a duplex system. In this case the reliability component $P_c = 1/r(1 + \frac{1}{2}) = 3/2r = 3m/2$. The maintainability component $M_c = \mu/2r^2 = \mu m^2/2$. The effective mean time between failures m_n is, therefore, the sum of P_c and M_c, viz.: $m_n = 3m/2 + \mu m^2/2$, which may be written in the form $m_n = 3m(1 + m\mu/3)/2$. Therefore, the improvement in reliability as a result of maintainability is $(1 + m\mu/3)$. Or, as we can see, the contribution of the maintainability component to the mean time between failures is $(1 + m\mu/3)$ times greater than that of the reliability component. Therefore, if we neglect the reliability component, approximate equations for availability can be written as follows:

Mission Availability. The approximate version of Eq. (9-31) which neglects the reliability component of availability is

$$\Lambda_M = \exp\left(\frac{-nr^nTe^{-n\mu t}}{\mu^{n-1}}\right) \tag{9-37}$$

Equipment Availability. The approximate version of Eq. (9-32) which also neglects the reliability component of availability is

$$\Lambda_E = 1 - e^{-n\mu t}\left[1 - \exp\left(\frac{-nr^nT}{\mu^{n-1}}\right)\right] \tag{9-38}$$

and if no time constraint is permitted, i.e., $t = 0$, the approximate equation for mission and equipment availability is identical, namely,

$$\Lambda_M = \exp\left(\frac{-nr^nT}{\mu^{n-1}}\right) \tag{9-39}$$

We will now solve some typical examples to illustrate the use of availability equations.

Example 9-4. (*a*) Derive the mission availability equation for a triplex system by substituting in Eq. (9-30).

(*b*) Calculate the availability of this system by using Eq. (9-34) and that derived from Eq. (9-30). Compare the results. (Assume $r = 0.01$ fph, $\mu = 2$, $T = 24$ hr.)

Solution. (*a*) Substituting $n = 3$ in Eq. (9-30),

$$\Lambda_M = \exp\left[\frac{-3r^3T}{3\left(\sum_1^3 1/i\right)r^2 + \mu^2}\right]$$

But
$$\sum_1^3 \frac{1}{i} = 1 + \frac{1}{2} + \frac{1}{3} = \frac{6+3+2}{6} = \frac{11}{6}$$

Substituting,
$$\Lambda_M = \exp\left[\frac{-3r^3T}{3(^{11}\!/_6)r^2 + \mu^2}\right] = \exp\left(\frac{-6r^3T}{11r^2 + 2\mu^2}\right)$$

which agrees with Eq. (9-34) if the cross products are neglected.

(*b*) Substituting $r = 0.01$, $\mu = 2$, and $T = 24$, the availability is

$$\Lambda_M = \exp\left[\frac{-6(0.000001)(24)}{11(0.0001) + 2(2)^2}\right]$$
$$= \exp\left(\frac{-0.000144}{8.0011}\right) = e^{-0.0000179}$$
$$= 0.9999821$$

Comparing with availability obtained by substituting in Eq. (9-34), we get

$$\Lambda_M = \exp\left[\frac{-6(0.000001)(24)}{11(0.0001) + 7(0.01)(2) + 2(2)^2}\right]$$
$$= \exp\left(\frac{-0.000144}{8.1411}\right) = e^{-0.0000177}$$
$$\Lambda_M = 0.9999823$$

As can be seen, there is no appreciable difference in the results and, therefore, for all practical purposes we can use Eq. (9-30) without any reservations.

Example 9-5. Two computers operate in redundancy to assure a high degree of availability. As long as one computer is functioning, the mission is successful. Failure is defined as both computers being inoperative for a period exceeding 5 min (i.e., $t = \frac{1}{12}$ hr). Calculate Λ_M if $\mu = 4$, $r = 0.05$ fph, $T = 24$ hr.

Solution. Substituting $n = 2$ in Eq. (9-31), the duplex availability expression is found to be

$$\Lambda_M = \exp\left(\frac{-2r^2Te^{-2\mu t}}{3r + \mu}\right)$$

Substituting $\mu = 4$, $r = 0.05$, $T = 24$, and $t = \frac{1}{12}$ hr in the above expression, we get

$$\Lambda_M = \exp\left[\frac{-2(0.05)^2(24)e^{-2(4)(\frac{1}{12})}}{3(0.05) + 4}\right] = \exp\left(\frac{-0.12e^{-0.66}}{4.15}\right)$$
$$= e^{-0.0149} = 0.9851 = 98.5 \text{ per cent}$$

which means that the duplex mission availability is 98.5 per cent, or, in other words, 98.5 per cent of the missions will be successful.

As an exercise for the reader, it is suggested that he solve Examples 9-4 and 9-5 by using the approximate equations (9-37) and (9-39) so that he may gain an appreciation of their simplicity and relative worth.

9-14. Various Types of Redundancy. There are two basic types of redundancy, namely, switchover and parallel. Switchover redundancy is used when the device under use fails to perform its function and the alternate unit is switched into service to take its place. If, before switchover, the alternate unit had been in standby in a constantly energized state so that it could be switched into service in a minimum of time, this is called standby switchover redundancy. If the redundant unit had been in readiness but not energized, we simply call this switchover redundancy.

In some instances, standby switchover redundancy is often referred to as priority redundancy. This applies when the standby unit is in use on a mission of less importance than the primary mission. Under these circumstances, the primary mission takes priority over the secondary mission, and therefore the secondary mission is temporarily suspended and its unit is switched over to sustain the primary mission. This is the reason for the use of the term priority redundancy.

In the case of switchover redundancy it can be shown that the expected length of time to reach an inoperative state is $(2r + \mu)/r^2$. This is the equivalent mean time between failures. If we take the reciprocal, we have the equivalent failure rate, which is $r^2/(2r + \mu)$. Figure 9-6 shows the relationship involved. In this case we assume unit A to be in use and unit B to be the standby unit. In the event that A fails, B takes its place by being switched into service and A goes into repair. The switch in this case is considered to be perfect with a failure rate of zero.

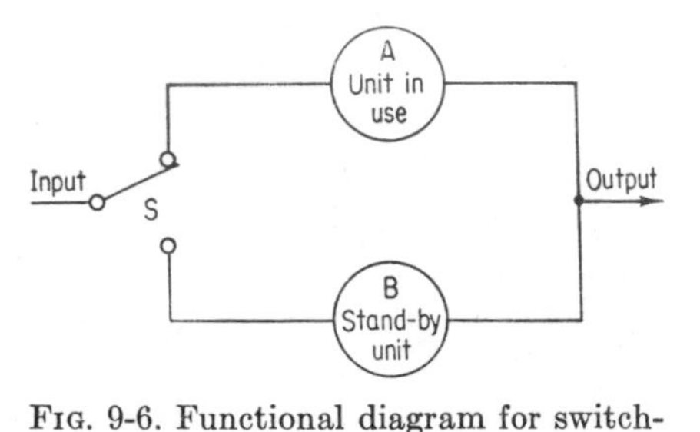

FIG. 9-6. Functional diagram for switchover redundancy.

The mission and equipment availability equation applicable to the system shown in Fig. 9-6 is ($\Lambda_E = \Lambda_M$ when $t = 0$)

$$\Lambda_M = \exp\left(\frac{-r^2T}{2r + \mu}\right) \tag{9-40}$$

and if we introduce a time constraint, the mission availability becomes

$$\Lambda_M = \exp\left(\frac{-r^2Te^{-\mu t}}{2r + \mu}\right) \tag{9-41}$$

and the equipment availability is

$$\Lambda_E = 1 - e^{-\mu t}\left[1 - \exp\left(\frac{-r^2T}{2r + \mu}\right)\right] \tag{9-42}$$

Let us now consider the parallel reliability. In this case, both units A and B are working at the outset. In the event of failure of one, it goes into repair while the other continues its function. When both fail the system is considered a failure. In this instance, from Eq. (9-33) and from the solution of Example 9-5, we find that the failure rate is $2r^2/3r + \mu$, and therefore the mission and equipment availability when $t = 0$ is

$$\Lambda = \exp\left(\frac{-2r^2T}{3r + \mu}\right) \tag{9-43}$$

and if we introduce a time constraint,

$$\Lambda_M = \exp\left(\frac{-2r^2Te^{-2\mu t}}{3r + \mu}\right) \tag{9-44}$$

and

$$\Lambda_E = 1 - e^{-2\mu t}\left[1 - \exp\left(\frac{-2r^2T}{3r + \mu}\right)\right] \tag{9-45}$$

In general, it is more realistic to consider the transfer from one redundant element to the other in terms of a switching action rather than in terms of a simple switch operation. The reason is that in a complex subsystem, the act of transferring from one device to another is more complicated than merely pressing a switch. Thus, suppose we have two redundant transmitters, one in service and the other in standby (unenergized). When the first one fails, in order to put the redundant unit into service, it is usually necessary to go through a series of steps such as switching the power on, tuning various circuits, and making certain adjustments. The more complex the component, the longer the required switching action time. A computer, for example, usually requires more time to switch into service than does a radio transmitter. In a sense we can consider the act of switching a complex redundant unit into service as a form of primary maintenance, since we are depending upon the redundant unit to sustain the operation in the event of a failure. In this case, the actual failure rate of the switch for all practical purposes can be assumed to be zero. Therefore, the switching action depends upon the various steps required to put the redundant unit into service. If the average time required to effect this switchover is determined, it can be considered as the mean time to repair for primary maintenance and, therefore, its reciprocal is μ, the maintenance-action rate. On this basis this value of μ can be used in Eqs. (9-40) to (9-42), which are applicable for switchover redundancy.

9-15. Determination of Spares Requirements. The number of spares required to ensure that parts are always on hand to maintain the equipment properly and effectively is of major importance. Therefore, the determination of the quantity of spares should be made in a scientific manner. If too few are on hand when urgently needed, this lack of spares may adversely affect the mission. On the other hand, if too many are provided, this represents an unnecessary expenditure of funds, storage space, and, in some instances, critical material.

In the early days of World War II there was no scientific method of determining spares requirements. Consequently, some blanket rule was adopted, such as requiring that an additional quantity of 20 per cent of all parts be ordered as spares. Another practice for major equipments was to order spare-parts kits with at least one of each part contained therein as a minimum quantity. Both of these practices were very inefficient and costly. In the first instance, the blanket 20 per cent spares rule led to many ridiculous situations. For example, in one instance an additional quantity of 20 per cent spare chassis were ordered for a complex radio receiver. It should have been apparent that it was ridiculous to order spare chassis, since these could not be replaced without rebuilding the entire radio set. Hence, none of the spare chassis were ever used, and therefore they represented a waste of critical materials which were on the wartime priority list.

In the second case, while many spare-parts kits assigned to specific equipment reposed in depots for years containing parts which were urgently needed in critical equipment, these parts were not available for issue because they had been predesignated for other equipment. Regulations at that time would not permit the dismemberment of spares kits to supply shortages for other equipment. Consequently, many of these kits found themselves in the surplus property disposal category at the conclusion of the war. This unscientific spares practice resulted in heavy monetary loss to the government.

Today, with our knowledge of reliability principles it is possible to forecast spare-parts requirements in a more scientific manner. Basically, the simplest method is to divide the expected life or mission requirement by the mean time between failures. This should provide us with the average number of spares required. However, the average itself is not always adequate, because there is a definite probability that more than the average number of spares might be required for a specific period. Therefore, more sophisticated methods should be used to calculate spares requirements. The ultimate aim is to provide a sufficient quantity of spares to ensure against shortages at critical times. This means that our estimates must be realistic but not overly optimistic. As a matter of fact, in the interests of economy we might be willing to take some risk of "stockout" at very infrequent periods. Therefore, we should always associate a confidence level with our estimate of the required number of parts. There are many sophisticated mathematical techniques which could be used to make this determination of the proper number of spares to satisfy the confidence level, but since the author prefers simplified methods when they are practical and sound, we can resort to Eq. (1) of App. 2 for a simple, rapid method for solving this type of problem. Let us solve a hypothetical example to see how this method works. (This method is applicable only to spare parts which are not repairable after failure and which are discarded.)

Example 9-6. A hypothetical subsystem consists of 10 tube types and 500 other parts. Assume for each of the tubes an equal failure rate of $r_T = 0.001$ failures/hr, and for simplicity assign to each of the other parts an equal failure rate r_p of 0.0005 failures/hr. The equipment is in use for 23 hr per day with a 1-hr scheduled maintenance. What quantity of depot spares is necessary for tubes and other parts to satisfy a 99.73 confidence level that there will be no stockout over a 2-year period? What quantity is necessary as running spares for a 30-day supply?

Solution. The total number of operating hours in 2 years is

$$2 \times 365 \times 23 = 16{,}790$$

The average number of tube failures per type is

$$r_T T = 0.001 \times 16{,}790 = 16.790$$

and for the other parts we get

$$r_P T = 0.0005 \times 16{,}790 = 8.395$$

From Eq. (1), App. 2, for a 99.73 per cent confidence level $Z = 3$, we get

$$A_w = r_T T + 3\sqrt{r_T T} = 16.790 + 3\sqrt{16.790} = 29.06$$

Thus, we must provide 29 tubes of each type as depot spares, or a total of $29 \times 10 = 290$ tubes of all types.

Similarly, to determine the number of spare parts, we have

$$A_w = r_P T + 3\sqrt{r_P T} = 8.395 + 3\sqrt{8.395} = 17.095$$

Therefore, the number of spares per part is 17, but since there are 500 parts of equal failure rate, the total number of spare parts $= 500 \times 17 = 8{,}500$.

The running spares are calculated in a similar manner. In this case, for 30 days, T in hours is

$$30 \times 23 = 690 \text{ hr}$$
$$r_T T = 0.001 \times 690 = 0.690$$
$$r_P T = 0.0005 \times 690 = 0.345$$

A_w for tubes is equal to

$$0.690 + 3\sqrt{0.690} = 0.690 + 2.48 = 3.17$$

This means that we must provide 3 tubes per type as spares for each 30-day period, or a total of $3 \times 10 = 30$ tubes.

A_w for parts equals

$$0.345 + 3\sqrt{0.345} = 2.09 \qquad \text{(use 2)}$$

Therefore the total number of running miscellaneous spares is

$$500 \times 2 = 1{,}000 \text{ parts}$$

Discussion of Example 9-6. The problem could have been solved by the use of Fig. 6-1 or by means of the Poisson exponential. However, the author's approximate equation is ideal, since it is fast and sufficiently accurate for most applications.

Note that the quantity of running spares required for a 30-day period is not a direct linear relationship of the quantity of depot spares. The reason is that we are considering probabilities which are intended to assure sufficient spares to maintain the equipment. Hence, the figures indicating the number of spares per type should not be misconstrued as meaning that exactly three tubes per type are needed for every 30-day period. Actually some 30-day periods might consume zero tubes, others one tube, and in 99.73 per cent of the cases we can expect that three or less will be consumed. On this basis, we must maintain three on hand per type per 30-day period simply as a safety factor. Hence, at the end of each 30-day period, it is necessary to replace only that quantity which is needed to maintain the complement of tubes and parts which were calculated as being required.

PRACTICE PROBLEMS

9-1. The data from a reliability test are shown in the table below. Calculate ϕ, μ, and M for 1 hr, 5 hr.

f = frequency of occurrence	Duration of each maintenance action in hours
1	1
2	2
3	3
5	4
7	5
10	6
8	7
4	8
3	9
1	10

9-2. An analysis of the reliability test data of a prototype indicated that $r = 0.0015$ failures/hr. $\mu = 2$ maintenance actions per hour. Calculate M_Δ when $T =$ 50 hr and $t = 1$ hr.

9-3. What is the mission availability of a device when $T = 100$ hr if $\mu = 1$, $r =$ 0.001 fph, $t = 0.5$ hr? What per cent improvement is the calculated availability over what would have been the reliability after 100 hr without maintenance action (i.e., $\mu = 0$)? Check your results by the use of Fig. 9-2.

9-4. From Fig. 9-5 determine the values of μ and r to assure that a device will have a Λ_E of 99.8 per cent when $t = 0.5$ and $T = 50$ hr. Are these the only possible values obtainable? Suppose the value of r was fixed by design considerations; would this fix the value of μ if t and T remain unchanged?

9-5. Calculate the availability of two devices functionally in parallel (duplex parallel redundancy) if the failure rate for each is 5 per cent per 1,000 hr when $t = 1$ hr, $\mu = 2$, and $T = 100$ hr. Do this for time and mission availability.

9-6. Two computers are connected to provide for switchover redundancy. Calculate the availability of the system if the failure rate r of the computers is 0.025 failures per 1,000 hr and $\mu = 0.5$, $t = 0.01$, $T = 24$ hr. Use Eq. (9-41).

9-7. Solve Prob. 9-6 by assuming duplex parallel redundancy. Compare the results with those obtained for Prob. 9-6. Use Eq. (9-31).

9-8. Derive the mission availability equation for a quadruplex system by substituting in Eq. (9-30). Calculate the availability if $\mu = 4$, $T = 100$, $r = 0.1$.

9-9. Solve Prob. 9-8 for a value of $t = 5$ sec.

9-10. Two computers are used as parallel redundant subsystems of a major system. Find the probability of at least one computer's being operable at all times when $m = 20$, $\mu = 3$, $T = 24$, $t = 0.2$. Disregard the effects of scheduled maintenance.

9-11. A subsystem consists of 50 tubes whose failure rate is 30 per cent failures/1,000 hr and 600 other parts, each having the same failure rate of 4 per cent failures/1,000 hr. Calculate the necessary number of depot spares to support requirements for an 18-month period. The equipment is used for 12 hr per day, 5 days per week, and no maintenance is performed after shutdown. (Use confidence level of 99.73 per cent.)

CHAPTER 10

SAMPLING METHODS

10-1. Introduction. The technique of attributes acceptance sampling is universally recognized by quality control engineers as an effective and economical method of evaluating the quality of a product. Similarly, reliability engineers use time sampling to verify the reliability of parts, components, or equipment. In general, the methods used and terms employed in reliability time sampling bear a striking similarity to those of attributes sampling. This is as it should be, since in reality, although we might be sampling for different parameters, the mathematical basis for both types of sampling plans is the same. The basic differences in nomenclature are due chiefly to the fact that attributes sampling is concerned with the number of defects in a physical unit of product, whereas time sampling is involved with the number of failures with respect to time. Let us look at some comparisons.

Attributes lot-quality protection plans specify an acceptable quality level (AQL) in terms of per cent defective or defects per hundred units. The symbol usually employed for the standard or acceptable level of quality is p'. In reliability time sampling the comparable term is called the acceptable reliability level (ARL), which is usually expressed in such terms as failures per hour, 100 hr, 1,000 hr, or per some desired standard of time. Or, in other words, the ARL is specified as a failure rate. The symbol we will use for the failure rate corresponding to the ARL is r_1.

Although the ARL may be expressed in terms of failures per any unit of time, there is an advantage to expressing it specifically in terms of failures per 100 hr, because if this is done, AQL attributes sampling tables can be used for reliability sampling as well as for attributes sampling. An example of an attributes AQL sampling plan which can be used in this dual role is MIL-STD-105A. However, in order to use this table for both quality and reliability measurement, one should be familiar with the characteristics of sampling plans, and, therefore, that subject will be briefly discussed in this chapter.

A simple means of converting ARL values to failures per 100 hr is by the use of Table 10-1. This is a simple conversion table which facilitates the change from one type of failure rate measure to another. It should be noted that the table also includes the *bit* as a unit of reliability measure. The bit is a relatively new type of unit for expressing reliability. Its chief advantage is that it reduces the labor and improves the accuracy of reliability calculations such as are used in reliability prediction studies. One failure bit equals one-thousandth of the per cent failures per thousand hours, or

$$\text{One failure bit} = 1/1{,}000 \text{ (per cent failures/1,000 hr)}$$

Table 10-1 gives simple conversion rules for converting from one type of failure rate measure to the other. To convert from column to row, multiply the column by the factor shown at the point of intersection of column and row. Thus, to convert per cent failures per 1,000 hr to bits, multiply by 10^3, because the point of intersection of column 2 and row 1 is 10^3.

TABLE 10-1. FAILURE RATE CONVERSION TABLE

Multiply by / To obtain	(1) Bits	(2) % failures per 1,000 hr	(3) % failures per 100 hr	(4) Failures per 100 hr	(5) Failure rate, failures per hour
(1) Bits	1	10^3	10^4	10^6	10^8
(2) % failures per 1,000 hr	10^{-3}	1	10	10^3	10^5
(3) % failures per 100 hr	10^{-4}	10^{-1}	1	10^2	10^4
(4) Failures per 100 hr	10^{-6}	10^{-3}	10^{-2}	1	10^2
(5) Failure rate, failures per hour	10^{-8}	10^{-5}	10^{-4}	10^{-2}	1

10-2. The Principle of Acceptance Sampling. Acceptance sampling is a method of evaluating the quality of a large lot of material by taking random samples from the lot and basing the decision to accept or reject the lot on the quality of the sample or samples so selected. The advantage of a sampling plan, therefore, is to reduce the expense which would have been incurred if every item in the lot had been inspected. There is another advantage. Some lots cannot be inspected 100 per cent because the tests are destructive in nature. An example is the tensile strength of a steel rod. The only logical way to test a lot of steel rods is to take random test samples from the lot of rods and, if these appear to exhibit a specified tensile strength, then conclude that the rods which have not been tested have a comparable strength. It would have been foolish to test all the rods because at the conclusion of the test, although the strength of the rods would have been established, there would have been no product left to ship to the customer.

The question now arises about the number of samples which should be randomly selected from a lot. If one takes too many samples, the process of testing may be prohibitive in terms of cost and time. Yet, if too few samples are selected, there may be a doubt about the authenticity of the results obtained on the basis of such a small sample. The ideal is to select a random sample of such a size that it will assure maximum confidence at minimum cost that the quality of the lot has been properly assessed. This is what a good sampling plan will do. It will specify the number and size of samples which should be selected and the maximum number of defects which can be tolerated in the sample to assure a certain confidence that the lot is equal to or better than the specified quality.

Similarly, in reliability time sampling a good time sampling plan will specify the number of failures which are tolerable in a given number of hours of testing if we are to be assured that the reliability requirement has been complied with. In this case, a sample may be thought of as consisting of so many hours of testing. This is why it is called a time sample. In this instance, there is no equivalent concept of a lot in terms of selecting so many random sample hours out of a specified lot of hours as is the case in attributes sampling, where the size of the lot is fixed by the number of units of product undergoing inspection. However, there is a comparison which we can draw between a time sample and an attributes sample if we think in terms of infinite populations. For example, if we were to take samples of the product from a continuous production process at various intervals, the process itself may be thought of as being equivalent to an infinite population of items which exhibit certain characteristics. Such would be the case if we cut samples of cloth from a weaving machine at specified intervals and inspected these samples to assure that the cloth met the specified standards of quality. We could consider this as taking samples in a continuum of time.

Likewise, if we consider the number of hours of a reliability test as testing in a continuum of time and select a number of test hours for a sample which will give us the desired confidence that the results obtained from this time sample give a measure of product reliability, we have a good time sampling plan.

From the foregoing discussion, it is obvious that what we are looking for in a sampling plan is economy and as high a degree of confidence in the results as is practicable. The operating characteristic of a sampling plan is designed to provide us with such knowledge, since the shape of this curve is a function of the sample size and acceptance number, the latter being the largest number of defects or failures which the sampling plan will tolerate as an acceptance criterion.

10-3. The Operating Characteristic Curve. Figure 10-1 shows a typical operating characteristic curve (OC curve) for an attributes sampling plan. The process average $\bar{p}$ (see Sec. 6-2) is shown on the abscissa, and the probability of acceptance of lots of particular process averages is shown on the ordinate. On this basis it should be logical to conclude that acceptance sampling plans must be based on probability theory. This is actually the case, as we shall see later in the chapter when we will explain the method of deriving sampling plans to fit specific requirements.

An outstanding advantage of the operating characteristic curve is that from it we can determine the probability of acceptance of samples randomly selected from lots whose process average is as shown on the abscissa. Thus, for the sampling plan described by the OC curve of Fig. 10-1, only 90 per cent of the samples selected from a lot whose process average is 2 defects per 100 units (2 dphu) are expected to be accepted and 10 per cent rejected. This means that for this sampling plan, if the AQL for a specific process were specified as 2 dphu, we could expect that even if the process were producing to the specified level of quality, we should still find that 10 per cent of the samples randomly selected would be rejected. This is what is referred to as the producer's risk. This will be discussed at greater length later in this chapter.

Ideally, we would expect that if our process average equaled or was better than the AQL, all items of product would be accepted. Conversely, if the process average $\bar{p}$ was worse than the AQL, a sharp cutoff would occur and all samples should be abruptly rejected. If this occurred, the ideal OC curve would look like the rectangle crosshatched in Fig. 10-1. In practice, this exact discrimination is not feasible, since the ideal OC curve could never occur, because of the effects of the chance variation which exists between samples selected from the same population. The result is that some samples tend to look better than others because less defects are found in them, while other samples might look worse because more defects exist in them. Therefore, some samples are bound to be accepted while others are rejected. The percentages which would be expected to be accepted and rejected are shown by the OC curve.

Thus, it is obvious that the shape of the OC curve of a sampling plan determines the degree of the plan's discrimination between good and poor lots. The steeper the curve, the greater is the discrimination.

A time sampling plan has an operating characteristic curve similar to the one for the attributes sampling plan just described. This is called the reliability operating characteristic curve and is labeled RO-C curve. Figure 10-2 shows such a curve. By way of analogy, in lieu of the process average, the RO-C curve is concerned with the reciprocal of the mean life $\bar{m}$, which we shall label $\bar{r}$ and call the mean reliability level. The MRL is the mean failure rate, which is calculated from successive time samples by dividing the total number of failures by the total number of hours of test. Therefore, in a manner similar to that described for the attributes case, the RO-C curve indicates the probability of acceptance for time samples taken from a device having a specific MRL. When the MRL equals or is better than the ARL, the probability of acceptance is very favorable.

10-4. Producer's and Consumer's Risk. Figure 10-1 illustrates that lots of varying process average have differing probabilities of acceptance. It also introduces some new terms in addition to the AQL, such as producer's risk α, consumer's risk β, and lot tolerance per cent defective LTPD.* The producer's risk α, expressed as a per cent, is the risk that the producer takes of having a lot rejected, because of a pessimistic sample, even if it is equal in quality to the AQL. In a typical sampling plan this probability is usually approximately 10 per cent, as is shown in Fig. 10-1, although it could be some other value. Likewise in reliability sampling the producer's reliability risk is R_α. This is the risk that a time sample might be rejected even though the product has an MRL equal to the ARL. This could occur because of the laws of chance, which permit a pessimistic grouping of random failures for a specific time sample.

The consumer's risk is the probability of acceptance when the quality or reliability is worse than the poorest tolerable standard because of an optimistic sample. In attributes sampling the poorest tolerable standard is called the lot tolerance per cent defective, LTPD, and the symbol used is p_2. In time sampling the comparable term is the lot tolerance fractional reliability deviation LTFRD and the symbol used is r_2.

Figures 10-1 and 10-2 show the various levels discussed and the risks involved, appropriately labeled.

As a general rule of thumb, in attributes sampling it is customary to make the producer's and consumer's risks approximately equal to each other. Any sampling plan which does not do this is considered to be biased in favor of either the producer or consumer. As a result, the quality engineer attempts to design sampling plans which tend to equalize the producer's and consumer's risk. These risks are usually made approximately equal to 10 per cent because, in most instances, as we shall appreciate later, it is not possible to maintain an exact equality for specific sample sizes. On the other hand, on some occasions, it may be desirable to

* α and β are also used in Chap. 8 to indicate level of significance.

emphasize either the producer's or consumer's risk to fit a specific situation. In this case, the OC curve of the sampling plan is useful in pointing this out. It is for this reason that MIL-STD-105A includes a series of OC curves for various sample sizes and AQL values. Thus, by selecting the proper OC curve for a specific AQL, one can determine the sample size and acceptance number from the tables included in MIL-STD-105A.

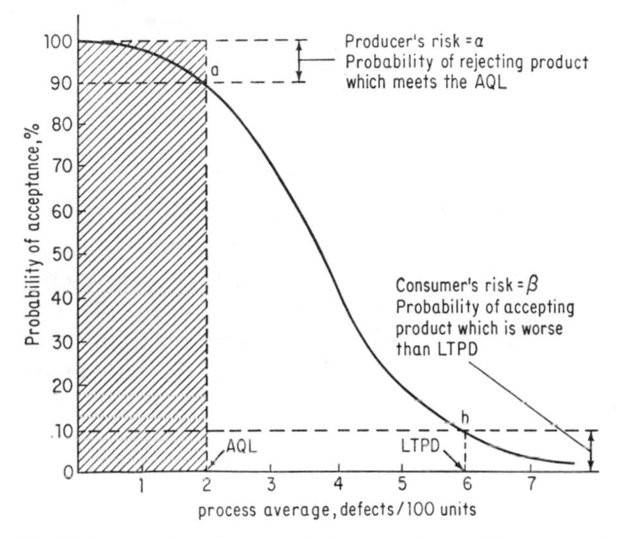

FIG. 10-1. Typical operating characteristic curve for attributes sampling plans.

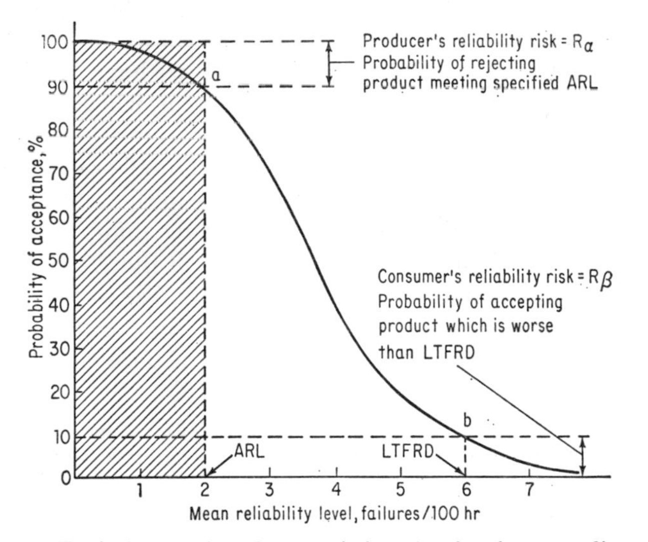

FIG. 10-2. Typical operating characteristic curve for time sampling plans.

Section 10-8 illustrates a typical method of designing sampling plans for required values of α and β.

The same reasoning can be applied to designing time sampling plans if it is remembered that in this instance the sample size is expressed as T hours and that in lieu of p_1 we use r_1.

To summarize, it should be clear that sampling involves some risk of accepting inferior product or rejecting superior product, but as long as these risks are reasonable, the sampling plan serves a useful purpose. Generally, when we sample, the procedure is to select a sample randomly, and if the number of defects or failures in the sample equals or is less than a stipulated number A, which we call the acceptance number, the lot from which the sample was selected is accepted. If the acceptance number is exceeded, the lot is rejected.

For a given process average $\bar{p}$ or mean reliability level $\bar{r}$ the probability of acceptance or rejection can be determined rapidly from the Poisson curves shown in Fig. 6-1. We shall see how these curves are applied later in our discussion, but first a word about the MRL.

10-5. Mean Reliability Level $\bar{r}$. The mean reliability level, expressed in terms of failures per hour, is calculated as the ratio of the total number of failures to total number of hours of test time.

$$\bar{r} = \text{MRL} = \frac{f}{T}$$

When the MRL is desired in failures per 100 hr, the fraction f/T is multiplied by 100. Similarly, failures per 1,000 hr are obtained by multiplying by 1,000.

To reiterate, it is apparent that in the MRL we have another analogy with a familiar quality control term *process average*, which is calculated as the ratio of the total number of defects to the total number of units in the samples, while the MRL is the ratio of the total number of failures to the total hours of reliability testing. Moreover, as in the case of the process average, the MRL is indicative of the actual reliability, as determined from test data, and should not be confused with the ARL, which is the desired or specified reliability. This, again, is similar to comparing process average with acceptable quality level (AQL) in quality control work.

The following example will illustrate the use of the Poisson curves and the effect of different sample sizes on R_α.

Example 10-1. As a result of repeated samples over an extended period of time, the mean reliability level (MRL) was calculated to be 10 failures/100 hr. What is R_α for time samples of 2 hr? For 4 hr, if the acceptance number $A = 0$ in each case? (*Note:* For time sampling, this acceptance number corresponds to a value of $f = 0$ in Fig. 6-1.)

Solution. Convert the MRL in failures per 100 hr to failures per hour. This gives a failure rate of 0.1 failures/hr $= \bar{r}$.

When $T = 2$, then $\bar{r}T = 0.1 \times 2 = 0.2$. From Fig. 6-1,

$$R_\alpha = 18 \text{ per cent}$$

When $T = 4$, then $\bar{r}T = 0.1 \times 4 = 0.4$ and $R_\alpha = 34$ per cent.

From the foregoing solution it is apparent that a definite relationship exists between sample size and the producer's reliability risk R_α for a given acceptance number and MRL. Similarly, it can be shown that the consumer's reliability risk R_β is a function of the time sample for a given LTFRD and acceptance number.

10-6. Basic Types of Sampling Protection. The basic types of sampling protection are lot quality protection and average quality protection. Lot quality protection is desired and utilized if an item is received from a supplier in homogeneous discrete lots, whereas when the lots are mixed and of random sizes, average quality protection is used.

As we have previously noted, in reliability sampling we do not consider lots as being composed of physical units, but rather we think of lots in terms of hours. However, since in the past most of us have been familiar with quality control work, it appears that it would be advantageous for the following sampling discussion to use quality control sampling language and later translate it into reliability parlance by analogy. This leads us to a discussion of lot quality and average quality protection plans.

1. *Lot Quality Protection Sampling Plans.* These are usually of two types: (*a*) AQL (acceptable quality level) plans and (*b*) LTPD (lot tolerance per cent defective) plans. An AQL plan is usually employed by the consumer without regard to the process average of the producer. The acceptable quality level is the quality level the consumer desires and prescribes without regard for the producer's ability to manufacture to that standard. The government usually uses AQL plans, because, as the consumer, it takes the attitude that if the product is good it will be accepted most of the time (usually 90 per cent) and if poor, it will be rejected most of the time.

This attitude puts the burden on the producer and forces him to improve his process average or face the risk of a high percentage of his lots being rejected. MIL-STD-105A sampling tables used by the government are AQL type plans.

Lot tolerance per cent defective plans are usually employed internally within and by industry. These plans define the sample size and acceptance number for a particular process average and specified lot tolerance per cent defective.

The plans also list the average outgoing quality which can be expected to result from each sampling plan. The amount of inspection for this type of plan varies in accordance with the process average. Thus, if the process average is poor, the amount of inspection required will be higher than if it were good. Under this type of plan, the consumer makes certain that he will not accept lots which are worse than the specified lot tolerance per cent defective, LTPD, more than a specified per cent of the time (usually 10 per cent), which, as we have seen, is called the consumer's risk. As the process average gets worse, the consumer is prepared to increase his inspection density to weed out the bad lots so that they can be detailed for inspection and then resubmitted. In this way, the consumer is assured of good outgoing quality despite increased inspection costs for initial poor quality of submitted product. This is a good method of inspection since the cost of inspection is always geared to the process average. This ensures that if the process average is good, the costs of inspection will be low. This type of plan is very effective in controlling the type of quality furnished the consumer, but it has the disadvantage of requiring a continuous knowledge or estimate of the process average to implement it properly.

2. *Average Quality Protection Plans.* This type of plan is usually employed when lots have lost their identity, as occurs when they are mixed in stores. In this case, the sampling plan must ensure that it will pass an average quality which is satisfactory but is no worse than a prescribed limiting value. The average quality is called the AOQ (average outgoing quality), and its limiting value is called the AOQL (average outgoing quality limit). In order to accomplish the sampling objective, a knowledge or an estimate of the process average should be available so that sample sizes and acceptance numbers can be determined. In general, when the process average is poor, the sample size must be greater than when the quality is relatively high. The Dodge-Romig "Single and Double Sampling Tables," which are commercially available, have worked out all the details. These tables also make provisions for 100 per cent screening of product whenever a lot is rejected by the sampling plan. It is this screening which achieves the desired AOQL.

From the foregoing discussion it is seen that plans which provide for screening or elimination of defective product cannot readily be used for reliability sampling, since screening out failures is not always feasible. Therefore, AQL plans are used for reliability sampling because they do not depend upon screening to assure a prescribed reliability level.

10-7. Sampling Methods. The outstanding methods of sampling are (1) single, (2) double, (3) sequential, (4) continuous.

1. *Single Sampling.* Single sampling is a method whereby a decision is made to accept or reject a lot on the basis of a single sample selected at

random from the lot. These sampling plans have some advantages and disadvantages. The chief advantages are that they are easy to design, explain, and administer. The disadvantage is that sample sizes are usually larger than those for equivalent double and multiple sampling plans. This is especially true for lots that are of unusually high or low quality, but for lots of intermediate quality, the sample size is usually smaller than the size for double sampling for plans having the same OC curve. Another disadvantage of single sampling is psychological. To the layman single sampling appears to be unfair. To him, the occurrence of a defect or two in a sample might have been just a coincidence and the lot should not have been rejected without its getting a second chance. This second chance is provided by double sampling as we shall see in the subsequent discussion.

2. *Double Sampling.* In general, as previously explained, except for intermediate quality, the sample size or amount of inspection is less for double sampling than for single sampling. Moreover, double sampling has the psychological advantage of appearing to give the lot a second chance of acceptance. Actually, this purported advantage is not factual, since the probability that the lot will pass the first sample is less for double than for single sampling.

This type of plan is implemented in the following manner. A sample of size n is selected from the lot. If the number of defects is less than A_1, the acceptance number for the first sample, the lot is accepted. If the number of defects is greater than A_2 for the first sample, the lot is rejected. If the number of defects is between A_1 and A_2, then a second sample is taken. If the sum of the defects in the first and second sample is less than A_2, the lot is accepted. If greater, it is rejected. See Fig. 10-3.

3. *Sequential Sampling.* Sequential sampling may be classified into two categories: (*a*) item-by-item analysis, (*b*) multiple sampling.

Originally, sequential sampling meant increasing the sample size one item at a time until a sample was large enough and with a sufficiently small or large number of defects upon which to base a decision to accept or reject. This procedure is also described as item-by-item analysis. Multiple sampling follows a similar pattern, except that instead of one item being taken at a time, groups of items are selected and the sampling is terminated at a specific point. This procedure is also referred to as sequential sampling, by groups, with truncation.

a. Item-by-item Analysis. In this type of sampling one item is inspected at a time, after which a decision is made to accept, reject, or continue inspection until a decision to accept or reject is reached. Usually, sampling is terminated at a particular point. This method of sampling has the advantage of requiring less inspection before a decision is reached than does single, double, or multiple sampling, regardless of how poor the quality of the submitted lots is. Thus, it does have good application in those instances where the tests might be destructive.

b. Multiple Sampling. Multiple sampling is a relatively economical sampling method. In general, it requires a smaller sample than does single or double sampling. However, in some instances when the quality is exceptionally poor, it may require a slightly larger sample than does double sampling. A good practical estimate for comparison's sake shows that the average sample size is about one-half to two-thirds as large as

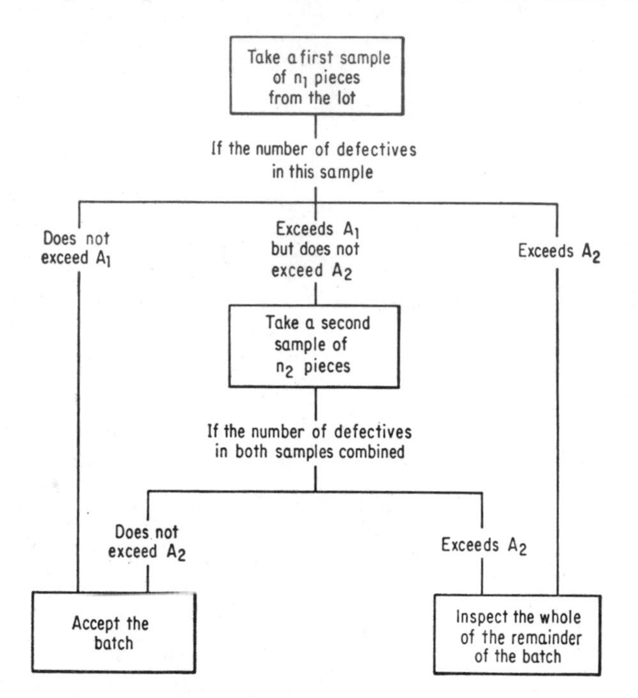

FIG. 10-3. Double sampling inspection procedure.

that required for single sampling for lots of good quality. Another advantage of multiple sampling is that it is convenient in that the items inspected need not necessarily be selected at random but can be inspected in the order that they are received or manufactured.

However, multiple sampling also has some disadvantages which are characteristic of any sequential sampling plan. Typical of these are the following: The inspector's relative knowledge of sampling must be greater than it is for single sampling. The administrative costs are high, and the clerical records and work are considerable. Moreover, sometimes it takes too long to come to a decision, resulting in a psychological resentment of the plan. Another disadvantage is that the amount of inspection is variable, and because of this, only the data from the first of several possible samples can be plotted on a control chart because of varying control limits which would have to be plotted if the data from subsequent samples were used.

4. *Continuous Sampling.* Continuous sampling is used when it is impractical to accumulate discrete lots. A typical example of the use of continuous sampling is its application to the output of parts from a production line which cannot be accumulated because other subassemblies are dependent upon them as a constant source of supply. Continuous sampling plans are also used when storage space is not available for the accumulation of lots. Generally, the method involves taking random samples at specific periods of time and inspecting them. If the number of defects exceeds the acceptance number, the defectives are isolated and inspected 100 per cent, and the process deficiency is corrected.

There are several versions of these plans in existence. The most common are the Dodge Continuous Sampling Plans. In each case the plans are designed to ensure a given AOQL regardless of the process average. This is accomplished by varying the amount of inspection as a function of quality. If the quality is good, the required inspection will be a minimum; if the quality is bad, 100 per cent inspection is instituted until a quality improvement results.

10-8. Design Methods for Single Sampling AQL Plans. As we have seen, the Poisson distribution is applicable for predicting the occurrence of events in a continuum of time, area, or volume, etc., when the expectation is constant. Therefore, it can be used as a basis for sampling tables which are concerned with the prediction of the number of defects per unit or 100 units, as desired, while the binomial distribution is utilized for the prediction of the per cent defective of populations of a finite size. However, it has been established that the Poisson is a good approximation for the binomial when the per cent defective is 10 per cent or less. Therefore, we can utilize the Poisson to play a dual role. It can be used for calculation of sampling plans wherein the per cent defective is less than 10 per cent as well as for plans less than or exceeding this value when expressed in terms of defects per unit. Military standard MIL-STD-105A makes this distinction.

Example 10-2. It is desired to construct a sampling plan for an AQL of 1 defect per 100 units (dphu) and an LTPD of 6 dphu. The required producer's and consumer's risks are approximately 10 per cent. Specify the sample size n and acceptance number A.

Solution. Let

$$\text{AQL} = p_1 = 0.01 \text{ dpu*}$$
$$\alpha = 10 \text{ per cent}$$
$$\text{LTPD} = p_2 = 0.06 \text{ dpu}$$
$$\beta = 10 \text{ per cent}$$

* In the past we have used p' to indicate AQL; however, in conjunction with the LTPD we use p_1 for AQL and p_2 for LTPD.

Step 1. (For producer's risk $\alpha = 10$ per cent and $c = 0$)

On the Poisson chart, Fig. 6-1, locate the intersection of the $c = 0$ curve with the horizontal line representing a probability of occurrence of c or less defects of 90 per cent (producer's risk of 10 per cent). This point corresponds to an np_1 value of 0.11 on the abscissa.

Divide the value of np_1 by p_1 to get the sample size n. Thus

$$n = \frac{np_1}{p_1} = \frac{0.11}{0.01} = 11$$

Step 2. (For consumer's risk $\beta = 10$ per cent and $c = 0$.) Locate the intersection of the curve $c = 0$ with the horizontal line representing a probability of occurrence of c or less defects of 10 per cent (consumer's risk). This point corresponds to an np_2 value of 2.3. The sample size is now found as

$$n = \frac{np_2}{p_2} = \frac{2.3}{0.06} = 38$$

Obviously this cannot be the proper sampling plan, since the value of n found in step 2 is not equal to the value found in step 1, and these values must be equal. Therefore, a similar procedure will be followed by trying the curve $c = 1$ to attempt to get sample size equality.

Step 3. (For producer's risk $\alpha = 10$ per cent and $c = 1$.) Using the same procedure as for steps 1 and 2, but using curve $c = 1$ instead of $c = 0$, we get

$$np_1 = 0.52$$

$$n = \frac{np_1}{p_1} = \frac{0.52}{0.01} = 52$$

Step 4. (For consumer's risk $\beta = 10$ per cent and $c = 1$.)

$$np_2 = 3.8$$

$$n = \frac{np_2}{p_2} = \frac{3.8}{0.06} = 63$$

Again, the sample sizes for steps 3 and 4 do not equal each other, although they are closer than in steps 1 and 2. At this point it is obvious that an exact producer's and consumer's risk cannot always be achieved if the sample size for α and β must be equal. Therefore, a decision must be made as to what these risks shall be. If it is desired to maintain the producer's risk at 10 per cent, then the sample size for step 3 would govern (i.e., $n = 52$). Under these conditions the consumer's risk is found by calculating np_2 corresponding to $n = 52$ and $p_2 = 0.06$, reading it off the Poisson chart. Thus $np_2 = 52(0.06) = 3.12$, and the corresponding consumer's risk is about 19 per cent. If the designer of the sampling plan is satisfied, the problem is solved. Otherwise, an adjustment must be made in the sample which will result in an approximation closer to his desires. Usually, the producer's and consumer's risks can be made almost equal by averaging the two sample sizes which come closest to satisfying the required conditions. Thus, in this case the average of 52 and 63 in round figures is 58. Therefore, the sampling plan would be $n = 58$ and $A = 1$, and the resulting producer's and consumer's risks (from the Poisson chart) are 12 per cent and 13 per cent, respectively, which are approximately equal to what is desired.

Example 10-3. Design a time sampling plan for an ARL of 2 failures/100 hr (fphh) and $R_\alpha = 5$ per cent; and an LTFRD of 10 fphh with $R_\beta = 10$ per cent. Specify the time sample T in hours and the acceptance number.

Solution. For a producer's reliability risk, $R_\alpha = 5$ per cent and $f = 0$, convert ARL and LTFRD to failures per hour, i.e.,

$$\text{ARL} = r_1 = 0.02 \text{ fph} \qquad \text{and} \qquad \text{LTFRD} = r_2 = 0.10 \text{ fph}$$
$$R_\alpha = 5 \text{ per cent} \qquad\qquad R_\beta = 10 \text{ per cent}$$

Using Fig. 6-1, locate the point of intersection of the $f = 0$ curve with the horizontal line representing a probability of f or less failures of 95 per cent ($R_\alpha = 5$ per cent). There is no such point of intersection. Next try $f = 1$ and $R_\alpha = 5$ per cent. This point of intersection corresponds to an r_1T value of 0.36 on the abscissa. Dividing,

$$T = \frac{r_1T}{r_1} = \frac{0.36}{0.02} = 18 \text{ hr}$$

For $R_\beta = 10$ per cent and $f = 1$, locate the point of intersection of curve $f = 1$ with the horizontal line for $R_\beta = 10$ per cent. This point corresponds to an r_2T value of 3.9. Therefore,

$$T = \frac{r_2T}{r_2} = \frac{3.9}{0.10} = 39 \text{ hr}$$

The two sample sizes are not equal. Therefore, we must try the same procedure for $f = 2$, and again there is no solution. Finally, for $f = 3$, $R_\alpha = 5$ per cent we get

$$T = \frac{r_1T}{r_1} = \frac{1.35}{0.02} = 67.5$$

and for $f = 3$, $R_\beta = 10$ per cent

$$T = \frac{r_2T}{r_2} = \frac{6.75}{0.10} = 67.5$$

Therefore, since the time sample for the producer's and consumer's reliability risk is the same, namely 67.5 hr, the time sampling plan to the nearest whole number which satisfies the given conditions is

$$T = 68 \text{ hr} \qquad A = 3$$

It should be noted that the symbol A is termed the acceptance number, which represents the limiting number of failures which are permitted for this sampling plan. It should not be confused with f, the maximum number of failures which are expected in time samples of size T hours for a given R_α or R_β. The symbol A is appended for that value of f which ultimately satisfies both the consumer's and producer's reliability risks.

10-9. AQL Sampling Tables. The United States government uses AQL plans for product acceptance in contractor's plants from whom it procures material. The reason for preferring AQL plans over other types is that these result in less inspection costs. An AQL plan encourages a manufacturer to keep his process average high because it minimizes his producer's risk. This results in less rejected lots, and, therefore, the government's reinspection costs, because of the resubmission of formerly rejected lots, are reduced. If AOQL plans were used by the government, there would be no incentive for the manufacturer to submit a high-quality product, since the government would screen all the rejects and the burden would in turn be its responsibility. As was previously mentioned, the AQL tables used by the government are MIL-STD-105A. This specification is readily obtainable from the Government Printing Office and, therefore, will not be included here. However, we will briefly describe its general layout and later show how we adapted certain of its features in our Table 10-2 and made this applicable for quality and reliability sampling. We will also describe the scope and limitations of Table 10-2, since it cannot be used as a sampling table for extremely high orders of reliability. In this instance, other methods are used in lieu of the table.

The numeral 3 represents a good reminder for sampling types applicable to MIL-STD-105A with regard to methods, levels, and inspection.

The three sampling methods are respectively *single*, *double*, and *multiple;* the three levels are *level I*, *level II*, and *level III;* and the three types of inspection are *normal*, *tightened*, and *reduced*. Since sampling methods were outlined in Sec. 10-7, we shall discuss only levels and types of sampling at this time.

There are three levels of inspection: I, II, and III. These levels determine the amount of sampling which will be conducted per given lot. Level II is usually specified and considered normal. Level I represents the least amount of inspection, while level III involves the greatest amount of inspection.

On the other hand, the three types of inspection are not concerned primarily with the amount of inspection, but rather with the degree of risk the consumer is willing to take. In general, when the process average is within control limits, normal inspection is used, whereas when it is better than the lower limit of the AQL, reduced inspection is utilized. However, when the process average is worse than the upper limit of the AQL, tightened inspection is used to reduce the probability of acceptance of poor lots.

These characteristics of MIL-STD-105A were described briefly because the author assumes a general understanding of this specification by the average reader, and since this is not a text on quality control, detail has been strictly avoided. This has been done because the reliability sampling tables which follow and their use will be explained in detail, and since these tables are based on MIL-STD-105A, an explanation of them will serve a dual purpose, i.e., an explanation of the use of MIL-STD-105A as well as the use and application of the reliability sampling tables.

10-10. ARL Reliability Sampling Plans. As was explained in Sec. 10-9, the reliability sampling table (Table 10-2) was patterned after MIL-STD-105A. As a matter of fact, the code letters, acceptance and rejection number, and AQL values are identical to those of MIL-STD-105A for as far as it goes. However, the reliability sampling table was extended beyond the scope of the MIL standard to include additional code letters as well as more AQL values. The reliability sampling table extends the code letters up to letter W, while the MIL standard goes only to code letter Q. Moreover, the MIL standard minimum AQL value is 0.015, whereas the reliability table has a minimum of AQL of 0.00010 dphu.

An additional feature of the reliability sampling table is that it introduces the concept of the ARL value, i.e., acceptable reliability level, in terms of failures per 100 hr. This was done despite the fact that much literature describes and measures ARL in terms of failures per 1,000 hr. Thus, the same table can be used to measure either reliability or quality by interpretation of the decimal fraction associated with either the AQL or ARL in terms of defects per 100 units or failures per 100 hr, as the case might be. In other words, the selection of 100 was deliberate in order to give the table a dual role.

The table also lists the value of the mean time between failures. This value may be determined from data gathered from a reliability test, in which case it is calculated as the reciprocal of the MRL. Or it may be as specified in a procurement contract or design specification and in this instance is the reciprocal of the ARL expressed in failures per hour, or the reciprocal of the ARL in failures per 100 hr multiplied by 100. Thus, for an ARL of 0.00010 failures/100 hr,

$$m = \frac{1}{0.00010} \times 100 = 1{,}000{,}000 \text{ hr}$$

Table 10-2 represents a single sampling ARL table which is similar to the single sampling AQL table of MIL-STD-105A. The extreme left column gives the sample size code letters. The second column represents the sample size in terms of number of units or number of aggregate hours of test in a sampling reliability test, while the remaining columns give the acceptance number A and rejection number R corresponding to a particular sample size.

The top row of the table lists the type of inspection and test, while the next to the top indicates the AQL in terms of defects per 100 units, or the ARL in terms of failures per 100 hr. It also indicates the acceptable mean time between failures m corresponding to each ARL value. The bottom row gives the values of AQL, ARL, and m corresponding to tightened inspection and reliability testing.

Although MIL-STD-105A includes tables for double and multiple sampling as well as for reduced inspection, these have not been included here because it is felt that single sampling is adequate for most purposes and that the sequential methods explained in Chap. 11 fulfill other requirements.

Table 10-2 is used in a manner similar to MIL-STD-105A, except that some judgment must be exercised in determining sample sizes which will satisfy specific values of R_α and R_β. This was accomplished in the MIL standard by the inclusion therein of a series of OC curves for each sample size code letter. We have not provided any such operating characteristic curves in this book. They were excluded because it was felt that most of the OC curves are readily available in the MIL standard and because the sample size can be readily calculated for specific values of R_α and R_β by other methods which we shall describe later in the chapter.

As we have mentioned, the operating characteristic curve of a sampling plan is a handy device for indicating the producer's and consumer's risks. This is one of the important reasons for their being included in MIL-STD-105A. For example, it is obvious that for an ARL of 1 failure/100 units one might select code letter F from Table 10-2, corresponding to a time sample of 15 hr and an acceptance number of zero. Or one might select another code letter, calling for 50 hr of test and having an acceptance number of 1. In the first sampling plan, i.e., $T = 15$, $A = 0$, since $r_1T = 0.01(15) = 0.15$, we find R_α (from Fig. 6-1) equal to 14 per cent. In the second sampling plan, i.e., $T = 50$, $A = 1$, since

$$r_1T = 50(0.1) = 0.5$$

R_α is read from the Poisson curves as equal to 9 per cent. Thus, we see that R_α is definitely a function of sample size and acceptance number.

MIL-STD-105A attempts to resolve all of these problems by providing OC curves and Table III. The OC curves permit one to tailor sampling

Table 10-2. Master Table for Normal and Tightened Inspection and Reliability Testing

Code letter	Sample size	Sample size is interpreted in units for AQL; Sample size is interpreted in hours for ARL																Normal inspection and reliability testing																
	AQL,ARL m	0.00010 1,000,000 A R	0.00015 667,000 A R	0.00035 286,000 A R	0.00065 154,000 A R	0.001 100,000 A R	0.0015 67,700 A R	0.0035 28,600 A R	0.0065 15,400 A R	0.010 10,000 A R	0.015 6670 A R	0.035 2860 A R	0.065 1540 A R	0.10 1000 A R	0.15 667 A R	0.25 400 A R	0.40 250 A R	0.65 154 A R	1.0 100 A R	1.5 67 A R	2.5 40 A R	4.0 25 A R	6.5 15 A R	10 10 A R	15 6.7 A R	25 4.0 A R	40 2.5 A R	65 1.54 A R	100 1.0 A R	150 0.67 A R	250 0.4 A R	400 0.25 A R	650 0.15 A R	1000 0.10 A R
A	2	↓	↓	↓	↓	↓	↓	↓	↓	↓	↓	↓	↓	↓	↓	↓	↓	↓	↓	↓	↓	↓	↓	0 1	↓	↓	1 2	2 3	3 4	5 6	8 9	12 13	19 20	28 29
B	3	↓	↓	↓	↓	↓	↓	↓	↓	↓	↓	↓	↓	↓	↓	↓	↓	↓	↓	↓	↓	↓	0 1	↑	↓	1 2	2 3	3 4	5 6	8 9	12 13	18 19	28 29	41 42
C	5	↓	↓	↓	↓	↓	↓	↓	↓	↓	↓	↓	↓	↓	↓	↓	↓	↓	↓	↓	↓	0 1	↑	↓	1 2	2 3	3 4	5 6	8 9	12 13	19 20	29 30	44 45	65 66
D	7	↓	↓	↓	↓	↓	↓	↓	↓	↓	↓	↓	↓	↓	↓	↓	↓	↓	↓	↓	0 1	↑	↓	1 2	2 3	3 4	5 6	7 8	11 12	17 18	26 27	39 40	60 61	89 90
E	10	↓	↓	↓	↓	↓	↓	↓	↓	↓	↓	↓	↓	↓	↓	↓	↓	↓	↓	0 1	↑	↓	1 2	2 3	3 4	5 6	7 8	10 11	15 16	23 24	36 37	54 55	83 84	123 124
F	15	↓	↓	↓	↓	↓	↓	↓	↓	↓	↓	↓	↓	↓	↓	↓	↓	↓	0 1	↑	↓	1 2	2 3	3 4	4 5	7 8	10 11	15 16	22 23	33 34	51 52	78 79	121 122	178 179
G	25	↓	↓	↓	↓	↓	↓	↓	↓	↓	↓	↓	↓	↓	↓	↓	↓	0 1	↑	↓	1 2	2 3	3 4	5 6	7 8	11 12	16 17	24 25	35 36	51 52	80 81	124 125	192 193	↑
H	35	↓	↓	↓	↓	↓	↓	↓	↓	↓	↓	↓	↓	↓	↓	↓	0 1	↑	↓	1 2	2 3	3 4	5 6	7 8	10 11	15 16	22 23	33 34	48 49	69 70	110 111	168 169	↑	↑
I	50	↓	↓	↓	↓	↓	↓	↓	↓	↓	↓	↓	↓	↓	↓	0 1	↑	↓	1 2	2 3	3 4	4 5	6 7	9 10	13 14	20 21	30 31	46 47	67 68	96 97	151 152	↑	↑	↑
J	75	↓	↓	↓	↓	↓	↓	↓	↓	↓	↓	↓	↓	↓	0 1	↑	↓	1 2	2 3	3 4	4 5	6 7	9 10	13 14	19 20	29 30	43 44	66 67	96 97	138 139	↑	↑	↑	↑
K	110	↓	↓	↓	↓	↓	↓	↓	↓	↓	↓	↓	↓	0 1	↑	↓	1 2	2 3	3 4	4 5	6 7	8 9	12 13	18 19	26 27	40 41	60 61	93 94	135 136	↑	↑	↑	↑	↑
L	150	↓	↓	↓	↓	↓	↓	↓	↓	↓	↓	↓	0 1	↑	↓	1 2	2 3	3 4	4 5	5 6	8 9	11 12	17 18	24 25	34 35	53 54	80 81	123 124	↑	↑	↑	↑	↑	↑
M	225	↓	↓	↓	↓	↓	↓	↓	↓	↓	↓	0 1	↑	↓	1 2	2 3	3 4	4 5	5 6	8 9	11 12	17 18	24 25	34 35	48 49	76 77	115 116	↑	↑	↑	↑	↑	↑	↑
N	300	↓	↓	↓	↓	↓	↓	↓	↓	↓	0 1	↑	↓	1 2	2 3	3 4	4 5	5 6	7 8	10 11	14 15	20 21	32 33	44 45	63 64	98 99	↑	↑	↑	↑	↑	↑	↑	↑
O	450	↓	↓	↓	↓	↓	↓	↓	↓	0 1	↑	↓	1 2	2 3	3 4	4 5	5 6	7 8	10 11	14 15	20 21	29 30	43 44	62 63	89 90	↑	↑	↑	↑	↑	↑	↑	↑	↑
P	750	↓	↓	↓	↓	↓	↓	↓	0 1	↑	↓	1 2	2 3	3 4	4 5	6 7	8 9	11 12	15 16	20 21	31 32	45 46	↑	↑	↑	↑	↑	↑	↑	↑	↑	↑	↑	↑
Q	1500	↓	↓	↓	↓	↓	↓	0 1	↑	↓	1 2	2 3	3 4	5 6	7 8	9 10	13 14	18 19	25 26	↑	↑	↑	↑	↑	↑	↑	↑	↑	↑	↑	↑	↑	↑	↑
R	3000	↓	↓	↓	↓	↓	0 1	↑	↓	1 2	2 3	3 4	5 6	8 9	11 12	↑	↑	↑	↑	↑	↑	↑	↑	↑	↑	↑	↑	↑	↑	↑	↑	↑	↑	↑
S	5000	↓	↓	↓	↓	0 1	↑	↓	1 2	2 3	↑	↑	↑	↑	↑	↑	↑	↑	↑	↑	↑	↑	↑	↑	↑	↑	↑	↑	↑	↑	↑	↑	↑	↑
T	8000	↓	↓	↓	0 1	↑	↓	1 2	2 3	↑	↑	↑	↑	↑	↑	↑	↑	↑	↑	↑	↑	↑	↑	↑	↑	↑	↑	↑	↑	↑	↑	↑	↑	↑
U	15,000	↓	↓	0 1	↑	↓	1 2	2 3	3 4	↑	↑	↑	↑	↑	↑	↑	↑	↑	↑	↑	↑	↑	↑	↑	↑	↑	↑	↑	↑	↑	↑	↑	↑	↑
V	35,000	↓	0 1	↑	↓	1 2	2 3	3 4	↑	↑	↑	↑	↑	↑	↑	↑	↑	↑	↑	↑	↑	↑	↑	↑	↑	↑	↑	↑	↑	↑	↑	↑	↑	↑
W	50,000	0 1	↑	↑	1 2	↑	↑	↑	↑	↑	↑	↑	↑	↑	↑	↑	↑	↑	↑	↑	↑	↑	↑	↑	↑	↑	↑	↑	↑	↑	↑	↑	↑	↑
	AQL,ARL m	0.00015 667,000	0.00035 286,000	0.00065 154,000	0.0010 100,000	0.0015 66,700	0.0035 28,600	0.0065 15,400	0.010 10,000	0.015 6670	0.035 2860	0.065 1540	0.10 1000	0.15 667	0.25 400	0.40 250	0.65 154	1.0 100	1.5 67	2.5 40	4.0 25	6.5 15	10 10	15 6.7	25 4	40 2.5	65 1.54	100 1.0	150 0.67	250 0.4	400 0.25	650 0.15	1000 0.10	
		Tightened inspection and reliability test																																

↓ = Use first sampling plan below arrow

↑ = Use first sampling plan above arrow

A = acceptance no.

R = rejection no.

AQL = acceptable quality level in defects per hundred units (dphu)

ARL = acceptable reliability level in failure per hundred hours (fphh)

m = mean time between failures

plans to the approximate producer's and consumer's risks which are desired, while Table III gives the relationship between lot size and sample size for inspection levels I, II, and III. For example, for any lot size between 2 and 8 units, for level II sample size code letter A is shown in Table III of the MIL standard. This code letter calls for a sample size of 2. For a lot size between 26 and 40 units, sample size code letter D is shown for level II and the corresponding sample size is 7 units. In attributes sampling, Table III helps considerably, since production lots are of a discrete size, and, therefore, the optimum sample must be selected to assure minimum sampling risks and inspection costs.

In time sampling, however, since lot sizes are measured in hours and not in discrete physical units, our only constraint on lot size (time sample) is the available test time and cost. Theoretically, if our only interest were a high degree of confidence in the results of a reliability test, the optimum course to follow would be to test for as long a period as possible selecting the corresponding time sample size and acceptance number from Table 10-2. However, we cannot always do this because time and cost are important considerations. Under these circumstances, it appears that the best course of action to follow is to select a sampling plan from Table 10-2 which is consistent with the available time and cost, remembering that the confidence level is a function of the sample size and acceptance number.

Generally, plans corresponding to the smallest possible time sample for a given ARL in Table 10-2 will give a value of R_α ranging from 5 to 20 per cent. The higher the code letter, the less are the producer's and consumer's reliability risks. If we are interested in both R_α and R_β, we may determine what they are, for sampling plans consisting of specific values of T and A, in a manner similar to that used in Example 10-3.

It is convenient to use Table 10-2 for values of m less than 1,000,000 hr. For values larger than this we have developed a statistic called kappa square which makes possible the designing of sampling plans to satisfy specific values of R_α and R_β. We shall discuss this later in this chapter and also in Chap. 11.

When the ARL or AQL value is different from those shown in the various columns of Table 10-2, we may use Table 10-3 as a guide. Column 1 of this table shows the range of AQL or ARL values expressed in terms of defects per 100 units and failures per 100 hr, respectively. The purpose of this column is to assist in determining which value of AQL or ARL will be used in Table 10-2 when the value specified is somewhere between the ranges shown in column 1 of Table 10-3. For example, if we calculated m from the data of a reliability test and found it to be equal to 0.00048 failures/100 hr, we would refer to column 1 for the range of values 0.0004 to 0.00069 because 0.00048 lies between these two values. From column 2 of Table 10-3 we would determine that the ARL value to use in Table 10-2 is 0.00065, and from column 3 we observe that this is equal to a mean time between failures of 154,000 hr.

As was mentioned previously, it can be shown that samples selected at random from a population of known process average will tend to

TABLE 10-3. AQL, ARL,* AND m VALUES

(1) For AQL, ARL values falling in these ranges—	(2) Use these AQL, ARL values	(3) Mean life equivalents of ARL values = m
0 – 0.000109	0.00010	1,000,000
0.00011– 0.00018	0.00015	667,000
0.00019 0.00039	0.00035	286,000
0.0004 – 0.00069	0.00065	154,000
0.0007 – 0.00109	0.0010	100,000
0.0011 – 0.0018	0.0015	66,700
0.0019 – 0.0039	0.0035	28,600
0.0040 – 0.0069	0.0065	15,400
0.007 – 0.0109	0.010	10,000
0.011 – 0.018	0.015	6,670
0.019 – 0.039	0.035	2,860
0.040 – 0.069	0.065	1,540
0.070 – 0.109	0.10	1,000
0.110 – 0.164	0.15	667
0.165 – 0.279	0.25	400
0.280 – 0.439	0.40	250
0.440 – 0.699	0.65	154
0.700 – 1.09	1.0	100
1.10 – 1.64	1.5	67
1.65 – 2.79	2.5	40
2.80 – 4.39	4.0	25
4.40 – 6.99	6.5	15
7.00 – 10.9	10.0	10
11.00 – 16.4	15.0	6.7
16.5 – 27.9	25.0	4.0
28.0 – 43.9	40.0	2.5
44.0 – 69.9	65.0	1.54
70.0 – 109.0	100.0	1.0
110.0 – 164.0	150.0	0.67
165.0 – 279.0	250.0	0.4
280.0 – 439.0	400.0	0.25
440.0 – 699.0	650.0	0.15
700.0 –1,090.0	1,000.0	0.10

* ARL values are expressed in failures per 100 hr; AQL values are expressed in defects per 100 units.

reflect a better or poorer process average than the parent population. This is due to the fact that, by chance, some samples will look better because they have less defects or failures than other random samples, while some will look poorer because they show more of either.

In general, if the results of several samples are considered in the aggregate, the estimate of the MRL from the total time sample will be better than that calculated from any individual time sample. A convenient term for the individual time samples comprising the total time sample is sub-time sample. Thus, in time sampling, the aggregate or total time sample is the sum of the sub-time samples in hours. Each subsample in turn is made up of the total hours of test on a number of physical units which make up the subsample. To clarify this terminology further, in the future the term time sample will refer to the total or aggregate time sample consisting of the sum of all sub-time samples. (See Chap. 11.)

From the foregoing discussion, it should be obvious that a time sample of a specific number of hours may be made up of test times on a small number of units, each undergoing test for a large number of hours; or conversely a large number of units may be tested for a small number of hours.

In any event, if we calculated the failure rate for each subsample by dividing the number of failures in each by the number of hours of test comprising the subsample, we would find that the variation in the failure rates, so calculated from each subsample, would vary about the MRL. The degree of variation from subsample to subsample would be greatest for small subsamples and least for large subsamples.

It is often desirable to determine the 3-sigma limits of the ARL so we can establish the upper and lower control limits for the MRL calculated from various subsamples. The reason for this is to assure ourselves that the calculated value of the MRL for a specific time sample does not exceed the control limits so established. For example, if the upper control limit of the ARL were to be exceeded by the calculated MRL for a time sample, this would be a very strong indication that the mean life, expressed as mean time between failures, is worse than what it should be. This situation can occur even though some subsamples may be considered as conforming under the prescribed time sampling plan, since the MRL is usually calculated from the data of a series of subsamples.

The 99.73 confidence limits, i.e., the 3-sigma limits, may be determined from Table II of MIL-STD-105A, "Limits of the Process Average." This table shows the upper and lower confidence limits of the AQL for various sample sizes. The table may be used for ARL values by considering the sample size in terms of hours instead of in terms of units.

These confidence limits may also be determined from Eqs. (10-1) and (10-2) when the ARL is expressed in terms of failures per 100 hr (see pages 186 and 187).

$$U_{ARL} = ARL + 30\sqrt{\frac{ARL}{T}} \tag{10-1}$$

$$L_{ARL} = ARL - 30\sqrt{\frac{ARL}{T}} \tag{10-2}$$

where ARL = failures/100 hr
T = total test time, hr (sample size)
U_{ARL} = upper control limit of ARL
L_{ARL} = lower control limit of ARL

To understand the use of these equations, we shall find the 3-sigma control limits for an ARL of 0.015 failures/100 hr when $T = 1{,}200$ hr. Substituting,

$$U_{ARL} = 0.015 + 30\sqrt{\frac{0.015}{1{,}200}} = 0.015 + 0.1056 = 0.121 \text{ fphh}$$

$$L_{ARL} = 0.015 - 30\sqrt{\frac{0.015}{1{,}200}} = 0$$

The above formulas are probably recognized as being similar to those for calculating the upper and lower 3-sigma limits of the AQL in the field of quality control, except that in this case we are concerned with the acceptable reliability level instead of the acceptable quality level. These limits give us the boundary lines, so to speak, between which it is expected that the mean reliability level (MRL) will fluctuate from sample to sample. This is an indication that the reliability of the product is actually equal to the ARL value within the confidence limits that these boundaries represent.

To summarize, the ARL is a specification value, whereas the MRL is an actual computed value based on data which have been obtained from several samples. However, since the aim is to have the MRL equal to or better than the ARL, we must have a method of determining from these data whether this is being achieved. Equations (10-1) and (10-2) are used for this purpose, since they give the limiting values we have been discussing. The reason for these limiting values is to assure us that if they are not exceeded, we must be operating with an MRL equal to or approximately equal to the ARL. If we did not have such limits, there would be no way of determining the value of the MRL on the basis of a single sample with any degree of confidence, because it might be either an optimistic or pessimistic sample. However, we can obtain a good estimate of an average or mean value of the reliability level from several subsamples. We also know that the MRL itself has a standard error, i.e., it varies inversely as the square root of the sample size T in hours. Thus, the larger the time sample, the closer the estimate we can make of the MRL. When T becomes very large, that is, approaches infinity, the MRL does not vary appreciably and, therefore, becomes synonymous with the constant failure rate explained in Chap. 6.

10-11. Determination of Sample Size T. Before selection of a sampling plan from Table 10-2, it is important to determine the required ARL and the desired R_α and R_β. In attributes sampling, if one decides to use

MIL-STD-105A for a particular AQL, the decision regarding sample size is made for him, since this table shows a definite relationship between lot sizes and sample sizes. In this case, the producer's risk α and consumer's risk β for the indicated plan corresponding to a particular code letter are described by the operating characteristic curve shown for the particular plan. Moreover, the lot size is discrete and physically discernible and available, and it is not a difficult task to select a random sample from it. However, in reliability sampling the lot is not discrete, and, therefore, we cannot use the same technique for determining sample size as is used in the case of attributes. This makes our choice more difficult, since, in most instances, we are torn between a desire for a particular R_α and R_β and economy. The economy aspect is most important, since, in general, reliability testing is relatively more expensive than inspection for attributes. Therefore, we cannot arbitrarily look at Table 10-2 and select a sample size indiscriminately without consulting applicable RO-C curves for that code letter to make certain that we are satisfied with the R_α and R_β corresponding to the sampling plan we choose. However, since we have not included RO-C curves in this volume, we must resort to other means.

As an illustration, if we wanted to determine the best time sampling plan for an ARL of 1 failure/100 hr and an LTFRD of 5 failures/100 hr, we would consult Table 10-2, calculate R_α and R_β, and tabulate the results.

TABLE 10-4. POSSIBLE SAMPLING PLANS FOR AN ARL OF 1 fphh AND LTFRD OF 10 fphh

Sample size code letter	Sample size T	Acceptance no. A	R_α, %	R_β, %
F	15	0	14	42
I	50	1	9	28
J	75	2	4	23
K	110	3	3	19
L	150	4	2	12
M	225	5	3	4
N	300	7	4	2

An examination of the tabulated plans indicates that the most economical one to use is that corresponding to code letter F. However, in this case, since R_α is 14 per cent, the producer runs a risk of rejection 14 per cent of the time. This high rejection rate may well offset the initial economy because of the costs incurred. If plan I is selected, the testing costs are increased over plan F, but R_α is decreased. However, R_α does not differ appreciably for plans J, K, L, M, and N, so that any of these could be selected if R_β were not a factor. If that were the case, then it would be logical to select plan J, since this represents the minimum test time for a relatively small value of R_α which does not differ appreciably from smaller R_α values which call for significantly greater test time. From the producer's viewpoint, it is always desirable to select a sampling plan with a small value of R_α because it is not always practical to rework rejected lots and resubmit them for acceptance, as can be done in attributes sampling. However, from the consumer's viewpoint, a small R_β is desirable, since the consumer wants to ensure that he is protected from accepting a poor level of reliability. The safe course to follow is to select a plan which is satisfactory to both parties and which involves a minimum risk for both. In this case, as can be seen from Table 10-4, such a plan would be that corresponding to code letter M, where $T = 225$ and $A = 5$. This sample of $T = 225$ hr may be made up by testing one or more samples until an aggregate of 225 is accumulated. It need not necessarily come from testing only one unit for 225 hr.

The author has developed an equation which makes our problem of selecting sample size relatively easy. This is Eq. (8) of App. 2, which is repeated below for ready reference.[1]

$$T = \left[\frac{Z_\alpha - Z_\beta \sqrt{k}}{\sqrt{r_1}\,(k - 1)}\right]^2 \qquad (10\text{-}3)$$

The equation expresses the sample size T for a two-tail confidence level. This eliminates the necessity for going through the procedure of tabulation as shown in Table 10-4, since it gives the required sample size T in terms of hours in one fast calculation. Example 10-4 will illustrate its use.

Example 10-4. In the solution of Example 8-5 we found that

$$r_1 = \frac{1}{135.2} = 0.00745$$

and $r_2 = 1/55.6 = 0.018$ for a two-tailed confidence level of 95 per cent. (The data from which these values were obtained were $T = 1{,}650$ hr and $f = 20$ failures.) Find the value of T [using Eq. (10-3)] for r_1 and r_2 and $R_\alpha = 2.5$ per cent and $R_\beta = 2.5$ per cent. How does this compare with the value of T of Example 8-5? Find the value of A_w and A by using Eqs. (1) and (2) of App. 2.

Solution

$$Z_\alpha = +1.96 \qquad r_1 = 0.00745 \qquad k = \frac{r_2}{r_1} = \frac{0.018}{0.00745} = 2.42$$

$$Z_\beta = -1.96 \qquad r_2 = 0.018 \qquad \sqrt{r_1} = \sqrt{0.00745} = 0.086$$

[1] This equation is the basis for the kappa square statistic, which we shall introduce in Chap. 11.

Substituting in Eq. (10-3),

$$T = \left[\frac{1.96 + 1.96 \sqrt{2.42}}{\sqrt{0.00745}\,(2.42 - 1)} \right]^2 = \left[\frac{4.96}{0.122} \right]^2 = 1{,}650 \text{ hr}$$

Thus, we see that the result obtained by the use of Eq. (10-3) is equal to the known value of T which was given in Example 8-5.

To find the value of A we first substitute in Eq. (1) of App. 2 and obtain A_w and then obtain the value of A by substituting in Eq. (2) of App. 2.

$$A_w = r_1T + Z_\alpha \sqrt{r_1T} \tag{10-4}$$

$$\begin{aligned} A_w &= (0.00745)(1{,}650) + 1.96 \sqrt{(0.00745)(1{,}650)} \\ &= 12.3 + 1.96 \sqrt{12.3} \\ &= 12.3 + 6.85 \\ &= 19.15 \\ A &= 19.15 - 0.5 = 18.65 \end{aligned}$$

Since 18.65 is greater than 18.5, the value of A in whole numbers is $A = 19$. Thus 19 is the acceptance number and 20 is the rejection number. A glance at the Poisson curves of Fig. 6-1 will show that for a value of r_1T of 12.3 and $f = 19$, $R_\alpha = 2.5$ per cent. This means that if r_1 is equal to the ARL, the probability of rejection is only 2.5 per cent or, expressing it another way, $R_\alpha = 2.5$ per cent. Likewise, if r_2 is equal to the LTFRD, we would get a probability of acceptance, called the consumer's reliability risk R_β, of 2.5 per cent. We calculate this as follows:

$$\begin{aligned} A_w &= r_2T - Z_\beta \sqrt{r_2T} \\ &= (0.018)(1{,}650) - 1.96 \sqrt{(0.018)(1{,}650)} \\ &= 29.7 - 10.6 \\ &= 19.1 \end{aligned}$$

$A = A_w - 0.5 = 19.1 - 0.5 = 18.6$, and therefore A to the nearest whole number is 19.

Thus, we see that A is the same for both R_α and R_β, which is as it should be. Therefore the calculation performed with Eq. (10-4) would have sufficed.

Suppose we were concerned with finding T for only a one-tail confidence level corresponding to either R_α or R_β for a fixed value of A_w. We would use Eqs. (10-5) and (10-6), respectively, which are based on Eq. (6) of App. 2. These equations express T as a function of A_w and Z.

$$T_1 = \frac{2A_w + Z_\alpha^2 - \sqrt{4A_wZ_\alpha^2 + Z_\alpha^4}}{2r_1} \tag{10-5}$$

$$T_2 = \frac{2A_w^2 + Z_\beta^2 + \sqrt{4A_wZ_\beta^2 + Z_\beta^2}}{2r_2} \tag{10-6}$$

We shall illustrate the use of these equations by means of an example.

Example 10-5. What should the sample size T be in order to assure that 97.5 per cent of the reliability samples will be accepted ($R_\alpha = 2.5$ per cent) when the acceptance number is 10 and the ARL is specified as $r_1 = 1$ failure/100 hr?

Solution

$$A_w = A + 0.5 = 10 + 0.5 = 10.5$$

Substituting in Eq. (10-4),

$$T_1 = \frac{2(10.5) + 1.96^2 - \sqrt{4(10.5)(1.96)^2 + 1.96^4}}{2(0.01)}$$

$$= \frac{21 + 3.84 - \sqrt{161.28 + 14.74}}{0.02}$$

$$= \frac{24.84 - 13.27}{0.02} = 578.5 \text{ (which is essentially } T = 579 \text{ hr)}$$

To determine R_α we use Eq. (4) of App. 2 as a basis:

$$Z_\alpha = \frac{A_w - r_1 T}{\sqrt{r_1 T}} \tag{10-7}$$

and since $r_1 T = (0.01)(579) = 5.79$, by substitution, we get

$$Z_\alpha = \frac{10.5 - 5.79}{\sqrt{5.79}} = \frac{4.71}{2.40} = 1.96$$

and this value of Z_α corresponds to the required R_α value of 2.5 per cent. It should be remembered that the equations of App. 2 are approximate, but despite this, the results obtained are extremely good.

10-12. Comparison of Table 10-2 and Equations for Determining Sample Sizes. From the foregoing discussion it has been shown that sampling plans can be determined from Table 10-2 or from Eqs. (10-3) and (10-4). When using the table it is necessary to use the Poisson curves of Fig. 6-1 if one is interested in specific values of R_α and R_β, since RO-C curves are not available. Each of these methods has specific advantages. The first is faster if one has the necessary curves and tables available. The advantage of the equations lies in the fact that once having memorized certain key values of Z corresponding to frequently used confidence levels, one can rapidly derive relatively accurate sampling plans.

A further refinement of Eq. (10-3) is made in Sec. 11-3, in which we derive a statistic called kappa square, whose values are tabulated in Table 8 of App. 7 and plotted in Fig. 11-1 for various confidence levels. This technique further simplifies and facilitates the derivation of a time sampling plan. The technique will be illustrated in Chap. 11.

A word of caution is in order at this time with regard to the interpretation of Table 10-2. Let us assume that we select a sampling plan $T = 50$,

$A = 1$. In this case we are actually stating that for a sample test time of 50 hr we are allowing one failure. At first glance this might appear to be an inconsistency, since, by definition, if we divide the total test time by the number of failures, we should get the mean time between failures. Thus, if we divide 50 by 1 the quotient should give us a value $m = 50$ hr. However, from Table 10-4 this sampling plan is supposed to assure a value of m of 100 hr at least 91 per cent of the time. This means that 91 per cent of the time we can expect that samples of 50 hr will have 1 or less failures, but 9 per cent of the time they will have 1 or more failures. On this basis, it should be obvious that before we know what m actually is, we must take several samples of 50 hr, since no one sample will tell us the complete story. We can synthesize the procedure in the following manner. The respective probabilities are found by calculating $rT = 0.01 \times 50 = 0.5$ and reading from the curves of Fig. 6-1. Thus, in round numbers the probability of exactly zero failures is 60 per cent; of exactly 1 failure, 31 per cent; of exactly 2 failures, 7 per cent; and of exactly 3 failures, 2 per cent. Thus, if 100 individual or subsamples were each tested for 50 hr, it would be expected that 60 samples would be expected to have zero failures, 31 would have 1 failure, 7 would have 2 failures, and 2 would have 3 failures. The total hours of test T would equal 100×50, or 5,000 hr. The total number of failures would be $60(0) + 31(1) + 7(2) + 2(3) = 51$ failures. Actually this figure should theoretically equal 50 failures, but since we rounded off the actual probabilities instead of using decimal fractions because we could not have fractional failures, we get 51 instead of 50. Thus, $m = 5{,}000/51 = 100$ hr (approximately). In general, the larger the number of samples, the closer would be the approximation to the true value of m.

From the above discussion it should be apparent that the sampling tables do work, provided it is understood that they reflect probabilities with relation to particular samples, but that no one sample by itself can provide the whole picture unless the degree of confidence is considered. For example, if by chance the first sample happened to have 3 failures, we would have an erroneous estimate of m; or if it had no failures in 50 hr, we could not have calculated m, because dividing by zero would give us an m equal to infinity, which is a ridiculous result.

Table 10-2 is also useful in those cases involving the number of operations or actuations of a device with respect to time. Thus, if the number of operations is known, it is a simple matter to translate these units into hours and apply the table.

Example 10-6. A device operates at a frequency of 100 operations per hour. Suppose that the specified mean life in terms of operations between failures is 100,000. What sample size in terms of number of operations would be used when $A = 0$?

Solution

$$m = \frac{\text{number of operations between failures}}{\text{operations per hour}} = \frac{100{,}000}{100} = 1{,}000 \text{ hr}$$

From Table 10-2 for $m = 1{,}000$ and $A = 0$, we find that $T = 110$ hr, corresponding to code letter K.

But 110 hr is equal to 110×100, or 11,000 operations. Therefore, the sample in terms of operations is 11,000 and the acceptance number is $A = 0$.

PRACTICE PROBLEMS

10-1. Convert 1,000 per cent failures per 1,000 hr to (*a*) failures per hour, (*b*) failures per 100 hr, (*c*) failures per 1,000 hr, and (*d*) bits.

10-2. The total number of failures from 10 samples was 5. The total time of test was 5,000 hr. What is the MRL expressed in terms of failures per hour? What would the acceptance number for each sample have to be to assure an R_α of 10 per cent? (Use MRL for ARL.)

10-3. Using the Poisson curves of Fig. 6-1, construct a sampling plan for an ARL of 1.5 failures/100 hr and an LTFRD of 6 failures/100 hr and maintain R_α and R_β at approximately 10 per cent.

10-4. Which sampling plan would you select from Table 10-2 to satisfy an ARL of 0.40 and $R_\alpha = 8.0$ per cent and an LTFRD of 4.0 and $R_\beta = 8.0$ per cent? Use Fig. 6-1 to verify conclusions.

10-5. Design a reliability sampling plan by using Eqs. (1) and (8) of App. 2 to satisfy the following conditions. ARL $= r_1 = 1$ fphh = LTFRD $= r_2 =$ 4 fphh; $R_\alpha = 10$ per cent, $R_\beta = 10$ per cent. Can you get a plan with R_α and R_β exactly equal, or must you arbitrate and compromise exact equality?

10-6. Find the sample size T for an ARL of 1.5 and $A = 0$ when $R_\alpha = 5$ per cent. What will this sample size become if $A = 5$ and the other conditions remain the same? (ARL expressed in dphu.)

10-7. How many sampling plans can you select from Table 10-2 to assure a mean time between failures of 6,670 hr? Which would you favor if you considered such elements as cost, the shape of the RO-C curve, other factors?

10-8. A telephone switch must be tested to assure a mean time between failures of 1,000,000 operations. It has been estimated that the switch operates at the rate of about 20 operations per hour. What sampling plans would you select from Table 10-2? Calculate the value of R_α that this sampling plan assures.

10-9. Design your own sampling plan for Prob. 10-8 for R_α and R_β equal to 2.5 per cent. Consider an ARL of 1,000,000 operations and an LTFRD of 250,000 operations.

CHAPTER 11

RELIABILITY SAMPLING AND CONTROL CHARTING—TIME SAMPLES

11-1. Relationship between Sampling and Time Samples. In most texts on quality control, sampling and control charting are treated as separate topics. The usual distinction between the two is that sampling is used to classify a finished product as defective or acceptable, whereas control charting concerns itself with the control of the production process. This is accomplished by plotting periodic representative data on control charts to determine whether or not the manufacturing method is producing an acceptable product. If the control chart indicates that defective items are being produced, the manufacturing process is adjusted accordingly. Thus, the basic difference is that sampling is fundamentally concerned with the finished items, whereas control charting is used to assess and control the manufacturing process. However, there are instances during production where data obtained from a sampling operation can also be used for control charting. A typical illustration of where such an application is possible is the sampling inspection following a final manufacturing or test operation. The data which are gathered here primarily as a sampling effort can also be used for control-chart purposes.

In the science of reliability, we can use the control-chart technique to assess reliability just as effectively as quality control engineers use it to control quality. However, as we mentioned previously, in the reliability case our samples are expressed in units of time, whereas in quality control the sample size is specified in terms of a number of physical units.

In Sec. 10-10 we discussed the ARL reliability sampling plans shown in Table 10-2. It was pointed out that this table is based on MIL-STD-105A, as extended and modified, and that the reason for altering it was to give us a table capable of being used for both the AQL and ARL. We also explained that from an RO-C curve we could determine the R_α associated with the ARL for each sampling plan. This means that the reliability producer's risk R_α in Table 10-2 is not the same for all sampling plans. It usually varies from approximately 2 to 20 per cent. In general, the larger the sample size, the smaller will be the value of R_α for the sampling plans specified in Table 10-2. However, a small value of R_α, although very desirable, is much more expensive, since the larger the sample size, the more it costs to conduct the testing program. Thus, there appears to be an optimum point at which increased testing costs are offset by the value of R_α, since, if it is relatively small, the risk of nonconformance is reduced. This balance, or optimum point, must be carefully worked out, because in reliability sampling it is sometimes impossible to rework a nonconforming lot since the nature of the required rework is not always known; or another reason might be that the tests are of a destructive nature. On the other hand, if R_α is large, the reliability might be unjustifiably adjudged as nonconforming on the basis of evidence gleaned from a small subsample. Therefore, there appears to be no alternative but to keep R_α as small as possible, consistent with reliability testing costs. This is why it is necessary to refer to RO-C curves when they are available, or resort to the means explained in Sec. 10-11, when they are not, to determine the value of R_α which satisfies the sampling plan.

11-2. Basic Requirements. In quality control work it is always emphasized that one does not sample unless he has a reasonable assurance that the lot from which the sample is taken is homogeneous, i.e., the product must come from a controlled manufacturing process. Moreover, the sample must be selected at random from the lot in such a manner as to assure that each item has an equal chance of being selected. When this is done, then the sample is considered to be representative of the lot; therefore, its quality is a true measure of the lot quality within specified confidence limits.

Likewise, in reliability sampling the existence of a certain degree of maturity of design must be demonstrated before sampling is undertaken. This means that there must be an indication that one component does not differ appreciably from another and that each was built to conform to the same engineering requirements and was manufactured under the same conditions using similar methods and processes. In brief, there must be an assurance that each component comes from the same or equivalent family. Another way of summarizing this situation is to say that all of the elements which are included in the component from the initial design phase to final production must be in a state of statistical control. This indication of statistical control usually manifests itself in terms of a constant failure rate. Chapter 6 discusses the subject of constant failure rate thoroughly. Actually, the constant rate is the limit of the MRL (usually specified as ARL), as T approaches infinity. Recalling that the symbols for MRL and ARL are respectively $\bar{r}$ and r_1, we see that on this basis, $\bar{r}$ would have to be between the upper and lower control limits of r_1 [whose value can be determined by the use of Eqs. (10-1) and (10-2) if

3-sigma limits are desired] to assure that it approximates or approaches r_1. This should be obvious, since as T increases in size, the control limits approach r_1 as a limit, at which time $\bar{r}$ approaches r_1, the specified constant failure rate (ARL). This was the reason for using such a large time sample in Table 6-3, which was used to illustrate a constant failure rate.

From the foregoing discussion, it is apparent that before we institute sampling, we must be certain that adequate maturity of design and manufacturing process control exist. We must also be prepared with an adequate sampling plan which, among other things, describes the steps to be taken in the event that the results for a time sample produce evidence of substandard or nonconforming reliability. During production this is not difficult to do, because sufficient units or test time is usually available. However, during the engineering phase, when only one or two prototypes have been built, they must of necessity provide the data which will be the basis for determining design maturity. These data are accumulated by the simple expedient of testing a unit to failure, then repairing and retesting it until sufficient information is available to demonstrate that a constant failure rate has been achieved. The theoretical basis for this procedure depends on the fact that the probability of failure of a device is the same after repair as it was before, provided that the wear-out phase has not yet been reached and that the repair operation does not cause damage to the unit. If the experiment shows that a constant failure rate has not been achieved, it is prima facie evidence that a mature design has not been demonstrated and that a statistically controlled process is not in being. Therefore, there must be influences other than chance causing the failures. In this case, time sampling should not be used until additional engineering effort succeeds in finding and eliminating the causes of failure. When data show that a constant failure rate exists, then sampling may be instituted in accordance with a plan selected from Table 10-2. However, we must not forget the fact that the R_α of any one of these plans will permit an occasional rejection of a sample even if it is from a population whose MRL equals the ARL.

This is why it is important to secure additional evidence before a decision of nonconformance is made. The method used is explained in subsequent discussions, particularly in Secs. 11-3 and 11-4. It depends upon accumulating data from sub-time samples which are plotted on a control chart from which it is possible to study trends and arrive at various conclusions.

A sub-time sample is a fraction of the total sample which is needed to satisfy certain values of R_α and R_β. Henceforth, we shall refer to the total time sample merely as the time sample. We shall also eliminate the word "time" from the term "sub-time sample" and merely refer to it as a subsample. This elimination is considered justified, since it should be understood that we are referring to a sub-time sample in a time sampling scheme.

Plotting data representative of subsamples on a control chart is a powerful and effective technique. This control chart permits a continuous surveillance of reliability testing from subsample to subsample. Moreover, it provides an excellent medium for making a positive decision with respect to conformance or nonconformance of the reliability test.

Figure 11-3 shows subsamples plotted from the data of Example 11-2. This chart will be discussed in detail when we explain Example 11-2. In the meantime, it serves to illustrate how subsamples are plotted. It also demonstrates the method used to record the type and frequency of failures. This ensures the availability of data which can be used for analysis and corrective action. Thus, by combining the technique of acceptance sampling and control charting, we are assured that decisions are based on indications of trends from subsample to subsample, rather than solely on the results from one time sample. In any event, no decision regarding conformance or nonconformance should be made on the basis of any one subsample unless good evidence exists for making such a decision. The decision of conformance or nonconformance should be made only after all the subsamples required for the time sample have been tested.

Another reason for plotting the results of subsamples is to assure that sufficient evidence is available from a wide random cross section of the population to substantiate a decision of nonconformance, should this appear evident. The reason for exercising such caution is that an adverse verdict is so costly that it should be made only after all the facts have been carefully evaluated. In quality control work the rejection of a lot merely involves reworking and resubmission, whereas in reliability such a rejection affects not only the lot, but the entire production process and design.

Another reason for resorting to subsamples is that standard sequential sampling plans currently appearing in literature require excessively large test times before a decision is reached. This usually involves a long, costly reliability test cycle. The advantage of a subsample is that it provides some confidence within a relatively short time about the degree of reliability obtained. Thus, a very poor product will show up almost immediately as having poor reliability, while a very good product will also be evident. In those cases where a doubt exists about the ultimate results, the subsample technique makes possible the early study of trends and assignable causes, which facilitates timely corrective action. The details will be discussed in Sec. 11-4.

11-3. Determination of Sample Size and Acceptance and Rejection Numbers. In this section we shall discuss the technique of determining the size of time samples and subsamples and their associated acceptance and rejection numbers.

Time Samples. One of the methods of determining the size of a time sample is to use the techniques described in Sec. 10-11 until a time sampling plan is found which will satisfy the desired R_α and R_β. As we have seen, this is a lengthy and annoying procedure, particularly when we desire a two-tailed confidence level, i.e., one involving R_α and R_β. It is relatively simple for a one-tail confidence level with either R_α or R_β specified.

A simplified method of calculating the required time sample is to use tabulated values of the statistic K^2 (kappa square)[1] appearing in Table 8 of App. 7.

This is done by deriving Eq. (11-1) from Eq. (10-3). In this case $m = 1/r_1$; therefore, we can rewrite Eq. (10-3) as

$$T = \frac{(Z_\alpha - Z_\beta \sqrt{k})^2}{r_1(k-1)^2}$$

or

$$T = m\left(\frac{Z_\alpha - Z_\beta \sqrt{k}}{k-1}\right)^2 \tag{11-1}$$

If we call the squared parenthetic expression of the above equation kappa square (K^2), we have

$$T = mK^2 \tag{11-2}$$

It should be recalled that the value of k in Eq. (11-1) is the ratio of r_2/r_1 or LTFRD/ARL.

The critical values of K^2 for various values of k and various confidence levels[2] were calculated by substitution of various values of k, Z_α, and Z_β in the squared parenthetic expression of Eqs. (11-1).

For convenience, the values of K^2 have also been shown as a family of curves in Fig. 11-1.

The advantage of using K^2 is that it provides a rapid method of calculating a time sample for a two-tailed confidence level. In this manner, proper cognizance is taken of the producer's and consumer's reliability risks for the specified ARL and LTFRD without the necessity for much laborious trial-and-error mathematical computations.

Subsamples. As we have seen, a subsample is a fractional part of the time sample. The size of the time sample is calculated for the optimum desired confidence level (usually 90 or 95 per cent). The size of the sub-

[1] A statistic developed by S. R. Calabro.

[2] Strictly speaking, the term *confidence level* as used in conjunction with K^2 is not exactly as defined in Chap. 8. In this case we mean that γ per cent of the time the value of R_α will lie between the minimum value corresponding to an expectation of r_1T and a maximum value equal to r_2T. The unity complement of this latter value is R_β. Thus if $\gamma = 90$ per cent, $R_\alpha = 5$ per cent and $R_\beta = 5$ per cent; or, in general, $R_\alpha = R_\beta = \frac{100 - \gamma}{2}$ in per cent. It is probably better to call γ the *confidence factor*.

sample is calculated on the basis of a lower confidence level. In any case, the sample size is expressed in hours and consists of the aggregate test time on one or more physical units.

There are other considerations which one should weigh when determining sample size. These are (1) the availability, cost, and required amount of test equipment; (2) the availability of space and skilled technical personnel; (3) the ratio of the number of hours of reliability testing per article to the expected number of hours of useful life; (4) the cost of necessary refurbishing of the article after extensive reliability testing; (5) the ratio of the combined total testing time per article, consisting of such tests as debugging, production, reliability, and special tests, to the expected number of hours of useful life.

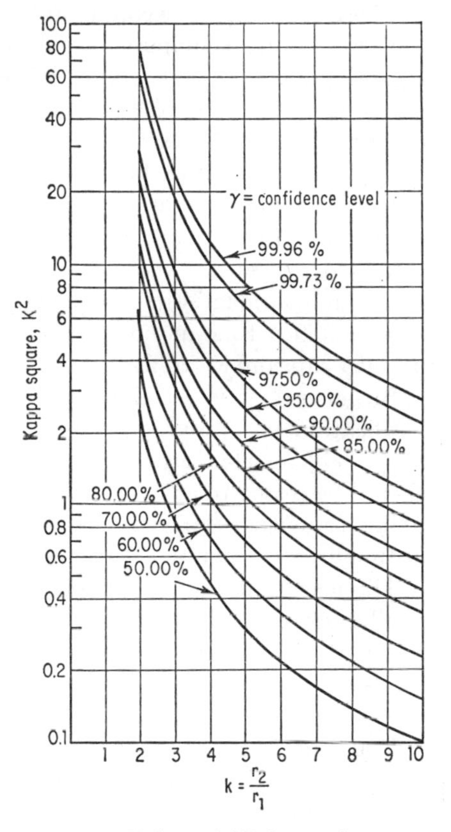

FIG. 11-1. Values of K^2 for various confidence levels. Note: Multiply K^2 value by mean time between failures for required time sample.

The above considerations are very basic, particularly with reference to points 1 and 2. The reason is that in many instances reliability testing requires much expensive test equipment, space, and skilled testers. Therefore, in forming a sub-time sample it might be more desirable at times to select a lesser number of articles and test each of them for a longer period than to test a greater number of articles each for a lesser time. This is a very important consideration, because usually the design and development of test equipment for complex units is just as involved as the design of the device being tested.

The above considerations also affect the cost of reliability sampling programs. Therefore, whenever possible, it is advisable to select the best combination of factors to keep the cost of testing low without affecting the confidence level of the reliability estimate. In doing this, one should be careful not to neglect points 3, 4, and 5, as they relate to the operating period. In general, it is not desirable to consume more than 10 per cent of the operating or useful life of equipment in testing without requiring refurbishing of the articles which comprise the subsample. In electronic equipment, in particular, specifications usually require extensive testing. These tests are of many varieties depending on the particular specification involved. Some are 100 per cent tests such as production and debugging tests, while others are of a sampling nature such as preproduction tests, type tests, or general periodic tests. One should consider the duration of all these tests before deciding on how much test time to assign to each of the articles which compose the subsample. As was previously mentioned, a good empirical rule to follow is to make certain that not more than 10 per cent of the predicted useful life of each article of the subsamples is consumed in reliability testing. This is a most important consideration of reliability sampling, because too much testing of any one article may reduce its remaining useful life.

There is one advantage, however, in testing some articles for an extended period of time when it is desired to gain some knowledge of wear-out characteristics. In this case, some subsamples may actually represent test time on only one article whose total test time was broken up into subsamples to facilitate plotting the data on control charts.

In general, it is recommended that at the beginning of a reliability verification test program, randomness of subsamples be used as widely as possible. This means that the greatest possible mixture of units should be tested to compose various subsamples. Moreover, some of the units of the subsamples should periodically be tested for extended periods of time to get a figure of merit of failure rate with respect to time.

The acceptance and rejection numbers associated with a time sample can be determined from Eq. (1) of App. 2 or from the Poisson curves of Fig. 6-1. They can also be calculated from K^2 when two-tail confidence levels are desired, i.e., when R_α and R_β are specified. The method should be evident from the following discussion.

Equation (11-2) can be rewritten as $T = K^2/r_1$ or $K^2 = r_1 T$ where r_1 = ARL. Therefore, it is obvious that K^2 equals the expectation, or average number of failures which may be expected in a time sample of T hours when the failure rate equals the ARL. In this case, the very fact that K^2 is used is an advantage because it gives a two-tail confidence level and, therefore, takes cognizance of r_2, the LTFRD, and $k = r_2/r_1$.

From Eq. (1) of App. 2 we see that $A_w = r_1 T + Z_\alpha \sqrt{r_1 T}$, but since $K^2 = r_1 T$, we can write

$$A_w = K^2 + Z_\alpha \sqrt{K^2} \qquad \text{(working acceptance number)}$$

and since $$A = A_w - 0.5 \qquad \text{(acceptance number)}$$

we have $$A = K^2 + Z_\alpha \sqrt{K^2} - 0.5 \tag{11-3}$$

But the rejection number $R = A + 1$. Therefore

$$R = K^2 + Z_\alpha \sqrt{K^2} + 0.5 \tag{11-4}$$

If the value of A or R obtained from the above equations is a decimal fraction, the nearest whole number will be used.

The method of finding K^2 for a specific value of k is the same as that just explained above. The value of Z for various two-tailed confidence levels may be found from Table 3 of App. 7. Once these values are obtained, all that is required is to substitute in Eqs. (11-3) and (11-4) to get the acceptance and rejection numbers, respectively. The acceptance or rejection number, so determined, may be used as the upper control limit of a subsample f chart, depending on whether it is intended as the acceptance or rejection criterion. The method of plotting the data on such a chart and the use of upper control limits are illustrated in the solution of Example 11-2.

11-4. Types of Time Sample Control Charts. There are two types of time sample control charts which we shall discuss: the r chart and the f chart. The r chart is also called the failure rate chart, because it is a plot of failure rates calculated from various subsamples of varying sizes. The f chart, on the other hand, is a plot of the number of observed failures which occur per subsample. When these failures are plotted per each subsample, the resulting chart is called the *subsample f chart.* If cumulative failures for cumulative time (i.e., the sum of all subsamples) are plotted, the chart is called the *sequential f chart.* In all cases, the charts are characterized by upper and lower control limits for specified confidence levels.

The use of control charts results in many advantages. They are an economical means of assessing reliability because they can be used as a control and acceptance medium. They also highlight trends which are invaluable when one is making product improvement studies or making decisions concerning conformance or nonconformance of designs. The charts are also useful for calculating or rapidly estimating the MRL. In addition, they can be used to supplement a failure reporting system and to isolate the causes of failure. We shall highlight all these points as we discuss each type of chart in turn.

11-5. The r Chart. This type of chart plots the failure rate $\bar{r}$ as calculated from various subsamples of varying sizes.

The purpose of the chart is to determine whether the variation of the values of $\bar{r}$, so calculated, are within the confidence limits of the ARL. This is done by calculating the upper and lower control limits. In those instances where an ARL has been contractually specified in terms of failures per 100 hr, the upper control limit (UCL) and lower control limit may be calculated by using Eqs. (10-1) and (10-2).

When an ARL has not been specified, the MRL can be calculated by dividing the total failures for all subsamples by the total of all subsamples. Control limits of the MRL can then be established. In any case, if individual values of $\bar{r}$ exceed the UCL, this is an indication that we do not have statistical control. Hence, either the manufacturing process is to blame or else the design is not mature. At this point, an investigation should be initiated to determine what is termed the *assignable cause of variation.*

We look for an assignable cause of variation because it is evident that some such cause must be responsible for one or more points exceeding the control limits, since it is highly improbable that such variation (particularly for 3-sigma control limits) could have occurred because of a constant-cause chance system acting alone.

The upper and lower Z-sigma control limits of the ARL, specified in terms of failures per hour, can be determined as follows:

$$\text{UCL} = \text{ARL} + Z\sqrt{\frac{\text{ARL}}{T}} \tag{11-5}$$

$$\text{LCL} = \text{ARL} - Z\sqrt{\frac{\text{ARL}}{T}} \tag{11-6}$$

When $Z = 3$ we have the 3-sigma limits written as

$$\text{UCL} = \text{ARL} + 3\sqrt{\frac{\text{ARL}}{T}} \tag{11-7}$$

$$\text{LCL} = \text{ARL} - 3\sqrt{\frac{\text{ARL}}{T}} \tag{11-8}$$

and if ARL is specified in terms of failures per 100 hr, we can derive Eqs. (10-1) and (10-2) as follows: Multiplying by 100, we get

$$\begin{aligned} \text{UCL} &= 100\ (\text{ARL}) + 3(100)\sqrt{\frac{\text{ARL}}{T}} \\ &= 100\ (\text{ARL}) + 3\sqrt{\frac{100\ \text{ARL}}{T/100}} \end{aligned}$$

It is interesting to note that the $T/100$ term in the denominator of the radical gives us a new unit of time in terms of 100 hr. Thus, if T were 1,000 hr, we would have 10 units, each equal to 100 hr. Thus, whether we are measuring failure rate in terms of failures per hour, with the hour being the basic unit, or in terms of failures per 100 hr, with 100 hr being the basic unit, we note that the standard deviation of the subsample is inversely proportional to the square root of the number of units.

Rewriting the above equation, we get

$$\text{UCL} = 100\ \text{ARL} + 30\sqrt{\frac{100\ \text{ARL}}{T}}$$

and if we remember that 100 ARL is the ARL expressed in failures per 100 hours, we get

$$\text{UCL} = \text{ARL} + 3\sigma\sqrt{\frac{\text{ARL}}{T}} \qquad \text{in failures per 100 hr}$$

This is the same as Eq. (10-1).

Since the Poisson is a nonsymmetrical distribution, it might be argued that probability limits should be used in lieu of sigma limits. Although the argument is theoretically valid, from a practical viewpoint it means little. This is due to the fact that in reliability work we are dealing with small values of r, and, hence, in most instances the lower calculated control limit is negative; and since we cannot have a negative value of r, we arbitrarily call this lower limit zero.

The upper control limit, as calculated from the sigma control limits, is sufficiently accurate for our purposes, since any departures from the Poisson actually cause the true standard deviation to be greater than the value $\sqrt{\text{ARL}/T}$, and, therefore, our estimate is at best conservative. Moreover, the advantages of speed in setting up our control limits, as shown, far outweigh any disadvantages of slight mathematical inaccuracy.

The r* chart is a valuable tool which can be used to particular advantage during the production stage. Its chief advantage is that it facilitates the determination of $\bar{r}$ within confidence limits at an early stage in the program. This means that it is possible to determine within a specified degree of confidence whether or not the measured value of $\bar{r}$, during a production reliability test program, is within the limits of what is desired or specified. The knowledge so obtained is usually of paramount importance because it may be the basis for determining whether or not production should continue.

Thus, we see that the r chart tells us early in production whether we are producing to standard or whether corrective action is required. This is a decided advantage over acceptance sampling per se, since the latter usually requires a relatively large sample before a decision is made regarding conformance or nonconformance. Hence, if a condition of nonconformance is discovered late in time, this could represent untold high costs and delays.

Example 11-1 will be used to illustrate the method of plotting and analyzing the r chart.

Example 11-1. The ARL of a device was specified as 1 failure/100 hr. A reliability prediction study indicated that the design was such as to satisfy this requirement. Units were tested to verify whether or not the reliability requirement had been met. The data were recorded and tabulated as in Table 11-1.

* It should be recalled that $\bar{r}$ is used to indicate the calculated mean reliability level of a subsample or time sample. The symbol r is used as a general symbol meaning failure rate, while r_1 is used to indicate ARL.

TABLE 11-1. DATA OF RELIABILITY TEST

Subsample no.	No. of failures	Subsample size, hr	Failures per 100 hr $\bar{r}$	$30\sqrt{\frac{ARL}{T}}$	UCL = ARL $+ 30\sqrt{\frac{ARL}{T}}$	LCL = ARL $- 30\sqrt{\frac{ARL}{T}}$
1	1	100	1	3.00	4.00	0
2	2	210	0.952	2.07	3.07	0
3	3	120	2.50	2.74	3.74	0
4	3	300	1.00	1.73	2.73	0
5	1	75	1.33	3.46	4.46	0
6	6	200	3.00	2.12	3.12	0
7	3	275	1.09	1.81	2.81	0
8	6	300	2.00	1.73	2.73	0
9	1	125	0.800	2.68	3.68	0
10	1	80	1.25	3.35	4.35	0
11	1	95	1.05	3.08	4.08	0
12	13	375	3.47	1.55	2.55	0
13	1	110	0.91	2.86	3.86	0
Totals.......	42	2,365				

Discussion of Example 11-1. The r chart is plotted as Fig. 11-2. It should be noted that the upper control limit varies inversely with the sample size. The greater the subsample, the closer are the control limits. This is only logical, since the variation of the calculated value of $\bar{r}$ from subsample to subsample should be less as the size of subsamples increases.

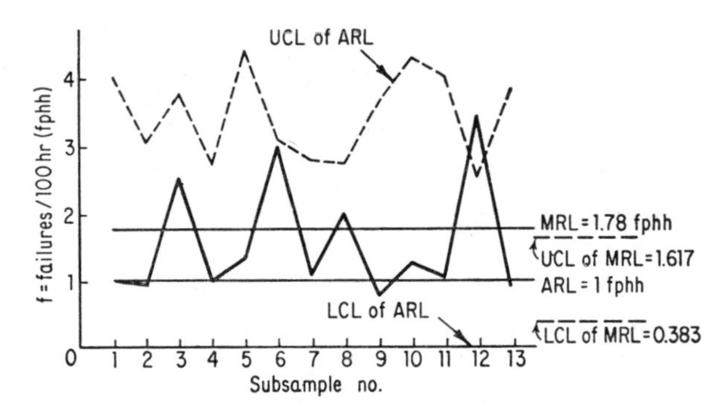

FIG. 11-2. r chart for data of Example 11-1.

Note that there is no lower control limit, since in each case the calculated value of the lower control limit is negative; therefore it is shown as zero.

At this point, a word of caution is in order, since much confusion often exists about the meaning of control limits. The control limits are established on the basis of the ARL value. Hence, they will not change as a function of $\bar{r}$. The only factor which affects them is the size of the subsample, since by specification the ARL is fixed, and from Eqs. (10-1) and (10-2) we see that the control limits are only a function of time. A common error is to add the computed value of $\bar{r}$ to the 3-sigma value. This is strictly to be avoided, since it will give an erroneous control limit. It is the ARL and not $\bar{r}$ which should be added to the 3-sigma value to get the UCL.

The reason is that we are measuring the variation of the plotted points against limits dictated by the specification. Therefore, if these plotted points do not exceed the control limit, we are certain that the failure rate of the device is reasonably close to the ARL.

In our example, it should be noted that the point corresponding to subsample 12 exceeded the upper control limit. This could be due to the fact that either the MRL exceeds the ARL or else there is an *assignable cause.* Let us first calculate the MRL.

$$\text{MRL} = \frac{\Sigma f(100)}{\Sigma T} = \text{fphh}$$

The MRL is the ratio of the total number of failures of all the subsamples to the sum of all subsamples. Thus, since there were 42 failures in 2,365 hr,

$$\text{MRL} = \frac{42(100)}{2{,}365} = 1.776 \text{ fphh}$$

Since we treated the sum of the subsamples as if it were one large sample (i.e., the time sample), we can calculate the UCL and LCL of the MRL, viz.:

$$3\sigma = 30\sqrt{\frac{\text{ARL}}{T}} = 30\sqrt{\frac{1}{2{,}365}} = 0.617$$

and therefore

$$\text{UCL} = 1 + 0.617 = 1.617$$
$$\text{LCL} = 1 - 0.617 = 0.383$$

Thus, we see that since the MRL = 1.776 failures/100 hr, it exceeds the UCL of 1.617, which indicates an out-of-control condition. Therefore, since subsample point 12 as well as the MRL which was calculated on the basis of 13 subsamples both show an out-of-control condition, an investigation is in order to determine the cause.

An obvious conclusion is that since all subsamples preceding number 12 were in good statistical control and so, apparently, was subsample 13, something peculiar must have occurred with subsample 12. This behavior might be attributed to a variety of things, but most likely is not due to design, because of the excellent statistical behavior of most of the subsamples. Therefore, the investigation should confine itself to human factors, test equipment, personnel, etc.

From Example 11-1 we can see that the r chart is a good tool for controlling the reliability during production as well as a medium of determining conformance or nonconformance. Its chief disadvantage is that the subsample size must vary, because failures occur at random times and we must have one or more failures in a subsample in order to calculate the failure rate. This requires additional calculations of varying control limits. It is obvious, then, that if we could keep our subsample size constant, we should be able to calculate the UCL only once, and the value so obtained would be applicable to all subsamples. The peculiarity of a constant subsample size is that in some subsamples no failures will be evident, and, therefore, the apparent failure rate is zero. In such instances we would have to plot the failure rate as if it were zero, despite the fact that we know this is not actually the case. In a sense, this might be an advantage, since with less test time we could still evaluate reliability from the standpoint of statistical control, since if no point exceeds the UCL in many subsamples, we are certain $\bar{r}$ is within the prescribed limits.

There is another variation of the r chart which is useful in the development stage when prototype testing is in progress. This involves testing to first failure, repairing, and continuing the test to the next failure, etc. This may be done on one unit or a number of units. In this instance, the number of failures per subsample will in all cases be unity, but the size of the subsample varies. The failure rate per subsample is, therefore, the reciprocal of the subsample size in hours. This calculated failure rate per subsample is the one which should be plotted on the r chart.

The chief advantage of this type of r chart is that it facilitates the determination of the time duration of the infant mortality period and operating period, because during the infant period the frequency of failure is greater than it is during the operating period. This results in a higher failure rate and a lower value of m. The r chart vividly points this out. It also highlights the point at which a constant failure rate starts to become evident, which is indicative of an approach to the operating period. On the other hand, in Example 11-1 there was no attempt to record the exact time of failure for each subsample; nor was a subsample restricted to only one failure, i.e., in this case data were accumulated from one or more units before being recorded and plotted.

Any of the above variations of the r chart is useful, depending upon the type of information or control which is desired. The reliability engineer should use his judgment, in each instance, in deciding on the specific method of plotting the r chart.

In general, when the subsample is kept constant in size, a better chart to use in lieu of the r chart is the subsample f chart, because the control limits are more facile to calculate, and, therefore, it is an easier chart to plot.

11-6. The Subsample f Chart. The subsample f chart is used to study trends during production as a result of data which are obtained and plotted from as wide a cross section of production units as is practicable. It is also used as an acceptance-rejection criterion or conformance-nonconformance vehicle.

The chief advantage of the subsample f chart is that it reduces reliability acceptance testing costs and also provides for product improvement. The testing costs are lower because reduced testing at lower confidence levels is permitted when chart trends indicate that the risk is justified. The total reduced testing time is the subsample.

The chart also provides for product improvement because it permits a rapid determination of high failure frequency and allows one to ascertain assignable causes of failure by referring to referenced failure report numbers appearing on the chart. Thus, we see that this type of chart is very useful as an integral part of a good failure reporting system.

In general, the function of an f chart is to determine whether statistical control exists. It is not intended to be used for making specific calculations of failure rate or for other purposes, even though the data exist to facilitate such calculations. Therefore, this chart must be used by those having the experience and maturity of judgment necessary to interpret chart results. For example, at the beginning of the reliability test cycle it might be desirable to test the first unit for a time sample which will give a 90 per cent confidence that the reliability levels have been achieved. Depending on the results obtained, the next time sample might call for a 70 per cent confidence level and subsequent time samples for a 50 per cent confidence level. In each instance, the points plotted on the chart should be indicative of whether the high risks involved with the subsample are justified or whether a return to a time sample corresponding to a high level of confidence (lower risk) should be resorted to.

Another advantage of the subsample f chart is that, if desired, it can also be used to plot the mean time to repair and reference it to specific failures and failure report form numbers. This is a decided advantage because availability calculations depend upon an accurate figure of ϕ. Moreover, since the repair or maintenance-action rate μ is the reciprocal of ϕ, it can also be easily calculated. These data are very important for logistics considerations, particularly for spare parts estimates.

Figure 11-3 shows a plot of ϕ on the lower part of the chart. These data could also have been plotted on the upper part of the chart by the addition of a time scale to the ordinate in addition to the number of failures scale. However, this might result in confusion due to the probable overlap of the plots of failures and mean time to repair. For this reason, it was considered more advantageous to plot ϕ at the bottom.

The mean time to repair is calculated by adding the individual times to isolate, locate, and repair each failure, the sum of which is considered the repair time, and dividing by the number of failures. This mean time to repair has also been referred to as the mean time of maintenance action to cover all general failure restoration activity, such as adjustments, module replacement, etc. It is permissible to use these terms interchangeably, provided that the speaker is aware of what is implied when he does so and that his interpretation is clear to other interested parties.

The control limits for the number of failures are established as explained in Sec. 11-3, Eqs. (11-3) and (11-4). If it is desired to use acceptance numbers as upper control limits, this means that if the number of failures equals or is less than A, the sample is considered as conforming for the confidence limit corresponding to the particular acceptance number. This confidence limit can also be shown on the chart.

If the rejection number is used as the upper control limit, this means that if R or more failures are observed in a sample, it is considered to be nonconforming and indicative of substandard reliability for the confidence level shown on the chart for the specific value of R.

The control limits for the mean time to repair may be determined in accordance with the methods explained in Secs. 8-6 and 8-7. For example, if the specified mean time of maintenance action for a device is 1 hr, this is the standard which must be achieved with prescribed confidence levels. If data from a time sample indicate that to repair 5 failures requires 2.5 hr, the value of ϕ as calculated from the time sample is $\frac{2.5}{5} = 0.5$ hr. The rejection number is 5 (i.e., equal to number of failures), and the degrees of freedom $2R = 2(5) = 10$. If we wanted a two-tailed confidence level of 95 per cent, then the level of significance in each tail would be 0.025 per cent. The control limits (refer to Secs. 8-6 and 8-7) are

$$L = \frac{2R\phi}{\chi^2_{10(0.025)}} = \frac{10(0.5)}{20.5} = 0.24 \text{ hr}$$

$$U = \frac{2R\phi}{\chi^2_{10(0.975)}} = \frac{10(0.5)}{3.25} = 1.53 \text{ hr}$$

The control limits could also have been obtained from Fig. 6-1 as explained in Sec. 8-7. In this case, we would have looked up the value for $f = 4$ and an α of 0.025 and found an expectation of 10.2. Therefore, the total repair time of 2.5 hr divided by 10.2 would give us the lower control limit. Similarly, if we looked up the corresponding value for a level of significance of 0.975, the expectation would be read as about 1.6, and if this were divided into 2.5, we would get the upper control limit. In this type of problem, our real interest is in the upper control limit, since any value of ϕ which is smaller than the lower control limit is better by far than what had been specified. The only advantage of plotting the upper and lower control limits is to give an indication of statistical control.

There is another advantage to the subsample f chart as it applies to failure reporting or product-improvement programs, because it charts all the failures during the time in which they occur. It also indicates, at a glance, the frequency of occurrence of specific failures. This is far superior to the failure reporting methods described in Chap. 7 because this type of chart visually highlights the causes of a failure on a timely basis.

A major deficiency of failure reporting systems is that they often result in too much paper work and too little analysis. They also wind up with reams of paper and IBM forms which defy analysis. The subsample f chart overcomes all these shortcomings. In addition to acting as a control chart, it also serves to highlight causes of failure in a clear visual and statistical sense and, therefore, represents a vital part of an equipment correction program.

From the foregoing discussion, it should be obvious that reliability verification based on the principles presented is much more economical than would have been possible under other current versions of sampling plans. This economy becomes particularly more meaningful when the control chart technique is used to monitor reliability at the higher levels such as a subsystem.

Example 11-2. Analysis of data from extensive testing of preproduction models of a microwave transmitter resulted in an estimate of the failure rate of 0.004 failures/hr and a mean time to repair (mttr) ϕ of 10 min. It was later decided that these values were satisfactory, and, therefore, they were specified in the procurement contract.

The infant mortality (early failure period) was established to be 25 hr. The specification read in part: "The ARL shall be 4 failures/1,000 hr and the LTFRD shall be 12 failures/1,000 hr." This corresponds to values of R_α and R_β equal to 5 per cent.

(*a*) Determine the required time sample to the nearest 100 hr and the applicable rejection numbers at the 90 per cent confidence level.

(*b*) Determine the subsample size to the nearest 100th hr and also find the rejection number R.

(*c*) Determine the confidence level for R_α, for cumulative subsamples.

(*d*) The data for a typical reliability run are shown in Table 11-2. Plot these data on a subsample f chart and discuss the subsequent analysis and conclusions.

TABLE 11-2. DATA OF EXAMPLE 11-2*

Time sample	Subsample number	Sample size	Number of failures	Mean time to repair for subgroups of 4
First	1	200	1	
	2	200	0	$\frac{24+20+8+16}{4} = 17$
	3	200	0	
	4	200	3	
	5	200	2	
	6	200	0	$\frac{12+4+20+16}{4} = 13$
Second	7	200	0	
	8	200	1	
	9	200	2(2)	
	10	200	1	
	11	200	0	$\frac{28+4+16+8}{4} = 14$
	12	200	1	
Part of third	13	200	0	
	14	200	2	
	15	200	3	$\frac{8+12+12+4}{4} = 9$
Total (for individual values)	...	...	16	212

* Grand average of mttr = 212/16 = 13.25 min. Specified mttr = 10 min.

Solution

(*a*) $r_1 = 0.004$ fph $R_\alpha = 5$ per cent $Z = 1.64$
$r_2 = 0.012$ fph $R_\beta = 5$ per cent $\gamma = 90$ per cent

$$k = \frac{r_2}{r_1} = 3$$

The value of K^2 when $k = 3$ and $\gamma = 90$ per cent is 5. Substituting in Eq. (11), App. 2, we get

$$T = mK^2 = 250(5) = 1{,}250 \text{ hr}$$
$$= 1{,}200 \text{ hr (to nearest 100 hr)}$$

The rejection number $R_{1,200}$ for a time sample of 1,200 hr and confidence level $\gamma = 90$ per cent is

$$R_{1,200} = K^2 + Z_\alpha \sqrt{K^2} + 0.5 = 5 + 1.64\sqrt{5} + 0.5 = 9.16$$
$$R_{1,200} = 9 \text{ failures (to the nearest whole number)}$$

(*b*) The subsample for $\gamma = 50$ per cent, $k = 3$ is

$$T = 250(0.84) = 210 \text{ hr}$$

Use $T = 200$ hr (to the nearest 100 hr) and

$$R_{200} = 0.84 + 0.68\sqrt{0.84} + 0.5 = 1.96$$
$$R_{200} = 2 \text{ failures (to the nearest whole number)}$$

(*c*) The confidence levels for the cumulative samples of 400, 600, 800, and 1,000 are determined in a manner similar to the illustration below for a sample of 400 hr.

$$K^2 = \frac{T}{m} = \frac{400}{250} = 1.6$$

Referring to Fig. 11-1 when $K^2 = 1.6$ and $k = 3$, one finds by visual interpolation that the confidence level γ is approximately 65 per cent.

The confidence levels for all samples, determined in a similar manner, are shown on Fig. 11-3.

Discussion of Example 11-2. Figure 11-3 shows a plot of the data of Example 11-2. Although most of the chart is self-explanatory, in the interest of clarity a few important matters will be highlighted.

The subsamples of 200 hr each are shown at the top of the chart directly below the subsample number. Directly below the subsample cells there are other cells which are subdivided into two sections. The top section shows the cumulative rejection number and the bottom section the actual cumulative number of failures for a group of subsamples. The actual number of failures per subsample is plotted on the control chart. For example, the numbers 5/1 for subsample 3 imply that in a cumulative sample of 600 hr (3 subsamples of 200 hr each) the cumulative rejection number is 5 and the actual cumulative number of failures is 1.

The ordinate of the graph of failures shows the number of failures. To the left are shown the rejection numbers R and confidence levels γ corresponding to the various sample sizes. Thus, $R_{1,200(0.90)} = 9$ means that the rejection number for a sample of 1,200 hr is 9 failures and the confidence level γ is 90 per cent. Thus R_α and R_β are each 5 per cent.

The reasons for failure are listed on the left side of the center of the chart. The number appearing at the point of intersection of a "reason for failure" row and a "failure" column is the failure report number. This identifies the failure report which lists the details regarding the nature and probable cause of failure. An advantage of this type of presentation is that it highlights the frequency of failure occurrence. For example, it is obvious that failure code 2 has a high frequency of failure, since it occurs respectively on failure report numbers 2, 7, 15, and 16.

The lower part of the chart shows a plot of the repair time for each failure in minutes. Each failure is numbered to correspond to the failure report covering it. The average or mean time to repair for subgroups of 4 failures was calculated and plotted as small triangles following each of 4 failures. Subgroups of 4 were arbitrarily selected to keep the upper control limit of the specified mean time to repair (mttr) constant. Any other subgroup size could have been used, if desired. For example, with subgroups of 4, the respective repair times for failures 1, 2, 3, 4 are 24, 20,

PRODUCT ASSURANCE DEPARTMENT

SHEET 1 OF 1

CHART NUMBER 1	SUBSAMPLE NO.:	1	2	3	4	5	6	7	8	9	10	11	12	13	14	15
ARL = 0.004 fph LTFRD = 0.012 fph k = 3	SUB - TIME SAMPLE, HR	200	200	200	200	200	200	200	200	200	200	200	200	200	200	200
	CUMULATIVE REJECTION NO. / CUMULATIVE FAILURES	2/1	3/1	5/1	6/4	7/6	9/6	2/0	3/1	5/3	6/4	7/4	9/5	2/1	3/2	5/5
	SERIAL NUMBERS – SAMPLE UNITS	(1)	(1)	(1)	(1)	(1)	(1)	(2)	(3)	(4)	(4)	(5)	(5)	(6)	(7)	(7)

Subsamples 1–6: Time sample 1; subsamples 7–12: Time sample 2.

TIME SAMPLE SIZE: 1,200 hr

SUB: 200 hr

MTTR = 10 min = ϕ'
$\mu = 4$

ITEM NAME: Microwave transmitter

ITEM DRAWING NUMBER: MWT-1380A

CODE NO.	REASON FOR FAILURE	1	2	3	4	5	6	7	8	9	10	11	12	13	14	15
1	Tube V-101 shorted	1														
2	Resistor R-203 open				2				7						15	16
3	Choke L-415 open				3											
4	Cold-solder joint				4											
5	Resistor R-301 open					5				8						
6	Resistor R-302 open					6				(9)						
7	Resistor R-303 open									(10)						
8	Unsoldered joint									11						17, 18
9	Short circuit pin 101 to pin 216										12					
10	Transformer T-1 shorted												13			
11	Capacitor C-502 shorted														14	
12																
13																
14																
15																

FIG. 11-3. Subsample f chart for Example 11-2.

8, and 16, and their average is $68/4 = 17 = \phi$. The first triangular point shows this value plotted on the control chart. To reiterate, the reason for selecting subgroups of constant size is to assure constant control limits for the standard mttr symbolized as ϕ.

The control limits may be calculated as explained in Secs. 8-6 and 8-7. However, since the average number of failures f equals the rejection number R, we can write the equations as shown below.

$$U = \frac{2f\phi}{\chi^2_{8(0.95)}} = \frac{2(4)(10)}{2.733} = 29.6 \text{ min}$$

$$L = \frac{2f\phi}{\chi^2_{8(0.05)}} = \frac{2(4)(10)}{15.507} = 5.12 \text{ min}$$

The lower control limit in this case has no importance other than statistical, since any value less than 5.12 is an indication of a better maintenance activity than that which has been specified. This is a desirable condition.

Let us now return to a discussion of the failure data which are based on sub- and time samples.

The first time sample consists of 6 subsamples of 200 hr each, all of which time was accumulated as a result of tests on the first article. This is shown on the chart as serial number 1 for each of the 6 subsamples. At the beginning of a reliability test it is not unusual to test the first article for the full time sample; as a matter of fact it is desirable in order to get an indication or verification of the infant mortality period. This is particularly important in those instances where it is not known. In this case, since the infant mortality period had been previously established as 25 hr and the actual production test time as an average of 30 hr per article, it was safe to assume that every article could be considered as debugged as a result of production testing. Therefore, no further debugging would be necessary before reliability testing commenced. However, despite these considerations a long reliability run on at least the first article is a good way of verifying preestablished data. Apparently, in this case the first sample proved to be very satisfactory on the basis of data obtained.

However, the second time sample, consisting of 1,200 hr of testing on articles 2, 3, 4, and 5, appeared to be satisfactory for its first two subsamples, and, therefore, on a calculated risk basis, articles 2 and 3 were prepared for shipment; but prior to their actual shipment article 4, representing the third subsample, had 4 failures. Based on the cumulative number of failures to this point there were 5 failures in total which were not permissible; therefore, a condition of nonconformance was apparent, and all shipments were held, pending an investigation of cause of failure. The results indicated that failures 9 and 10 were secondary because they were the direct result of an overload which resulted because of failure 8. Consequently, since these secondary failures were not chargeable, they were circled, and shipments were resumed and the test was permitted to continue. This is why the chart shows 2 failures, not 4.

Time sample 3 resulted in a rejection after its third subsample. Investigation revealed that the cause was due primarily to failures 15 and 16 (Code No. 2—Resistor R-203 open), which showed a high frequency of occurrence because the same failure occurred once in each of the previous subsamples. The rejection was also due to poor workmanship, as evidenced by the unsoldered joints shown in failure reports 17 and 18. The high frequency of failure code 2 was traced to a nonconservative use of this resistor, and, therefore, an engineering change was introduced and retrofit instructions were issued for back-fitting units which had already been shipped. Corrective action was also employed to make certain that the quality of the soldering operation was improved. Upon satisfactory evidence that the corrective action was adequate, the test was resumed.

The foregoing example was used to highlight the effectiveness of the subsample *f* chart. It can be seen that through its use, decisions can be made in a relatively short time on the basis of individual or cumulative subsamples. However, this entails a certain risk due to the low confidence levels associated with individual subsamples; but this risk can be minimized to a degree because of the dual protection of evaluation of results on the subsample basis and on a cumulative level as the testing continues. This gives further assurance that prior decisions were or were not justified. It also makes possible observance of trends which are very important in a product improvement program.

In general, the following rules are useful in interpreting a subsample *f* chart:

1. Ship the product if the number of failures per subsample is less than the corresponding rejection number.

2. Ship the product if the cumulative failures up to the time of shipment do not exceed the cumulative rejection number.

3. Hold shipment, but continue testing, if the number of failures for a subsample equals or exceeds its rejection number but the number of cumulative failures does not equal or exceed the cumulative rejection number of the cumulative sample up to the time in question. In this case, the final decision to accept or reject should be made at the completion of the full time sample.

4. Consider the product nonconforming and hold shipment for engineering investigation if the number of failures for a cumulative sample just equals the rejection number. Testing may continue on another sample to gain additional evidence if this is deemed desirable.

5. Reject if the number of failures for the full time sample equals or exceeds the rejection number.

11-7. The Sequential *f* Chart. The sequential f chart has been commonly prescribed as a vehicle to determine whether reliability is considered as conforming or nonconforming. The technique for setting up this type of chart is explained in various statistical texts. Basically, there are two methods used. The first depends upon formulas developed by Wald[1] and based on what he terms the probability ratio (PR). This ratio is explained in App. 2. This type of plan is called an item-by-item sequential plan. The second method is the multiple sampling plan, which depends on deriving various products of conditional probabilities for the subsamples which comprise the sample. It is much too complex and will not be discussed in this text; nor were these probabilities taken into consideration in the case of the subsample f chart.

We shall therefore focus our attention on the method developed by Wald for an item-by-item sequential sampling plan in which he uses the binomial probability ratio. From this ratio Wald developed certain logarithmic equations which he used to set the acceptance and rejection lines of his sampling plan. The author has used Wald's technique but substituted Poisson probabilities in lieu of binomial probabilities. This has resulted in a much simpler and faster method for setting up a sequential f chart. This development is shown in App. 2. The reason for this substitution of probabilities is that in reliability work we are always dealing with small failure rates. In this case, the binomial and Poisson probabilities are basically equivalent, and one may be used in lieu of the other. In this section we shall illustrate the method of developing a sequential f chart by means of Wald's binomial probability ratio equations and then compare the result with the method which utilizes the Poisson probability ratio.

The equations applicable to these data, namely, r_1, r_2, R_α, R_β, are

$$h_1 = \frac{\log\,[(1 - R_\alpha)/R_\beta]}{\log\,\{(r_2/r_1)[(1 - r_1)/(1 - r_2)]\}} \tag{11-9}$$

$$h_2 = \frac{\log\,1 - R_\beta/R_\alpha}{\log\,\{(r_2/r_1)[(1 - r_1)/(1 - r_2)]\}} \tag{11-10}$$

$$S = \frac{\log\,[(1 - r_1)/(1 - r_2)]}{\log\,\{(r_2/r_1)[(1 - r_1)/(1 - r_2)]\}} \tag{11-11}$$

In order to simplify the handling, we make

$$g_1 = \log\frac{r_2}{r_1} \qquad g_2 = \log\frac{1 - r_1}{1 - r_2} \qquad a = \log\frac{1 - R_\beta}{R_\alpha}$$

$$b = \log\frac{1 - R_\alpha}{R_\beta}$$

[1] Abraham Wald, "Sequential Analysis," John Wiley & Sons, Inc., New York, 1947.

By substituting in Eqs. (11-9) to (11-11), respectively, we get

$$h_1 = \frac{b}{g_1 + g_2} \qquad h_2 = \frac{a}{g_1 + g_2} \qquad S = \frac{g_2}{g_1 + g_2}$$

Moreover when $R_\alpha = R_\beta$, then $h_1 = h_2$. The values h_1, h_2, and S are the characteristic constants which make possible the construction of a sequential f chart. Such a chart would appear as shown in Fig. 11-4. This chart shows the characteristic constants and the rejection and acceptance lines. The parallel lines mark the limits which are used to determine when a sample is considered as conforming or nonconforming to the specified reliability for the required confidence level. Whenever

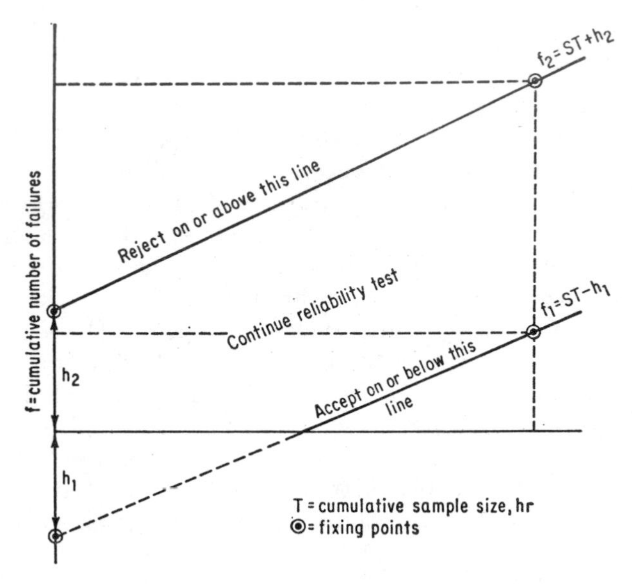

FIG. 11-4. Sequential f chart layout showing characteristic constants h_1, h_2, and fixing points.

the number of cumulative failures are on or above the rejection line, this is a condition of nonconformance. If the failures are between lines, no decision is indicated, and the test must continue. If the number of cumulative failures is on or below the acceptance line, it indicates a condition of conformance, and the requirements are considered to have been met. Moreover, a decision of nonconformance can be made at any time that the cumulative number of failures falls on or above the rejection line, but a decision of conformance is permitted only after a minimum time sample. This minimum time sample is determined by the point of intersection of the acceptance line and the abscissa.

Any point on the two parallel lines can be determined once we know the characteristic constants because each of these lines has an equation which applies to it. These are shown in Fig. 11-4. Thus, once two points on each of the lines are determined, the lines can be drawn. These fixing points, as they are called, are also shown in the figure as circled dots. The following example will illustrate the methods and procedures involved.

Example 11-3. Design a sequential f chart to assure an ARL of 1 failure/100 hr and an LTFRD of 5 failures/100 hr when $R_\alpha = 0.10$ and $R_\beta = 0.10$.

Solution

$$r_1 = 0.01 \qquad r_2 = 0.05 \qquad R_\alpha = 0.10 \qquad R_\beta = 0.10$$

but since $R_\alpha = R_\beta$, then $h_1 = h_2$.

$$g_1 = \log \frac{r_2}{r_1} = \log \frac{0.05}{0.01} = \log 5 = 0.6990$$

$$g_2 = \log \frac{1 - r_1}{1 - r_2} = \log \frac{0.99}{0.95} = \log 1.04 = 0.0170$$

$$a = \log \frac{1 - R_\beta}{R_\alpha} = \log \frac{0.90}{0.10} = \log 9 = 0.9542$$

But $R_\alpha = R_\beta$, and therefore $b = a = 0.9542$.

$$h_1 = \frac{b}{g_1 + g_2} = \frac{0.9542}{0.699 + 0.017} = \frac{0.9542}{0.716} = 1.32$$

$$h_2 = h_1 = 1.32$$

$$S = \frac{g_2}{g_1 + g_2} = \frac{0.0170}{0.716} = 0.0238$$

To fix the rejection line we need another point other than h_2. If we arbitrarily select a value of T, such as $T = 100$, we can determine this second point from the rejection-line equation

$$\begin{aligned} f_2 &= ST + h_2 \\ &= (0.0238)(100) + 1.32 \\ &= 3.7 \end{aligned}$$

Connecting the two points $h_2 = 1.32$ and $f_2 = 3.7$ gives us the rejection line.

Similarly, for the acceptance line we find the second point f_1.

$$\begin{aligned} f_1 &= ST - h_1 \\ &= (0.0238)(100) - 1.32 \\ &= 1.06 \end{aligned}$$

Connecting f_1 and h_1 we establish the acceptance line.

The resulting sequential sampling chart is shown in Fig. 11-5.

The probability ratio formulas which depend upon Poisson probabilities are shown below. These equations are derived in App. 2.

$$\frac{R_{\alpha_2}}{R_{\alpha_1}} = k^{R_T}e^{-d(k-1)} \tag{11-12}$$

$$\frac{R_{\beta_2}}{R_{\beta_1}} = k^{A_T}e^{-d(k-1)} \tag{11-13}$$

where $R_{\alpha_2}/R_{\alpha_1}$ = rejection probability ratio
R_{β_2}/R_{β_1} = acceptance probability ratio
A_T = acceptance number for time T
R_T = rejection number for time T
$k = r_2/r_1$
d = average number of failures (r_1T)

There are two probability ratios which will be used. The first one is used to determine points on the rejection line and is $R_{\alpha_2}/R_{\alpha_1}$; the second one determines points on the acceptance line and is R_{β_2}/R_{β_1}.

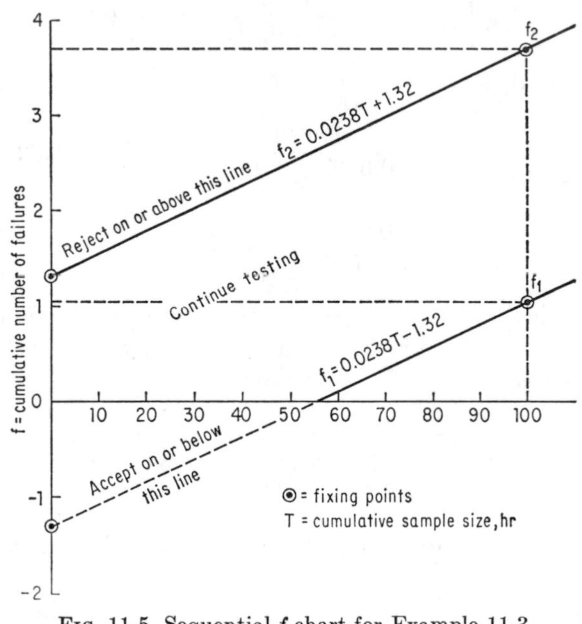

FIG. 11-5. Sequential f chart for Example 11-3.

A minimum of two fixing points for each is required to draw the rejection or acceptance lines. The time sample size for the required confidence level may be calculated from Eq. (8) of App. 2 or by using the product of the appropriate value of K^2 and m. The value of T so determined is not strictly required, since we could assume any two values of T and find the value of A_T or R_T corresponding thereto and still draw the acceptance or rejection line. The only value of locating T is to give a person an idea of how large a single sample would have to be in order to get the required confidence level. As will be realized from the solution of the problem, the number of failures corresponding to this value of T is more generous for a sequential f chart than it would have been for a single sample of T hours. This is due to the fact that in an item-by-item sequential plan we have products of conditional probabilities from item to item which apply, whereas in a single sampling plan this is not the case.

Example 11-4. Solve Example 11-3 by using Eqs. (11-12) and (11-13).

Solution. List the data:

$$r_1 = 0.01 \qquad R_{\alpha_1} = 0.10 \qquad R_{\beta_1} = 0.90 \qquad k = r_2/r_1 = 0.05/0.01 = 5$$
$$r_2 = 0.05 \qquad R_{\alpha_2} = 0.90 \qquad R_{\beta_2} = 0.10 \qquad Z_\alpha = 1.3 \qquad Z_\beta = -1.3$$

Solve for T:

$$T = m\left(\frac{Z_\alpha - Z_\beta\sqrt{k}}{k-1}\right)^2 = 100\left(\frac{1.3 + 1.3\sqrt{5}}{4}\right)^2 = 110$$
$$d = r_1 T = (0.01)(110) = 1.1$$

Substituting in Eq. (11-12) and solving for R_T when $T = 110$,

$$k^{R_T}e^{-d(k-1)} = \frac{R_{\alpha_2}}{R_{\alpha_1}} = 5^{R_T}e^{-(1.1)(4)} = \frac{0.90}{0.10} = 9$$
$$R_T \ln 5 - 4.4 \ln e = \ln 9 = R_T \ln 5 - 4.4 = \ln 9$$

but since $T = 110$ hr, the rejection number R_{110} is

$$R_{110} = \frac{\ln 9 + 4.4}{\ln 5} = \frac{2.197 + 4.4}{1.609} = \frac{6.597}{1.609} = 4.1$$

Substituting in Eq. (11-13),

$$\frac{R_{\beta_2}}{R_{\beta_1}} - k^{A_T}e^{-d(k-1)} = 5^{A_T}e^{-(1.1)(4)} = 0.111$$

but since $T = 110$ hr, the acceptance number A_{110} is

$$A_{110} = \frac{\ln 7.793 - 10 + 4.4}{\ln 5} = 1.36$$

When $T = 0$, $d = 0$; therefore, by substituting in Eq. (11-12), we get h_2.

$$5^{R_0} = 9 \qquad R_0 \ln 5 = \ln 9 \qquad R_0 = \frac{\ln 9}{\ln 5} = \frac{2.197}{1.609} = 1.37 = h_2$$

Similarly, we can get h_1 when $T = 0$ by substituting in Eq. (11-13). Thus

$$A_0 = \frac{\ln 0.111}{\ln 5} = \frac{7.793 - 10}{1.609} = -1.37 = h_1$$

Thus, since the four fixing points (A_{110}, R_{110}, h_1, h_2) have been determined, the sequential f chart can be constructed. This is essentially the same as that shown in Fig. 11-5 for Example 11-3.

The solution of Example 11-4 by using the author's Poisson probability ratio equation appears much simpler than the equations used in Example 11-3. This is a decided advantage because it reduces the amount of labor required to develop a sequential plan.

At this point, it might be interesting to develop a sequential sampling plan for the device described in Example 11-2. To make it more interesting, we shall change the value of γ, the confidence level, from 90 to 95 per cent. The data developed from a time sample are shown in Table 11-3.

TABLE 11-3. FAILURE DATA OF EXAMPLE 11-5

Failures	*Time interval in hours during which failures occurred*
0	0–100
1	101–200
0	201–300
0	301–400
0	401–500
1	501–600
0	601–700
4	701–800
0	801–900
0	901–1,000
1	1,001–1,100
0	1,101–1,200
0	1,201–1,300
0	1,301–1,400
2	1,401–1,500
1	1,501–1,600
1	1,601–1,700
2	1,701–1,800
13 total failures	

Example 11-5. Design a sequential sampling plan for data shown in Table 11-3. Plot the failures on the chart. Assume the following parameters:

$$r_1 = 0.004 \text{ fph} \quad R_{\alpha_1} = 2.5 \text{ per cent} \quad R_{\alpha_2} = 97.5 \text{ per cent} \quad Z_{\alpha_1} = 1.96$$
$$r_2 = 0.012 \text{ fph} \quad R_{\beta_1} = 97.5 \text{ per cent} \quad R_{\beta_2} = 2.5 \text{ per cent} \quad Z_{\beta_2} = -1.96$$

Solution. Obtain the value $K^2 = 7$ from Fig. 11-1 for $k = 3$ and $\gamma = 95$ per cent.

$$T = mK^2 = 250(7) = 1{,}750 \text{ hr}$$

Substituting and solving for a point on the rejection line $T = 1{,}750$ hr,

$$PR = \frac{R_{\alpha_2}}{R_{\alpha_1}} = \frac{0.975}{0.025} = 39 \qquad k = \frac{r_2}{r_1} = \frac{0.012}{0.004} = 3$$

$$d = r_1T = (0.004)(1{,}750) = 7 \qquad \text{(when } T = 1{,}750)$$

but

$$R_{\alpha_2}/R_{\alpha_1} = k^{R_T}e^{-d(k-1)}$$
$$39 = 3^Re^{-7(2)}$$
$$= 3^Re^{-14}$$
$$\ln 39 = R \ln 3 - 14 \ln e$$

Rejection number for $T = 1{,}750 = R_{1,750}$.

$$R_{1,750} = \frac{\ln 39 + 14}{\ln 3} = \frac{3.664 + 14}{1.099} = \frac{17.664}{1.099} = 16.2$$

Substituting and solving for a point on acceptance line when $T = 1{,}750$ hr,

$$PR = \frac{R_{\beta_2}}{R_{\beta_1}} = \frac{0.025}{0.975} = 0.0256$$

Acceptance number for $T = 1{,}750$.

$$A_{1,750} = \frac{\ln 0.0256 + 14}{\ln 3} = \frac{(6.290 - 10) + 14}{1.099} = 9.4$$

Since $R_{\alpha_1} = R_{\beta_2}$, then $h_1 = h_2$; to solve for h_2 make $T = 0$ in Eq. (11-12).

$$39 = 3^{R_0}e^{-0(2)}$$
$$= 3^{R_0}$$

$$R_0 = \frac{\ln 39}{\ln 3} = \frac{3.664}{1.099} = 3.4 = h_2$$

$$A_0 = \frac{\ln 0.0256}{\ln 3} = 3.4 = h_1$$

To find the minimum sample we must have before making a decision to accept, it is necessary to find the point of intersection of the acceptance line and the abscissa. This will occur when the acceptance number is zero.

Substituting in Eq. (11-13), we get

$$3^0e^{-0.004T(2)} = 0.00256$$

Solving for T,

$$-0.008T \ln e = \ln 0.0256$$

$$T = \frac{\ln 0.00256}{0.008} = 464$$

Therefore, a minimum of 464 hr of test time is necessary before a decision to accept is permitted. However, a decision to reject can be made at any time.

Discussion of Example 11-5. Figure 11-6 shows the data plotted on a sequential *f* chart. The actual numbers of failures are shown as points, while the chargeable failures are shown as small triangles. The chargeable failures are those which, after analysis, can actually be considered primary failures. It should be noted that in the solution of Example 11-2, the chargeable failures were designated with cross marks and the nonchargeable ones with circles. Any system can be used provided it is clear to those using it.

The analysis of what constitutes a chargeable failure was conducted in a manner similar to that described in Example 11-2 and will, therefore, not be amplified at this time. In other words, on the basis of an engineering analysis, a decision is made regarding the causes of failure and what corrective action, if any, is recommended. If it is found that the failure was secondary or that it can be corrected by redesign and retrofit as required, the failure is not charged with respect to making a decision to

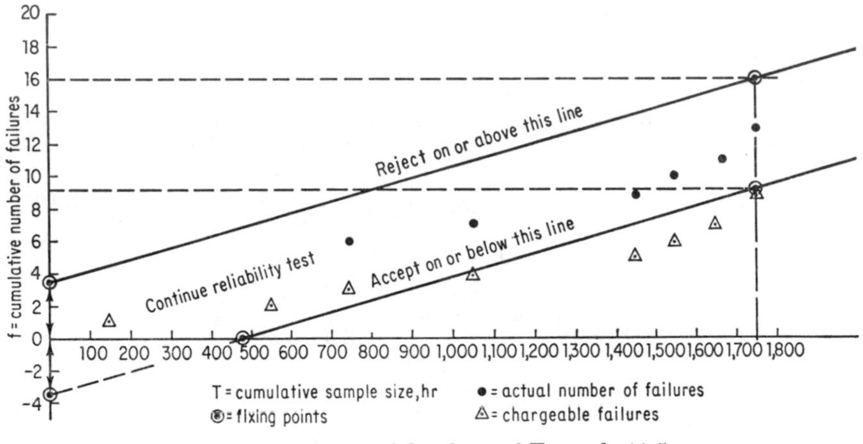

Fig. 11-6. Sequential f chart of Example 11-5.

accept or reject the measured reliability. An examination of Fig. 11-6 indicates that, in general, the trend is favorable, as shown by the triangles. As a matter of fact, during the interval 1,001–1,100, an acceptance could have been made because a triangle point is shown on the acceptance line. However, in order to get a higher degree of confidence, we agreed to test for the entire time sample and found, as a result, that the continued testing also indicated conformance.

It is interesting to note that data are recorded as if the failure occurred at the mid-point of the interval despite the fact that it may actually have occurred at its beginning or end. This is done as an economy measure in order to eliminate the necessity for monitoring the exact time of failure during reliability testing, since the additional degree of accuracy obtained by recording exact failure times is negligible. As an illustration, a failure occurring at 641 hr or 601 hr would be recorded on the sequential time sampling chart, Fig. 11-6, as the mid-point of the interval 600–700 hr.

TABLE 11-4. TABLE OF ACCEPTANCE AND REJECTION NUMBERS FOR EXAMPLE 11-5

Time interval	Acceptance no. A	Rejection no. R
0– 80	*	4
81– 220	*	5
221– 350	*	6
351– 484*	0	7
485– 620	1	8
621– 770	2	9
771– 900	3	10
901–1,035	4	11
1,036–1,180	5	12
1,181–1,270	6	13
1,271–1,450	7	14
1,451–1,580	8	15
1,581–1,750	9	16

* Acceptance is not permitted unless a minimum of 464 test hours have been accumulated.

PRACTICE PROBLEMS

11-1. Name four conditions which must be demonstrated before reliability sampling can be instituted.

11-2. What, in your opinion, is the chief advantage of a sequential f chart over a subsample f chart? What are the advantages of the subsample f chart?

11-3. What is the time sample and corresponding rejection number you would use to satisfy the following conditions: $R_\alpha = 5$ per cent; $R_\beta = 5$ per cent; $r_1 = 0.02$; $r_2 = 0.08$? What subsample size and rejection number would you use for a $\gamma = 50$ per cent? (Solve by use of Fig. 6-1.)

11-4. Solve Prob. 11-3 by the use of K^2.

11-5. Calculate the control limit for a subsample f chart when ARL $= 0.001$; LTFRD $= 0.005$; $R_\alpha = 10$ per cent; $R_\beta = 10$ per cent.

11-6. For one unit what is the minimum time sample you could use in Prob. 11-5 if the estimated production test time was 25 hr and the infant mortality period was 50 hr? Why? What actual time sample size would you recommend? Why?

11-7. Set up a sequential f chart for the same data as for Prob. 11-5.

11-8. Determine the MRL in terms of failures per hour from data given in the table. Using this estimate as the ARL, determine the sequential f chart when $k = 4$ and $R_\alpha = 2.5$ per cent and $R_\beta = 2.5$ per cent.

No. of failures	*Hours to failure*
1	970
1	2,500
1	3,300
1	3,905
1	4,625
1	5,785
1	6,085
1	6,925

CHAPTER 12

RELIABILITY AND AVAILABILITY PREDICTION METHODS

12-1. A Discussion of Reliability and Availability. Reliability prediction is a method of forecasting the probable reliability of a device by means of past experience or statistical methods. The principal advantage of this technique is that it assists the designer in making an early estimate of the expected reliability of the device during the design stage instead of waiting until the hardware is assembled, after which time it is too late to do much of anything to correct poor reliability. In other words, reliability prediction gives the designer a degree of assurance, from a reliability standpoint, that the basic design concept is sound. If he finds that it is sound, he can proceed with confidence that he is on the right track. Likewise, if the prediction technique indicates that the resultant reliability will probably be substandard, it also pinpoints the contributory factors of unreliability. In this manner, the designer can concentrate on the problem areas and refine the design accordingly in order to eliminate them.

Reliability prediction can be achieved in several ways. In some instances the only available method is the *guess*. This is usually resorted to when, because of urgent delivery requirements, sufficient time is not available to use more sophisticated techniques. The guess, however, should be based on a knowledge of comparable equipment with which the designer is familiar, or else it should not be used because of the risk of making an erroneous prediction.

Although the guess is the least desirable approach to reliability prediction, there are many applications, particularly at the early stage of development of large-scale systems, where the guess is the only method which can be applied. Since the guess involves eliciting judgments from people, the problem is to obtain the best possible responses from qualified individuals. Psychophysical methods are recommended for this purpose.[1]

Psychophysical methods which warrant particular attention are the

[1] J. P. Guilford, "Psychometric Methods," McGraw-Hill Book Company, Inc., New York, 1954.

methods of *paired comparisons* and *single stimuli.* These techniques are recommended to pinpoint possible sources of unreliability in a relative and absolute manner early in development, when design alternatives may be considered. The following description illustrates the step-by-step application of these methods to reliability analysis of large-scale systems.

1. *System Subdivision.* The system is subdivided into a manageable number (≤ 25) of physically equivalent components, e.g., modules, chassis, racks, or equipments.

2. *Judge Identification.* A group of engineers are identified who are familiar with each component and are otherwise qualified to judge.

3. *Rating of Components.* By use of file cards or IBM punched cards with component titles printed on opposite sides of the cards, each component is compared with every other component, one pair per card. Accordingly, $\frac{1}{2}n(n - 1)$ cards are prepared, where n equals the total number of components.

Each judge is provided with a set of cards with the following instructions: "For each pair ask yourself the question: 'Which component is more likely to perform its function consistently?' Make a check mark by the component you choose." The specific question posed, of course, should vary according to the nature of the problem at hand.

4. *Analysis of Rating.* Each judge's ratings are averaged and plotted for each component on the basis of the number of checks each component received. The plotted data are examined to provide a visual check of the variability of ratings assigned to each component.

Marked rater variation for a given component may be an indication of lack of knowledge concerning the component, or, more likely, it may indicate that one or more judges have information not possessed by the others. In any case, data characterized by high inter-rater variation are refined on the basis of judge interviews.

Over-all indices of inter-judge agreement and intra-judge consistency can be computed by means of the coefficients of agreement and consistency, respectively.[1]

5. *Judge Interviews and Refinement of Ratings.* Personal interviews are conducted separately with each judge, at which time the results of the analysis are fed back to him. He is questioned in general regarding the criteria he used in making his judgments. Specifically, he is questioned about those components for which his ratings were atypical.

From analysis of the interviews, ratings are refined and final relative component standings determined. In addition, criteria are assembled to be incorporated into a questionnaire to obtain absolute estimates of the problem potential of certain components.

[1] M. G. Kendall, "The Advanced Theory of Statistics," vol. I, Harper & Brothers, New York, 1952.

6. *Construction and Administration of Questionnaire.* Problem criteria are incorporated into a questionnaire and plotted against the low half of the components as ranked. The purpose of the questionnaire is to establish an absolute estimate of *how much* of a problem is presented in the lower-ranked components. The questionnaire is administered to the judges with these instructions: "Considering each component separately for each problem, indicate how you feel about the statement: 'The problem exists for this component and may lower its reliability.' Mark your answers in the appropriate squares, using the following code: (1) strongly agree, (2) agree, (3) undecided, (4) disagree, (5) strongly disagree, (X) cannot judge."

7. *Analysis of Questionnaire Responses.* The mean absolute rating for each component for each problem is computed. Components which average under three on any problem are regarded as critical and are subjected to extensive study to see what modifications can be made to improve the design.

Another method that is commonly used is *extrapolation.* This method constitutes more than a single guess, although it also depends upon comparisons with similar devices. The analysis, in this case, is a little more profound, because it considers the functions to be performed, conditions of use, and the similarity with other equipment. The assumption is then made that the reliability of the unknown equipment must be approximately equal to that which was experienced with the similar device.

A third method of predicting reliability is *mathematical calculation.* This is accomplished by using proved laws of reliability theory and applying them in accordance with past practice and by utilizing data which have been substantiated by past experiments. Later in the text, we shall see how this method is applied and how we handle the synthesis of the elements which comprise a component, subsystem, or system. In the case of a component we synthesize all parts which comprise it, whereas for a subsystem we synthesize the components. In the latter case the various components which comprise the system are considered individually as they affect the over-all operation, and their contribution is assessed accordingly. This is done usually through the medium of a functional diagram which describes whether the particular component is working in series or in parallel with other independent elements or components.

Another method of predicting reliability is *measurement.* This involves the accumulation of failure data in statistically significant amounts from tests conducted under simulated test conditions in the factory or laboratory. The test results can then be compared with similar results obtained from prototypes in the field to assure that correlation exists. The number of hours of testing required is a function of the degree of confidence which is expected. This subject was thoroughly discussed in Chap. 11.

The underlying reason for performing a reliability prediction is to determine whether the design configuration is sufficiently mature to assure that when the device is in service, it will have the necessary reliability to ensure the successful execution of the required mission. This is as it should be, since the successful execution of the mission is the underlying reason for the existence of any design configuration. However, in the interests of economy, there is always a minimal best configuration to achieve the desired result. The reliability prediction is one of the most effective tools which the engineer has at his disposal to assess his design achievement. Economy in design, consistent with achieving the mission, is also an inherent responsibility of every engineer because of the present-day strain on the national economy. The days when all the frills were tolerable have long since passed. The important thing to remember in our present-day world is that a military device is a weapon with a specific mission or purpose, and this is what must be achieved. In some instances, there may have to be trade-offs between cost and performance. These may be reflected in terms of reducing capability, defining the specific areas in which it is considered that the system will fail to perform its complete mission, or outlining those features which represent an embellishment but not a positive necessity. Another valuable feature of a reliability prediction is that, very early in the program, it gives us an awareness of those factors which it is necessary to take into account during design. However, it should be remembered that such a prediction represents a best estimate based on data which were available at the time and should not be considered in any other light. In other words, it should never be construed as depicting accuracy to the last decimal place. The real test of what has been achieved from the reliability viewpoint is possible only after the device has been constructed and placed into service under the actual conditions of use. However, past experience has established that there have been many reliability predictions made whose accuracy has been amazingly good when compared with actual data accumulated in the field.

Reliability prediction is essentially a design tool. It is the most effective means by which a designer can know whether his concept has any possible chance of achieving the objective he has in mind. Moreover, it provides a method of assurance that reliability is designed into the equipment and not tested into it. The old concept that one could test reliability into a device is as archaic as the thought that quality can be inspected into it. The use of special parts and conditioning tests for parts or components is not recommended, since experience has shown that this is a costly method which contributes very little to the end reliability and is not amenable to scientific prediction methods.

To supplement the use of the word *reliability*, in Chap. 9, we have introduced the concept of *availability*. Actually, in a sense, the two are synonymous from the operating point of view. As was mentioned before, when making a telephone call, if the subscriber is always successful in reaching the called party, he is not usually interested in knowing whether or not a failure of any equipment has occurred in the branch telephone exchange. Likewise, a person looking at a television program is concerned only with whether he has a picture constantly available to him and knows very little about the problems of the studio personnel, who might be coping with all kinds of technical situations. It appears, therefore, that there should be some compromise between availability and reliability from a practical viewpoint. Oftentimes, it is impossible to design a device with the reliability required to accomplish the mission, and, therefore, it becomes necessary to use a combination of reliability and maintainability which results in achievement of a certain specific availability. We have seen the derivations of the equations which were used to achieve this in Chap. 9. In a sense, reliability is a special case of availability, in which no maintenance activity is practiced. Therefore, it is applicable when the mission time is relatively very small; and because of this, if the failure rate is proper, the probability of failure for this small mission time is relatively insignificant. For example, in the case of a missile whose total flight time is measured in minutes, we must develop a certain degree of reliability in the design which will ensure that it will hit its target the preponderant number of times. In this instance, we cannot use the concept of availability unless we envision a number of little gremlins riding on the missile and making adjustments as each failure or malfunction occurs. As an alternative, we could design automatic methods of handling each situation, such as redundancy of certain elements, but this would add to the weight, complexity, and cost of the device and is usually done only within certain limitations.

On the other hand, if the device with which we are concerned must operate continuously, it is usually to our advantage to achieve this objective by means of the principle of availability instead of reliability. This is particularly true in a large installation where it is a relatively simple task to employ switchover redundancy or replacement of pluggable units as a standard method of operation.

As we have seen in Chap. 6, a mature design is usually characterized by a constant failure rate. This behavior has resulted in a specific advantage to the reliability engineer, because it has enabled him to predict the probability of success by means of mathematical equations. It should be clearly borne in mind that a constant failure rate is exhibited only for catastrophic failures and does not apply to wear-out failures, which are a function of time and usage. Moreover, reliability prediction depends upon the assumption that the probability of failure of a part or device which has not progressed to the point of wear-out does not change as a result of a repair or maintenance action. This concept is often difficult to understand from a practical viewpoint, although it has been proved mathematically in a theoretical sense as well as demonstrated through the analysis of data that have been gathered on a number of various equipments. However, from our everyday experience, we should be convinced that the principle of constant failure rate has some basis in fact. In our everyday lives, most of us drive automobiles equipped with tires which have a definite probability of failure. If one has been subjected to the inconvenience of fixing a flat tire, it should be obvious that as long as it has not worn out and the treads are in good condition, the probability of failure has not changed to any degree. The obvious reason is that the area in which the flat occurred was subjected to the same general conditions of use applicable to every other area. This is the random pattern which can be expected as the car is driven over the road. Thus, each specific area of the tire is completely independent and isolated from every other area, with no connection whatever between any of them. Similarly, in electronic equipment the failure of any part has no bearing or connection with other parts (exclusive of secondary failures) which comprise the equipment, and therefore failures are completely independent of each other. Thus, the fact that a failure has occurred in no way affects the probability of recurrence after the part has been repaired.

As we have seen from Chap. 6, the concept of constant failure rate has made possible the simple addition of failure rates for several parts or components and the subsequent determination of reliability on the basis of this summation.

On the basis of the foregoing discussions, it is obvious that a valid reliability prediction can be prepared if one is aware of the number of parts or components comprising the whole, their failure rates, and their relative functional relationships. These functional relationships are very important, since if we have series combinations, as we have seen in Chap. 6, the results are completely different from those which are possible with parallel combinations. Perhaps the best method of illustrating typical reliability calculations is to perform some typical calculations.

12-2. The Series Case. As we have previously stated, in order to perform a mathematical reliability prediction effectively, the designer must have a knowledge of the failure rates which he will employ. These failure rates may be obtained from published data or from the designer's own records; or else they may be calculated from data obtained from special tests. At this time, it must be reiterated that there is no such thing as an abstract figure which represents a failure rate, because the environment under which these failure rates have been determined is a controlling factor. Therefore, when using particular values of failure rates for specific reliability predictions, the designer must know that these failure rates were determined under conditions closely approximating the expected immediate part environment when in actual use in equipment. The closer the conditions of test to those of actual usage, the more valid will be the resulting reliability prediction. However, this close correlation is very difficult to achieve, because the immediate part environment is usually different from the environment which was used in the reliability test for determining the failure rate. It is for this reason that a reliability prediction at best is considered a broad estimate. It has value, however, because it gives the designer sufficient information to guide him in achieving his reliability objective.

In Sec. 6-10, we learned how to combine failure rates in families or groups. As examples, some common groupings are wire-wound resistors, mica capacitors, and chokes. If we assign the same failure rate to each member of a family, then the combined failure rate for the group is the product of the number of members in the family or group and the failure rate assigned to the member. The following example should illustrate the principles involved.

Example 12-1. A radio receiver consists of five pluggable units, namely, the power supply, the radio-frequency stage (r-f), the mixer, the intermediate-frequency stage (i-f), and the power output stage (PO). All parts for each of the pluggable units are to be considered as being functionally in series; therefore the failure of any one part will result in failure of the equipment. Make an estimate of the equipment failure rate r_e and the mean time between failure. Suppose the mission time T is 100 hr; what is the probability of survival P_s for this time?

Solution. Tabulate the data as shown in Table 12-1. The failure rates were determined by the design engineer on the basis of published data and special reliability tests. For this calculation, it is assumed that the same part failure rate will apply for the same type of part used in different components even though the immediate part environment might be different.

To get the component failure rate for each component, multiply the number of each type of part n by its failure rate r. This gives the value nr for that cluster of parts. Add the total nr values so found, and this will give the failure rate of the component. Thus, from Table 12-1 we see that for the power supply the sum of the nr values is 15.28 per cent failures/1,000 hr. Similarly, for the r-f, mixer, i-f, and power output stages the values of nr are respectively 40.28, 33.58, 54.34, and 32.89, while the grand total of all the nr values of the components is 176.3 per cent failures/1,000 hr. This is r_e, the equipment failure rate. If we convert this to failures per hour by taking the reciprocal, we get the mean time between failures m of the equipment, which equals 567 hr. The data are tabulated and the calculations are shown in Table 12-1.

TABLE 12-1. TABULATION OF DATA FOR ESTIMATE OF INHERENT RELIABILITY FOR A TYPICAL RADIO RECEIVER

Failure rates r are expressed in per cent per 1,000 hr

Part number	Parts nomenclature	Type	Failure rate r	Power supply		R-f stage		Mixer		I-f stage		Power output	
				Number of parts n	nr	Number of parts n	nr	Number of parts n	nr	Number of parts n	nr	Number of parts n	nr
1	Resistors	All	0.30	20	6.00	40	12.00	35	10.50	50	15.00	45	13.50
2	Capacitors	All	0.35	10	3.50	32	11.20	40	14.00	60	21.00	40	14.00
3	Capacitors	Feed-through	0.30	...	.....	10	3.00	10	3.00	15	4.50	8	2.40
4	Capacitors	Trimmer	0.20	...	.....	15	3.00	12	2.40	14	2.80	5	1.00
5	Electron tube	Dual triode	0.12	...	.....	3	0.36	2	0.24	6	0.72		
6	Electron tube	Pentode	0.12	...	.....	2	0.24	...	.....	2	0.24	2	0.24
7	Electron tube	Rectifier	0.18	1	0.18								
8	Electron tube	R-f power	0.14	...	.....	2	0.28	1	0.14	2	0.28		
9	Electron tube	Gas (volt. reg.)	0.10	1	0.10								
10	Transformer	R-f and i-f	0.30	...	.....	6	1.80	2	0.60	8	2.40		
11	Transformer	Audio	0.35	...	.....	...	.....	...	.....	...	.....	1	0.35
12	Transformer	Power	0.40	1	0.40								
13	Relays	General purpose	0.50	4	2.00	2	1.00	1	0.50	2	1.00	2	1.00
14	Coils	R-f	0.25	...	.....	10	2.50						
15	Chokes	R-f and i-f	0.30	...	.....	15	4.50	6	1.80	20	6.00		
16	Chokes	Power	0.35	6	2.10								
17	Switch	Toggle	0.60	1	0.60								
18	Connectors	All	0.20	2	0.40	2	0.40	2	0.40	2	0.40	2	0.40
Total number of parts per component..........				46	.....	139	.....	111	.....	181	.....	105	
Component failure rate, %/1,000 hr...........				...	15.28	...	40.28	...	33.58	...	54.34	...	32.89

Note: Equipment failure rate equals sum of component failure rates, viz.: 15.28 + 40.28 + 33.58 + 54.34 + 32.89 = 176.37% failures/1,000 hr = r_e, and the mean time between failures $m = \frac{1}{r_e} = \frac{1}{0.0017637} = 567$ hr (to nearest whole hour).

The probability of survival for a mission time $T = 100$ hr is calculated as follows:

$$P_s = e^{-rT} = e^{-0.0017637(100)} = 0.838 \text{ or } 83.8 \text{ per cent}$$

12-3. The Parallel Case—Redundancy. In the solution of Example 12-1, we considered all parts and components as functionally in series. We also predicted that the probability of survival P_s for a mission time of 100 hr would be 83.8 per cent. Suppose that this value of P_s is inadequate because the least value of P_s which is tolerable is 96 per cent; it would mean that we would have to improve the equipment failure rate r_e from a value of 0.00176 to 0.0004 failures/hr. In other words, the mean time to failure m of the equipment would have to improve from a value of 567 to 2,500 hr. This improvement could be effected by improvement of the radio set design or, as explained in Chaps. 3 and 9, by use of two sets redundantly or by improvement of the maintainability to provide an availability of better than 96 per cent when $T = 100$ hr. The choice of methods would be a function of time and cost. If the design of the prototype has been completed, it is usually impractical to effect a complete redesign, because the limitations of time generally do not permit this type of solution. Therefore, the most logical method is to use two equipments redundantly if they will provide the required reliability, since this is the fastest and probably the cheapest method of achieving the desired results. In Chap. 3 we studied the methods of solution for elements which are functionally in parallel or redundant. At that time we assumed inherent redundancy without considering the effects of maintainability, which were explained in Chap. 9.

Let us solve Example 12-1 for the duplex parallel redundancy case to determine whether this will satisfy the minimum reliability requirements. If the resulting reliability prediction is equal to or better than 96 per cent, we shall have established one method of satisfying the mission requirements. If not, we shall have to resort to methods which were explained in Chap. 9 and which will be further amplified in the next section.

The unreliability of the radio receiver when $T = 100$ hr equals $1 - 0.838 = 0.162$. The product of the unreliabilities of two receivers which are inherently redundant is $(0.162)^2 = 0.028$. Therefore, the reliability $P = 1 - 0.028 = .972$ or 97.2 per cent. The solution indicates that the two radio sets, when used redundantly, will have a predicted reliability of 97.2 per cent when $T = 100$ hr; and, therefore, since this value exceeds the desired 96 per cent, this satisfies the requirements of the mission, and the problem is solved.

12-4. Availability and Maintainability Prediction. The nomograph of Fig. 9-5 can be used to predict the maintenance-action rate and reliability which will be necessary to achieve a given availability. For example, suppose we required a minimum availability of 99.6 per cent in 10 hr of operation; we could attain this by some combination of r and μ. If we went to one extreme, we could design our equipment so that the availability would be wholly a function of the reliability r, for which case $\mu = 0$. If we refer to Fig. 9-3 when $\Lambda_E = 99.6$ and $b = 0, d = 0.004$ failures and r will equal 0.0004 failures/hr (since $r = d/T = 0.004/10 = 0.0004$). This would satisfy the requirements, since $P_s - e^{-0.004} = 99.6$ per cent.

However, if a reliability of 0.0004 failures/hr cannot be achieved because of the complexity of design, the deficiency can be overcome by improving the maintenance-action rate. Thus, suppose that various reliability predictions indicated that the best possible intrinsic reliability is 0.01 failures/hr; then $d = 0.01(10) = 0.1$ and from the nomograph, Fig. 9-5, when $\Lambda_E = 99.6$ and $d = 0.1$, we get $b = 3.25$. This means that in order to obtain an availability of 99.6 per cent when $T = 10$ hr, the product of the maintenance-action rate μ and time constraint t must be equal to 3.25. If a value of $t = \frac{1}{4}$ hr is tolerable, then $\mu = 3.25/\frac{1}{4} = 13$, and thus $\phi = 1/\mu = 1/13 \times 60 = 4.6$ min. The designer must now decide if he can achieve the required values of r, μ, or ϕ. He may do this either by using redundancy or by designing a device with interchangeable and replaceable modules with built-in fault-location and isolation devices. In this manner, when the trouble is isolated to a particular pluggable unit, if it can be readily replaced with a spare within the required time constraint, enabling the device to resume operation in t or less time, the required availability will have been achieved.

In general, when r is high, μ must also be large in order to achieve a given Λ_E in time T. This necessitates a larger complement of spares than might be considered desirable, because of the resultant higher costs. In this case, the designer should study the feasibility of reducing the required value of μ by design improvements which will provide a better reliability. For example, for the case in point, if a value of $r = 0.005$ could be achieved instead of 0.01, the value of d for $T = 10$ hr would be

$$T = 10 \text{ hr} = (0.005)(10) = 0.05$$

and for the required availability Λ_E of 99.6 per cent (from Fig. 9-5) we find a value of $b = 2.50$. This represents a reduction from the previous value of 3.25 and corresponds to a new μ value of 10. Thus, we see that in order to obtain a required availability for a time T, the designer must weigh the relative merits of r and μ in terms of practicality and cost. The value of Fig. 9-5 lies in the fact that by its use he can do just this. In this manner, if for some reason the designer finds that the state of the art restricts the value of r which he can design into the equipment, he now has a tool in the nomograph which makes it possible for him to predict the required maintenance-action rate which will assure the required mission availability. This knowledge thus provides a goal in terms of μ which must be achieved in order to satisfy the availability requirements.

There are many textbooks available in time and motion study which the designer can use as a source of data for predicting the theoretical value of μ. The best method is to simulate failures and then rate each individual element of the maintenance action in terms of time. For example, suppose that one assumed the failure of a pluggable unit. The value of μ could be determined by summing the standard times for each element comprising the maintenance action. This would include such details as the times required to observe or locate the failure, remove the defective unit and place it in its proper place, and substitute a workable unit and snap in place. In actual practice, as is done in modern time and motion study, each of the maintenance functions would be broken down into those subelements to which a standard time unit can be allocated.

By following the principles of time and motion study, the designer can thus predict μ for various levels of maintainability, such as part or pluggable unit, and in this manner determine the most practical and economical course to follow. These levels should be studied from the basic part level to the component subassembly or subsystem level or any other combination which the designer considers most feasible and advantageous. Generally speaking, in electronic equipment, maintainability at the part level is not recommended. The reason is that the accessibility of parts is usually poor. Moreover, the time required to isolate part troubles and replace them is usually prohibitive. Therefore, in general, it is recommended that maintainability be approached from the pluggable unit level with provisions for automatic fault location and isolation for each pluggable unit. This represents a form of primary maintenance, the secondary maintenance being applicable to those defective pluggable units which were removed for repair on the bench at a later date. The values of μ are also different for each type of maintenance. Thus, the value of μ associated with primary maintenance differs from that of secondary maintenance. The basic reason for these differences in μ values consists in the fact that in the first instance the μ is associated with maintainability at the pluggable unit level, whereas in the second case we are concerned with the part level. The part level usually requires a greater mean time of maintenance action, and for this reason the corresponding value of μ is usually much smaller in order of magnitude than the value of μ associated with primary maintenance.

There are certain basic steps which must be followed in order to make an effective availability prediction. The first step is to perform a reliability prediction study similar to the one shown in Example 12-1. Having gone through this exercise, one is then cognizant of the effect of the failure rate of parts, components, and equipment.

The next step is to determine the value of μ for primary maintenance. Thus, if an equipment consists of six pluggable components or drawers, the time of isolating the failure to a specific component and removing and replacing it can be estimated either through simulated time and motion study or by using accepted time measurement standards. If the time of replacement for each component is equal, then the primary maintenance-action rate must be the same for each pluggable drawer or component and hence for the entire equipment, since the very act of replacing a component constitutes the maintenance action.

In order to predict the estimated mean time of maintenance action of equipment whose components are functionally in series ϕ_e when the replacement time for each component is not the same, we must calculate the relative frequency of failure f for each component and multiply it by the estimated mean replacement time (estimated mean time of maintenance action) ϕ_c for each component and divide the product by the total number of maintenance actions N_a. The equation is[1]

$$\phi_e = \sum \frac{f\phi_c}{N_a} \tag{12-1}$$

The following example will illustrate the principles involved.

Example 12-2. In Example 12-1 we estimated the failure rate of the receiver as 176.37 per cent failures/1,000 hr.

(*a*) Estimate the value of m and μ for the receiver if the replacement times for each component are equal and are 0.1 hr.

(*b*) Estimate the value of ϕ_e and μ if the estimated replacement times in hours for each component are respectively 0.1, 0.15, 0.2, 0.3, and 0.4 (use estimated component failure rates shown in Table 12-1) when $T = 1{,}000$ hr.

(*c*) Estimate the radio set Λ_M when $T = 100$ hr and $t = 0.2$ hr.

Solution. (*a*) The mean time to failure of the receiver is

$$m = \frac{1}{0.0017637} = 567 \text{ hr}$$

Since the replacement times for each component are equal, their relative frequency of failure is not considered and, therefore, the replacement time of 0.1 hr is the mean time of maintenance action of the receiver ϕ_e.* The maintenance-action rate of the equipment μ_e is

$$\mu_e = \frac{1}{\phi_e} = \frac{1}{0.1} = 10 \text{ maintenance actions per hr}$$

(*b*) When the replacement times are not equal for each component, we must first tabulate the data as shown in Table 12-2. Assume $T = 1{,}000$ hr.

[1] When relative frequency is used, ΣN_a is always unity, and therefore Eq. (12-1) could be written simply as $\phi_e = \Sigma f\phi_c$.

* For components functionally in parallel, $\phi_e = \phi_c/n$, see Sec. 9-11.

TABLE 12-2. DATA OF EXAMPLE 12-2

(1) Nomenclature	(2) Probability of failure of each component $P_f = 1 - e^{-rT}$	(3) Relative failure frequency f	(4) Estimated component replacement time ϕ_c, hr	(5) Product of columns 3 and 4
PS	0.148	$\frac{0.148}{1.463} = 0.101$	0.1	0.0101
R-f	0.330	$\frac{0.330}{1.463} = 0.225$	0.15	0.0337
Mixer	0.286	$\frac{0.286}{1.463} = 0.196$	0.20	0.0392
I-f	0.417	$\frac{0.417}{1.463} = 0.285$	0.30	0.0855
PO	0.282	$\frac{0.282}{1.463} = 0.193$	0.40	0.0772
	$\Sigma P_f = 1.463$	$\Sigma f\phi_c = 0.2457$		$\Sigma f\phi_c = 0.2457$

Substituting in Eq. (12-1),

$$\phi_e = \sum \frac{f\phi_c}{N_a} = \frac{0.2457}{1} = 0.2457$$

$$\mu = \frac{1}{\phi_e} = \frac{1}{0.2457} = 4.48 \text{ maintenance actions per hour}$$

(*c*) To calculate the availability when $T = 100$ hr, we can resort to the nomograph of Fig. 9-5, but first we must find the values of d and b.

$$d = r_eT = 0.0017637(100) = 0.177$$
$$b = \mu t = 4.48(0.2) = 0.896$$

Setting a straightedge on these values of b and d in Fig. 9-5, we read Λ_M as equal to approximately 94 per cent.[1]

A similar method can be used to make predictions at the part level.

It is interesting to note that the availability of the radio receiver according to the maintainability schedule of Example 12-2 is almost as effective as the parallel redundancy described in Sec. 12-3. In the latter case, after a time of 100 hr we obtained a reliability of 97.2 per cent, as compared with our present calculation of 94 per cent. We can also see that if it were possible to improve μ, we could easily equal or exceed the 97.2 per cent figure. Therefore, if an analysis indicates that the cost of maintenance, i.e., the complement of spare parts, components, and personnel,

[1] See restrictions regarding use of nomograph in footnote in Sec. 9-9.

is justified, the availability may be obtained by the use of maintainability instead of redundancy. In each instance the method which is most beneficial to the user should be selected.

Moreover, as a general rule of thumb it is always more advantageous to think in terms of availability obtained through the medium of maintainability than it is to talk of reliability. On the other hand, if the mission time is short, it might be more effective to design a relatively high reliability into the equipment without regard for any maintainability whatsoever. This course of action is justified because the probability of failure is very small for a small mission time. In a sense, we are operating at the very top of the P_s curve, and therefore a preponderant number of our missions will be successful.

Oftentimes it is possible to improve the failure rate immeasurably by means of derating the parts. This implies subjecting parts to lesser stresses than they can normally withstand. By doing this the incidence of failure, or in other words, the failure rate, is reduced with the consequent improvement in reliability. There are many published tables which furnish derating curves for such parts as tubes, resistors, condensers, etc.

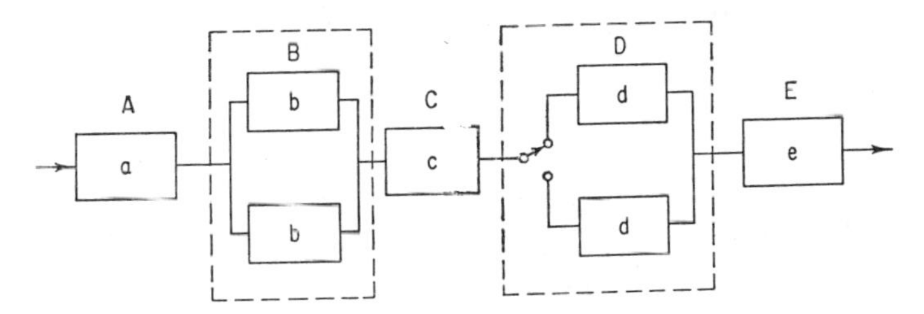

Fig. 12-1. Functional diagram of a typical system.

12-5. The Functional Diagram. Figure 12-1 shows a functional diagram of a frequency selector circuit. This type of diagram differs from a schematic one because it stresses the functional relationships which exist between the associated parts or components. In this manner, it is possible to observe whether certain functions are in parallel or series with each other, or whether certain switching provisions are provided for on a priority basis, giving precedence to more important missions.

A functional diagram need not be restricted to hardware alone but may also include manpower or facilities when they comprise part of an operational system. Although this type of integration of dissimilar elements may appear to be illogical, a little thought should convince the reader that such combinations are not only rational but necessary. As an example, if two military facilities had identical electronic installations but different weapons, it is obvious that neither could perform the other's mission. Therefore, if one facility were destroyed, the other could not take its place because of the lack of proper weapons. As an example, such a situation could occur if a bomber base were destroyed and the remaining fighter base had no bombers available to complete the mission. This is an illustration of two military facilities that would be considered as being functionally in series with respect to completing the specific mission, although they might be functionally in parallel with respect to communications facilities.

The following illustrative example was especially selected in order to highlight the effectiveness of a functional diagram in making availability predictions.

Example 12-3. Calculate the availability of the system shown in Fig. 12-1. The system consists of five subsystems operating functionally as shown. The upper-case letters are used in reference to the subsystems, while the lower-case letters refer to the components of a subsystem.

The applicable data are shown in Table 12-3.

TABLE 12-3. TABULATED DATA FOR FIG. 12-1

Component	r = failure rate, fph	μ = maintenance-action rate	T = mission time	t = time constraint
a	0.0055	5.0	24	0.2
b	0.0425	2.5	24	0.2
c	0.0003	7.5	24	0.2
d	0.0125	4.0	24	0.2
e	0.0009	7.5	24	0.2

Solution. 1. The availability of the system is equal to the product of the availabilities of all the series elements comprising the system. Therefore, the system has been subdivided into five series elements which are labeled A through E. The availability of the system of Fig. 12-1 is

$$\Lambda_S = \Lambda_A \times \Lambda_B \times \Lambda_C \times \Lambda_D \times \Lambda_E$$

2. The availabilities of the series elements Λ_A through Λ_E are determined individually as follows. All subscripts refer to the equipment of Fig. 12-1.

(*a*) $\Lambda_A = \exp(-r_A T e^{-\mu_a t})$

where $\quad r_A = r_a = 0.0055$

Substituting,
$$\begin{aligned} \Lambda_A &= \exp[-(0.0055)(24)e^{-5(0.2)}] \\ &= \exp(-0.1320e^{-1.0}) \\ &= 0.9526 \end{aligned}$$

(*b*) $\Lambda_B = \exp(-r_B T e^{-2\mu_b t})$

where

$$r_B = \frac{2r_b^2}{3r_b + \mu_b} = \frac{2(0.0425)^2}{3(0.0425) + 2.5} = 0.00137 \qquad \text{[see Eq. (9-18)]}$$

$$e^{-2\mu_b t} = e^{-2(2.5)(0.2)} = e^{-1} = 0.367$$

Substituting,

$$\begin{aligned} \Lambda_B &= e^{-(0.00137)(24)(0.367)} \\ &= e^{-0.01207} \\ &= 0.98793 \end{aligned}$$

(*c*) $\Lambda_C = \exp(-r_C T e^{-\mu_c t})$

where

$$r_C = r_c = 0.0003$$

$$e^{-\mu_c t} = e^{-7.5(0.2)} = e^{-1.5} = 0.223$$

Substituting,

$$\begin{aligned} \Lambda_C &= e^{-(0.0003)24(0.223)} \\ &= e^{-(0.0016)} \\ &= 0.9984 \end{aligned}$$

(*d*) $\Lambda_D = \exp(-r_D T e^{-2\mu_d t})$

where

$$r_D = \frac{r_d^2}{2r_d + \mu_d} \qquad \text{[see Eq. (9-41)]}$$

$$= \frac{(0.0125)^2}{2(0.0125) + 4} = \frac{(0.000156)}{0.0250 + 4} = 0.0003$$

$$e^{-2\mu_d t} = e^{-2(4)(0.2)} = 0.202$$

$$\Lambda_D = e^{-0.0003(24)(0.202)} = 0.99855$$

Hence

$$\Lambda_D = 0.99855$$

(*e*) $\Lambda_E = \exp(-r_E T e^{-\mu_e t})$

where

$$\begin{aligned} r_E &= r_e = 0.0009 \\ \Lambda_E &= \exp[-(0.0009)(24)e^{-7.5(0.2)}] \\ &= \exp(-0.0216e^{-1.5}) \\ &= 0.9952 \end{aligned}$$

The availability of the system is therefore

$$\begin{aligned} \Lambda_S &= \Lambda_A \times \Lambda_B \times \Lambda_C \times \Lambda_D \times \Lambda_E \\ &= 0.9526 \times 0.98793 \times 0.9984 \times 0.99855 \times 0.9952 \\ &= 0.93373 \end{aligned}$$

PRACTICE PROBLEMS

12-1. One of the requirements in the specification for a radio receiver is a mean time between failures of 1,000 hr. Data for a typical design are shown below. Calculate the reliability under the following conditions: (*a*) Assume that any part failure will cause complete failure of the receiver. (*b*) Most of the failure rates and derating factors were taken from RCA Report No. TR-1100, "Reliability Stress Analysis for Electronic Equipment." Does the prediction indicate that the receiver has a good or poor chance of meeting a value of $m =$ 1,000 hr?

Quantity of part	Description of part	Average derating	Failure rate, %/1,000 hr	
	Capacitors:			
29	Fixed, mica-silver	0.16	RCA	0.001
2	Fixed, glass	0.003	Assumed	0.0001
5	Dry polarized electrolytic	0.72	Assumed	0.010
2	Paper	0.38	RCA	0.010
4	Button, silver-mica	0.15	RCA	0.0001
34	Disk, ceramic	0.13	RCA	0.001
3	Variable, glass	0.07	RCA	0.001
7	Transformers and chokes	.....	RCA	0.100
	Resistors:			
80	Fixed composition	0.12	RCA	0.052
19	Fixed, wire-wound	0.10	RCA	0.140
11	Diodes, silicon	0.50	RCA	0.026
	Tube:			
7	Pentode, reliable	0.50	RCA	2.500
1	Double diode, reliable	0.50	RCA	1.500
4	Duotriode, miniature	0.50	RCA	9.000
1	Pentode, miniature	0.50	RCA	7.500
1	Thyratron, miniature	0.50	RCA	9.000
2	Diode, miniature	0.50	RCA	3.750
3	Duotriode	0.50	RCA	3.000

12-2. Estimate the value of A_M for the radio receiver of Example 12-1 for a mission time of 100 hr and a maintenance time constraint of 5 min when $\mu = 5$.

12-3. Estimate the time and mission availability of each of the configurations *a* and *b* shown in Fig. 12-2. Which has the better availability, and why? The values of component availabilities and other data are shown on the functional diagram. (Assume switchover redundancy and a perfect switching action for configuration *b*.*)

* Hint: Refer to Secs. 9-11 through 9-13 for applicable equations. Also assume outputs as independent and that both are required for successful operation.

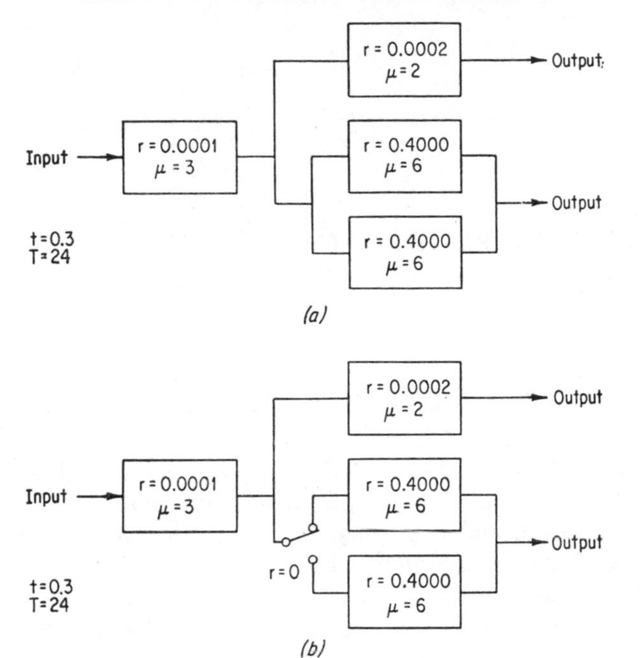

FIG. 12-2. Configurations of Prob. 12-3.

12-4. (*a*) For Example 12-2 shown in the text, find the value of ϕ_e and μ if the estimated replacement time for each component is 0.2 hr. (*b*) Estimate the value of ϕ_e and μ if the estimated replacement times in hours for each component are respectively 0.15, 0.2, 0.1, 0.25, and 0.3. Assume $T = 100$ hr.

12-5. Estimate the mission availability of two identical subsystems connected functionally in switchover redundancy. The data are as follows: $T = 100$ hr, $\mu = 2$, $r = 0.001$ failures/hr, $t = 0.2$ hr. (Refer to Eq. 9-41.)

CHAPTER 13

PRINCIPLES OF RELIABILITY DESIGN

13-1. General Design Objectives. The general objective of reliable design is to achieve and maintain the specified operational reliability at minimum cost. This implies that the reliability goals set by the designer must be achieved for all stages of product evolution such as design, development production, and field operation. These goals, although important guide posts, cannot be determined until the designer has first established some sort of concept of what is expected of the product, which he can determine only after analyzing the operational requirements.

Whether it be a component or system, it is necessary that the objectives which are established include the achievement of reliability, performance, accuracy, maintainability, operability, producibility, procurability, human factors, expandability, versatility, and vulnerability. The designer must also weigh the practicality of trade-offs with some of these or other design parameters in order to achieve optimum design. Such trade-offs may be with respect to accuracy, cost, weight, or related considerations.

Most of the factors mentioned above are self-explanatory or familiar to the reader as a result of past chapters, but there are some which it might be advantageous to discuss at this time.

Operability is the ability of the device to be operated in the manner intended, as installed in the final operational system configuration. This means that all controls should be accessible and all other factors, such as environment and accessibility to parts and components, must be conducive to good operation.

Producibility may be defined as the ease with which the device may be manufactured without the use of complex tools, jigs, fixtures, and test equipment.

Procurability is concerned with time. If, because of the complexity of the design, little consideration is given to the time required to fully develop and procure the final product, we can assume a poor degree of procurability. In general, we can say that procurability is producibility measured as a function of time.

Expandability is a measure of the capability of the product to perform more functions or more of the same functions by the addition of extra parts, modules, or facilities or by changing certain basic elements consisting of parts, modules, or components.

Versatility is the ability of the device to be used in several different systems for similar or related applications. For example, a radio receiver might be used in a military application as a simple receiver, or it might be used as part of a triple diversity receiving system.

Vulnerability is a measure of the ease with which a device may be disabled. The jamming of a radio transmitter by the enemy is a good example of vulnerability.

13-2. Design Factors to Be Considered in Order to Achieve Design Objectives. The design objectives mentioned in Sec. 13-1 can be realized only if several factors are considered which individually and collectively affect each of them. For example, the initial design concept should stress simplification without sacrifice of specification and contractual requirements. It should also weigh the relative merit of manual versus automatic operation or control, as well as speed of operation, programming, and other factors as they affect reliability, interchangeability and replaceability, and maintainability. Moreover, sound reliability engineering must consider the aptitudes of individuals as well as the capabilities of machines in order to ensure the greatest over-all reliability of the combination. Thus, human engineering definitely takes its place beside design engineering, since, in a sense, the human machine is an integral part of system reliability. Therefore, the engineer must consider whether by virtue of his design the human is capable of performing the things demanded of him. For example, is the human going to be required to exceed his limits of physical strength, perform too many functions simultaneously, or perceive and handle information at faster rates than his capabilities will permit? Is he going to be asked to maintain monitoring vigils without adequate informational feedback or perform meticulous tasks under difficult environmental conditions and for longer periods than he can function at a reliable level of performance? Or is he going to be required to work at peak performance, yet make judgments beyond the range of human capability without adequate illumination? Is he going to be required to perform visual motor functions as efficiently as any other person or work with tools in surroundings which are not compatible with his physical dimensions; or is he expected to have the same physical stature, motivational level, training, experience, and intellectual background and be of the same size and physical characteristics as any other person?

In addition to the human engineering aspects, the design must consider the technical and environmental factors involved in the initial design concept. There must be a consideration of the limitations of weight, complexity, and size on accessibility and replaceability of subassemblies. The advantage of using replaceable modules in lieu of unit construction must also be evaluated. The design should also consider the effects of heat dissipation, vibration, shock, and the mounting of parts, as well as specify the proper mechanical structures which should be used. The engineer should also weigh the practicality of the design with regard to standard factory production techniques, as well as weigh the probable effects of varying environmental conditions on heat dissipation and deterioration of parts and devise means for overcoming these effects. In this regard, he must evaluate whether the most appropriate techniques are being used in the most effective manner. Moreover, he must decide whether required performance will be achieved under all specified conditions of imposed environmental stress, power supply fluctuations, contaminants, tolerance accumulation of circuit elements, as well as temporary and permanent shifts in value imposed by variation of operating conditions and by age. He should also evaluate whether the desired function is accomplished in the most effective manner from the viewpoint of reliability, with a minimum of "series" elements, and proper and adequate application of redundancy techniques. Another consideration is to determine if the interaction between electrical and mechanical functions is optimum. This involves all mechanical aspects necessary for proper functional utilization of electrical and electronic circuit elements, such as structural support, control of electromagnetic pickup by physical orientation or isolation, heat source proximity, and lead dress.

The factors mentioned so far are by no means all that are involved. The designer must also consider past experience with similar equipment to assure that any shortcomings which were then apparent are corrected in his present design. For example, if the equipment he is designing has the potential of producing electromagnetic radiation hazards, he should take positive steps to shield or correct the problem in some other manner. In this respect, it is known that X rays produce cataracts of the eye, cancer of the skin and blood stream, sterility, and shortened life. Usually a high-powered klystron tube will produce X rays, and therefore a wise precaution is to insist that personnel using this equipment be equipped with dosimeters, to assure that they have not been overexposed to radiation. Similar precautions should be taken to protect operating personnel from microwave radiation whose wavelength is greater than or approximately equal to 30 centimeters, since this could also cause cataracts. Moreover, when the power density is greater than 0.01 watt per square centimeter, this is very dangerous.

The designer should also weigh the possible effects of heat and heat transfer on the operation of the equipment. Therefore, he should study methods of conduction, insulation, radiation, convection, and cooling to overcome deleterious effects.

Other factors worthy of consideration and summarization involve the effects of biological growth, sand and dust, shock mounting, part fasteners, wiring, lead-mounted component parts, transportation and storage problems, packing, marking, and installation.

13-3. Methods of Achieving Design Factors. In order to cope with the design factors of Sec. 13-2, the electronics engineer must be a capable, mature individual of wide experience and a thorough, sound technical training. He must be capable of balancing items of cost, manufacture, and installation versus his desires for incorporating pet experiments or circuits which are still in the experimental stage. A good engineer who is interested in achieving reliability will select basic circuits which have been proven and tried and will resist pressures of his own or of subordinates to experiment with the untried or purely hypothetical concepts. This does not mean that engineers should not be encouraged to be creative. There is much room for creativity in any field, and certainly nobody would be guilty of advocating stagnation in electronics, which is so crucial to the defense and development of the country. However, we are suggesting that new circuits be fully developed and tested in less critical or commercial devices before being considered for incorporation in highly reliable gear, particularly if it is destined for military service. To sum it up, a good designer should use simple circuits of proven reliability in his designs. These circuits either are usually available as a result of his own experience or, in some instances, are offered to the industry as prepackaged modules of proven characteristics and reliability by reputable vendors.

The designer should also be assured that the circuits he selects are stable without the necessity for excessive shielding or voltage regulation, or that they are not critical or marginal to the extent that special parts must be selected to assure the proper function. Such a condition is undesirable, because the replacement of a random spare part is not feasible unless the circuit in which it is to be used is retuned. It is not always possible to perform this retuning under field conditions, and therefore this type of unstable circuit should be strictly avoided. The foregoing is a good argument for designing or using circuits which will accommodate parts of wide tolerance without impairing the operation of the circuit. Moreover, as in the case of individuals who specialize in specific fields in order to attain a high degree of proficiency, so must we expect circuits to specialize. Dual-purpose circuits, although generally more economical because they use less parts, are usually less reliable. This is one of the major reasons for parts manufacturers' complaints that the alleged lack of reliability of their parts is not due to the parts at all but rather due to misapplication of the parts in circuitry. Dual circuits are usually blamed for such a situation. In summary, as a general rule it is a good principle to avoid the use of dual-purpose circuits when high reliability is desired.

Another argument in favor of standard circuits is that they are usually familiar to maintenance men, who have seen them used before in other equipment. Therefore, this type of circuit can be packaged in a module which is interchangeable and easily replaced, resulting in a high degree of maintainability.

A reminder for the electronics engineer with respect to tolerances is that he must consider the products of all manufacturers for the same type of approved part if he is to achieve a high degree of interchangeability. It is not unusual in electronics equipment for a part of a particular manufacturer not to function in a circuit while a part of another manufacturer does, even though each of the parts complies with the applicable specifications. It is apparent that in this case the interaction between part and circuit is different for each manufacturer. A good designer will try a cross section of parts from various approved manufacturers in his circuits and check their response by simulating tolerances through marginal checking techniques. This means that specific or critical voltages are varied and the outputs for particular inputs are studied in order that all possible effects can be assessed. If the study reveals that the only method of using parts of all approved manufacturers is to specify tight tolerances such as plus or minus one per cent, then the circuit should be redesigned because parts cannot be expected to maintain these tight tolerances under various environments or extended periods even though they have been so advertised.

Marginal checking may also be used on major subassemblies to determine their degradation characteristics and to pinpoint the reason for the effect. This includes the use of programming techniques and automatic read-out to plot subassembly degradation and thereby establish the proper preventive maintenance cycles.

Another variation of the marginal checking idea is called testing to destruction. In this case, various loads are imposed upon a part or device and the outputs monitored. The loads are varied by being gradually increased until failure occurs. After sufficient data are accumulated, a statistical analysis is made to determine the safe load limits which can be imposed on the part. A deficiency of this method is that the data are not obtained under the conditions of actual usage. In other words, the immediate part environment for each application is not the same as that under which the part was tested, nor are the electrical applications the same. The only advantage, therefore, of the test-to-destruction technique is that it provides a "ball park" estimate of what stresses the part can withstand; and as long as the results are regarded in this light, they are useful to the design engineer.

Another technique that is useful in improving reliability is called derating. This is a method of parts application which applies stresses on the part which are less than those specified by the manufacturer. Various published derating data and curves are available to the designer who is considering derating.

In those instances where the required degree of reliability is considered in excess of what is conceivably practical after consideration of all the techniques we have been discussing, the designer invariably resorts to redundancy. This is a technique which utilizes one or more additional parts, components, or equipments which are intended to sustain the functions if the originals fail. Most of the time the use of redundancy is not justified, because of additional cost and space, but it is more than worthwhile if the advantages which accrue from the increased resulting operational reliability are considered. However, it should be used only when no other more economical approach is possible, because it is usually the most costly method.

As we have seen in Chap. 9, maintainability is another technique which, if properly planned, can result in a high degree of operational reliability despite the fact that the inherent reliability might be relatively low. The objective in designing for a high degree of maintainability is to be able to locate the cause of trouble rapidly and eliminate it in order to realize a minimum maintenance-action rate. Since the subject was covered in great detail in Chap. 9, we will not repeat the concepts here.

On the other side of the scale, if the maintainability is relatively low, the designer should make every effort to achieve a high inherent reliability. In order to do this, he must not only concentrate on the techniques we have already discussed, but must select parts or components of the lowest possible failure rate which are available. There are many sources of published data which give failure rates for most standard parts, but when data are not available, it is necessary to conduct tests in the environmental laboratory under simulated conditions to determine the failure rate of a part being considered for use. If the tests indicate that the failure rates for the immediate part or general environment are too high, it may be expedient to design controlled environments into the equipment. This is not an uncommon practice. Many oscillator circuits are enclosed in thermal ovens which control the temperature closely to reduce frequency drift, while some airborne equipment is sealed in dry nitrogen in order to maintain a stable environment. Moreover, it is not uncommon these days to resort to hermetic sealing and encapsulation of parts. As a matter of fact, nowadays this is a common and accepted practice.

The foregoing outline of design methods and techniques is illustrative of what a good designer considers in formulating his concepts, but it is by no means all-inclusive. To cover all the techniques in a discussion of this sort is practically an impossibility, since the science of reliability is still in the process of evolution and we are learning more and more with each passing day. Nor are we attempting at this time to explain exactly how the designer is to apply the techniques described herein, since this is a function of his own skill and professional attainment which it is expected he will apply wisely to the problems at hand. Our purpose here is to outline, from the past experience of others, what has been successfully accomplished in the past and what it is expected can be accomplished in the future. Therefore, the designer should accept the suggestions embodied herein as a sort of check list which he can use to assure that he is availing himself of every possible approach to designing more reliable equipment. Accordingly, we shall proceed with additional design techniques for his consideration.

When modules of similar exterior configuration, but dissimilar electrical characteristics, are used for rapid interchangeability in case of failure of the module in the equipment, it is imperative that methods be provided to prevent the insertion of the wrong module. An effective method of accomplishing this objective is to provide a keying device so that it is impossible to insert the wrong module in a cavity. This is usually accomplished by providing pins spaced in such a manner that only one cavity will receive that type of module. Any other equivalent method to accomplish the same purpose is also satisfactory. It is also suggested that during production each module be checked with interchangeability gages to assure a good fit. Accordingly, the designer should weigh the system he uses for keying to assure that simple gauges can be designed to satisfy the interchangeability requirement. He should also provide a design which will assure quick removal without damage to connectors or other components and of a weight which is not outside the limitations of using personnel.

The design should also incorporate built-in test equipment which can be utilized to detect failures and the standardization of test points and test connectors to ensure rapid isolation of the trouble or to provide for calibration as required. Moreover, the color coding of wiring from each of these connectors and throughout the equipment should be standardized to assist in rapid and efficient trouble shooting. The signals at particular pins of the connectors should also be standardized in order to simplify maintenance and calibration.

The number of controls for operation and adjustment should be kept at a minimum, and their design and arrangement should be consistent with requirements dictated by good human engineering practices, which means that, among other things, they should be easily accessible and simple to manipulate.

13-4. Systems Analysis. An experienced designer will get all the facts before attempting the design of a system or subsystem. This is an analytical process and is therefore referred to as systems analysis. Actually, it is nothing more or less than a process of gathering technical and operational information in order to achieve the most economical and practical design. Proper perspective is a most important part of the analytical method, and therefore the designer must ascertain why the proposed system is considered necessary as well as define its basic purpose, i.e., the nature of the mission and the reason for considering it important. In other words, his impression of what is desired, and why, must be crystal clear.

To perform the analysis properly, the designer must discard all preconceived notions of his own and proceed to get the facts from whatever source they may be available, keeping an open mind and acute ear to absorb the suggestions of others. Bias on his part, or pet theories, will only result in an inadequate design. The fact-finding process must be well planned but not limited to any specific area or activity. For example, if the customer purchasing the system happens to be one of the military services, the designer will doubtless want to talk to military officers of the using arm who, it is expected, should have a thorough knowledge of the military characteristics which they desire, as well as talk with research and development engineers who are interested in the technical aspects. Of no less importance, however, are the operating personnel in the field who have had experience with similar equipment and whose ideas might conceivably result in designs for the projected equipment which would not have the same faults as were being experienced with current similar equipment.

In his quest for facts, therefore, the designer will find himself visiting a variety of customer installations and discussing mutual problems with their personnel in order to secure data which will be useful in his design endeavor. He will also doubtless attend many meetings and read many technical reports on all phases of the problem. However, regardless of the source of information, the designer must eventually get all the facts if he is expected to do an efficient design job. To this end, he should want to know what similar facilities are presently available and how adequate their performance is, and, if they are unsatisfactory, whether this is due to technical deficiencies of the equipment or lack of skill and knowledge of the using personnel. These factors are most important because they provide a relative measure, for planning purposes, of what must be designed into the new equipment to avoid similar pitfalls.

The operational requirements must also be explored in order to obtain an appreciation of the military use, the range of environments, and the expected reliability for a given duty cycle. Moreover, it is important to know if any limitations of size and weight have been imposed on the equipment in order to assess their effects.

Present facilities and regulations should also be studied to determine available power, buildings, and space, as well as frequency allocations, existing codes and regulations, and equipment duty cycles. Moreover, it is important to determine the type of location or site at which it is planned to install the device. If a hardened site is being planned to assure protection against atomic attack, it is important to list the probable environmental conditions which must be provided by means of air conditioning or heating plants, as the case might require.

It is also important to determine what other components, subsystems, or systems must operate with the proposed system or what it is intended to replace in whole or in part. In instances where replacement is required, it is necessary to gather sufficient information to permit its being done with the least interruption to service.

The requirements for expandability as envisioned by the customer should also form an important part of the fact-finding process. In this manner, the designer is in a better position to provide features to assure the expansion of the system facilities with time.

In conclusion, after all the required data have been accumulated, it is necessary to determine what the customer considers a satisfactory manner of demonstrating operational capability. This is an important consideration, because this is the time during which the system is operated to demonstrate that operational requirements have been complied with. The customer's plans about how he intends to participate in this demonstration should also be clearly understood. On the basis of the facts elicited at this time, the designer can then prepare the specification for demonstrating operational capability and submit it to the customer for approval. This is very important, since failure to do so might result in much litigation later because all the parameters which should have been demonstrated were not clearly defined. The specification should simulate, as nearly as possible, the actual operating conditions which will probably be encountered in service and provide a program of events which are indicative of those which will actually be used in operation. This type of activity, usually called programming, outlines the scheduled maintenance activity, which includes preventive maintenance checks such as marginal checking and recording of significant data for specific equipment features, as well as operational checks to assure that the capability of the equipment to perform its mission has not been impaired. The kind of operational checks is a function of the type of equipment or system being checked. In an instrument landing system these checks would include actual or simulated testing to assure the safe landing of an airplane, while monitoring such important patterns as the glide slope and localizer. On the other hand, a complex communication system might require operational checks to assess the effectiveness of particular programs such as war games; intelligence; logistics; data transmission and recording either visually, aurally, or on hard copy; availability of war material or personnel, etc.

13-5. Systems Engineering. The design of a weapons system represents a complex man-machine problem which requires a critical analysis of the characteristics of the man and machine components which comprise it, as well as the methods of interconnecting them to form an integrated operational system. This is the task of systems engineering. Using the facts available from the systems-analysis studies, the systems engineer initially attempts to establish requirements for components, subsystems, and the system which will optimize the reliability of the system to a sufficient degree to meet the contractual and mission requirements. These must be realistic, achievable requirements, based on consideration of factors other than technical ones, such as survivability under combat conditions, flight safety, maximum allowable maintenance load, and training missions.

The general approach to the problem is to calculate the intrinsic reliability of the system. This is accomplished by preparing a preliminary line-flow functional diagram and calculating the reliability in accordance with the techniques described in Chap. 9. If the resultant intrinsic reliability is not satisfactory, then necessary design changes or trade-offs with other mission requirements must be made in order to resolve the situation. At this point, either the line-flow functional diagram must be reassessed to see if any changes are possible, or else components of higher reliability must be considered if they are available. If they are not available, then other techniques, such as redundancy or derating, must be considered in order to bolster the reliability. In any event, regardless of the technique used, the engineer must eventually come up with a line-flow functional diagram which indicates the reliability which has been allocated to each of the components and subsystems which comprise the system. This allocation should also consider the effects of interconnections between components or components and subsystems, often called interface relationships. In other words, the allocation of an abstract reliability figure for a component, because it had demonstrated its capability to meet this figure by past performance in other equipment, should be strictly avoided. The reason is that the component must always be considered in relation to the system or subsystem configuration of which it is a part, because it will have the same reliability with varying usage.

The interface relationships are the major factors which delineate the governing component specifications, because they determine the component inputs and outputs with respect to specific environmental conditions. This is the basis for preparing component performance specifications for the guidance of the components design engineer. These component specifications must be carefully prepared to assure that they are reconcilable with subsystem specifications, and the latter must in turn be compatible with the systems requirement. These relationships are usually shown in schematic diagrams which indicate the interconnection concept. Other diagrams, such as ground schematics and analysis diagrams, show the ground connections and analysis of specific circuits. This is usually done by means of specifying continuity, voltage, current, or response tests from point to point in order to test the interconnection concept and also to evaluate the functional relationships of the system within itself as well as its compatibility with other systems with which it must work. The support equipment and facilities necessary to achieve this integration of operation must also be clearly specified. This will involve the preparation of specifications for the type and size of buildings, environments, and associated test and maintenance equipment.

13-6. Design and Development Practices. Each designer accepts as gospel the general specification for the component or subsystem he is charged with designing. It is his job to design, within the limitations imposed by the performance specification, a component which will meet all the performance requirements. In order to do this, he must eventually come up with a set of detailed engineering drawings and specifications which are peculiar to the component he is designing and which are suitable for reprocurement purposes. This means that, from the component engineer's point of view, any manufacturer could make the component by merely following the applicable specifications and drawings which he has generated, without knowing any specific details of the system as a whole.

In a sense the component engineer follows the same basic principles as the system engineer, except that he concentrates on the application of parts and works on the component level. However, he has the distinct advantage of proving his design by testing models of the breadboard or prototype variety before finalizing on the design. The systems engineer is denied this advantage, because he must wait for component and subsystem prototypes before he can perform system tests for evaluation purposes.

The methods available to the designer for building reliable equipment were amply described in Sec. 13-3 and therefore will not be repeated here in detail; but the steps which he must pursue will be discussed in order to explain the design and development procedure.

The first step is to conceive the design concept and to record the failure rate for each part used in the equipment. These failure rates may be obtained from published data, by reliability testing, or from the reliability engineer assigned to the designer to render consulting services. In any case, the failure rates assigned must be recorded with relation to the conditions of usage. A good practice to follow is to record the failure rates right on the schematic. Usually this also indicates the voltages and currents which are impressed on the part. The next step is to calculate the intrinsic reliability as described in Chap. 9. This first calculation may be made by assuming all parts functionally in series, although this may not actually be the case; but since it is the worst possible condition, if the resulting reliability prediction meets the requirements, then one is certain that a series-parallel combination would more than satisfy the conditions. On the other hand, if the intrinsic reliability falls far short of the mark, then another calculation should be made by use of the actual functional relationships of the parts to determine whether they are being used functionally in series or parallel. Again if we fail to meet the reliability objective, we must proceed with other measures to assure our achieving the required goal. As we have seen, we might resort to derating if our input-output and space requirements permit. This means that for those parts which the reliability prediction shows contribute the least reliability, we shall impress lesser stresses in order to improve their life. There are many published derating curves available which furnish the stresses which should be impressed on a part in order to improve its reliability. If we find that the improvement is not sufficient even if we derate the part, we should investigate the availability of parts of greater reliability from other manufacturers. If none are available, we would probably resort to redundancy, space, weight, and cost not being a factor, and evaluate the results on this basis. Care should be exercised when using redundancy to assure that there will be proper switching in the event of failure and that no reactions deleterious to the equipment function will be evident as a result.

If the designer finds that using all these techniques and others which we formerly described is ineffectual, then he must decide on what trade-offs he can make with the system engineer to achieve the reliability objective. These trade-offs may involve sacrificing weight, volume, power output, or similar parameters in favor of achieving the prescribed reliability.

When the eventual reliability prediction shows that the intrinsic reliability conforms to what was prescribed, the designer proceeds to the construction of breadboard, design, or prototype models, depending on the requirements of the contract and applicable schedules. Upon completion, this model is subjected to reliability tests to determine if the calculated value of intrinsic reliability and the value determined by testing agree. At this point, it is important to mention that it is the exception rather than the rule if the calculated and measured reliability agree. The purpose of the prediction technique is not to get an exact figure but rather to determine approximately in what area the true reliability figure might be. At most, it is but an estimate and must be treated accordingly. A major advantage of the prediction technique is that it is a powerful tool which can be used to isolate those parts which are expected to give the most trouble.

Whenever possible it is always recommended that the model be as much like the production unit as feasible, because reliability tests performed on these will carry a lot more weight than those on breadboards. A breadboard should be used only as a tool in the initial stages merely to test the function of some circuits or the over-all circuitry. It is for this reason that prototypes are suggested. A prototype, on the other hand, can be used to demonstrate operational capability to the customer. Moreover, during reliability testing, the data which are accumulated have a greater degree of confidence associated with them than if breadboards were tested. Therefore, any resulting engineering changes which appear evident will also be made with a high degree of certainty that they are necessary. Another argument in favor of the prototype is that since it is practically equivalent to production equipment, the effects of stray capacity and poor grounding as well as the interconnecting techniques may be adequately evaluated.

The availability of prototype models also enhances the study of the human engineering aspects, because the man-machine relationships can actually be studied on these models. This makes possible the reevaluation of human factors and the preparation of required specifications to assure the optimum man-machine relationship.

After the prototypes have been thoroughly tested and found to be satisfactory, it is then possible to prepare the applicable specifications and drawings for the particular components.

CHAPTER 14

RELIABILITY SPECIFICATIONS

14-1. General Discussion. It was mentioned in Sec. 1-7 that a complete reliability specification should provide means for *measurement*, *evaluation*, *improvement*, and *prediction*. It was also emphasized that each specification must in turn define the *purpose*, *place*, *methods*, *instruments*, *personnel*, *circumstances*, and *procedures* to be employed.

A complete reliability specification usually consists of subsidiary specifications which in turn concentrate on a special requirement. For example, one of these subsidiary specifications might be devoted exclusively to the technique of reliability measurement, while another might concern itself with prediction methods.

In order to be effective, a reliability specification must avoid meaningless adjectives, such as "equipment must have the highest order of reliability," and must specify measurable characteristics which are described in a clear, lucid, and intelligible manner. This has not always been easy to do. As a matter of fact, the lack of standards of measurement has been one of the reasons for inadequate reliability specifications. As a result, in the past various government agencies prepared and issued many inadequate specifications which were intended to assure reliability. These early efforts were as expensive as they were worthless because they had no basis in fact for their existence. Usually these specifications assumed that compliance with all requirements of the applicable MIL specifications for all parts ensured the required reliability. Consequently, they specified 100 per cent testing of all parts under rigid environmental conditions. In some extreme cases, these specifications required the microscopic inspection of every vacuum tube to determine the quality of welds, seal of glass envelopes, and general tube construction. If these attributes were found to be substandard, the tube would be rejected even though it might have passed all electrical and environmental tests. This was a very expensive program, since in many cases the rejection percentages for such items as tubes and relays were as high as 60 per cent. Moreover, despite all precautions and rigid control, some of the parts which passed the gamut of reliability testing still failed in the equipment. To compensate for these equipment failures, other specifications were developed for the purpose of detecting and eliminating early failures. One such type is the debugging or burn-in procedure. It is based on the theory described by Fig. 6-2, namely, that early failures occur during the infant period and that they therefore can be eliminated by the burn-in procedures. This type of specification, although by no means a cure-all, has resulted in an appreciable improvement in reliability.

However, all these methods should be considered rudimentary stopgaps, since they do not strike at the heart of the problem. Reliability practices begin with the initial concept of the design and follow it each step of the way until the equipment is shipped. On a comparable basis, just as it is said that quality cannot be inspected into equipment but must be built into it, so we can state that reliability cannot be tested or inspected into equipment but must be designed into it. It was this early lack of recognition of the importance of design for reliability that led to trouble. This was partly due to lack of appreciation of the design contribution to reliability, as well as a lack of knowledge of the parameters, methods, techniques, and devices necessary to assure more reliable equipment.

As a greater appreciation of the complexity of the problem developed, it became apparent that reliability could not be specified as an abstract number without being related to a particular environment. In other words, it is meaningless to say that a component has a reliability of 99 per cent or a mean time between failures of 100 hr. In each case, there is no significant relationship to any parameter or factor. Ninety-nine per cent of what? A mean time between failures of 100 hr under what operating conditions?

Moreover, it was realized that the type and duration of the mission was also important. This led to the development of a series of reliability specifications which were applicable to various categories of equipment and their use. Table 14-1 lists the various government and associated documents establishing reliability requirements.

TABLE 14-1. GOVERNMENT AND ASSOCIATED DOCUMENTS ESTABLISHING RELIABILITY REQUIREMENTS

I. Major reliability documents
   A. Missiles and space systems—AFBM 58-10 (USAF)
   B. Weapons system—MIL-R-26674 (USAF)
   C. Fleet ballistic missile—REL. ASSUR. PROGRAM
   D. Military electronic equipment—MIL-STD-441 (DOD)

II. Reliability in design, development, and production of equipment and subsystems
   A. Production electronic equipment—MIL-R-19610 (AER)
   B. Development systems and subsystems—MIL-R-26484A (USAF)
   C. Design equipment and systems—MIL-R-22256 (AER)
   D. Ground checkout equipment—MIL-R-27173 (USAF)
   E. Production ground equipment—MIL-R-26474 (USAF)

TABLE 14-1. GOVERNMENT AND ASSOCIATED DOCUMENTS ESTABLISHING RELIABILITY REQUIREMENTS (*Continued*)

III. Reliability organization, monitoring, assurance, demonstration, and quality control documents
- *A*. Quality control—MIL-Q-9858 (DOD)
- *B*. Maintainability—MIL-M-26512 (USAF)
- *C*. Assurance program—MIL-R-25717C (USAF)
- *D*. Demonstration requirements—MIL-R-26667A (USAF)
- *E*. Organization—USAF SPEC BLTN 510
- *F*. Monitoring—USAF SPEC BLTN 506

IV. Detail requirements
- *A*. Interference
  1. MIL-I-006051B (USAF)
  2. MIL-I-6181D (USAF)
  3. MIL-I-26600(2) (USAF)
  4. PD-R-186 (ABMA)
- *B*. Test reports—MIL-T-9107(1) (USAF)
- *C*. Design
  1. MIL-E-5400D(1) (ASG)
  2. MIL-E-4158B(1) (USAF)
  3. MIL-E-8189B(1) (ASG)
  4. MIL-W-9411A(2) (USAF)
  5. MIL-E-16400C (SHIPS)
  6. MIL-E-19100A (SHIPS)
  7. MIL-E-19600A (WEP)
  8. ANA BLTN 444
  9. AD 114274 (ASTIA)
  10. AD 148556 (ASTIA)
  11. RADC EXHIBIT 2629
- *D*. Provisioning
  1. MIL-B-5005A (DOD)
  2. DOD INSTR. 3232.7
  3. MCP 71-673
  4. PP-SIG-SE-IA
  5. SAR-400
  6. SAR-398
  7. MIL-E-17362C(1) (SHIPS)
- *E*. Training
  1. MIL-T-27382 (USAF)
  2. MIL-T-4860C (USAF)
  3. MIL-T-9344 (USAF)
  4. MIL-I-26036 (USAF)
  5. MIL-T-26046 (USAF)
  6. MIL-D-26239 (USAF)
- *F*. Preservation and packaging
  1. MIL-P-116C(1) (DOD)
  2. USAF SPEC BLTN 56AF
  3. MIL-P-9024B (USAF)
- *G*. Sampling
  1. MIL-STD-105B
  2. MIL-STD-414
  3. DOD-HDBK-H-106

TABLE 14-1. GOVERNMENT AND ASSOCIATED DOCUMENTS ESTABLISHING RELIABILITY REQUIREMENTS (*Continued*)

H. Test methods
  1. MIL-STD-446
  2. MIL-E-5272C(1) (USAF)
  3. MIL-T-5422E(1) (ASG)
  4. MIL-E-4970A
  5. MIL-STD-202B
  6. MIL-T-18303 (AER)
  7. MIL-T-4807A (USAF)

I. Installation
  1. MIL-I-8700 (ASG)
  2. MIL-E-0025366B (USAF)

J. Data
  1. MIL-D-70327A
  2. MIL-D-9310B(2) (USAF)
  3. MIL-D-9412C(2) (USAF)

K. Environmental factors
  1. USAF SPEC BLTN 106A
  2. USAF SPEC BLTN 115(1)
  3. ANC BLTN 22
  4. MIL-STD-201A
  5. MIL-T-152A (DOD)

L. Enclosures
  1. MIL-STD-108C
  2. MIL-C-172C-ID (DOD)
  3. MIL-E-2036C(2) (NAVY)

M. Human factors
  1. WADC-TR-56-488
  2. MIL-H-25946 (USAF)
  3. MIL-D-26207 (USAF)

N. Wiring
  1. MIL-W-5088B (ASG)
  2. PD-E-53 (ABMA)
  3. MIL-W-008160C (USAF)
  4. MIL-T-713A(2) (DOD)

O. Test equipment
  1. MIL-T-21200A (ASG)
  2. MIL-T-8191(1) (USAF)
  3. MIL-T-945A(2) (DOD)
  4. MIL-STD-415A
  5. MIL-T-18306A (AER)

V. Reference documents

AGREE Report: Reliability of Military Electronic Equipment
PB-121839: NEL Reliability Design Handbook
NAV-AER-16-1-519: Handbook of Preferred Circuits
Woodson, "Human Engineering Guide for Equipment Design," University of California Press, Berkeley
NEL: Suggestions for Designers of Electronic Equipment
PSMR-I: Parts Specification Management for Reliability

TABLE 14-2. TABULAR COMPARISON OF RELIABILITY SPECIFICATIONS

Subject	Specification								
	MIL-R-25717C (USAF)	MIL-R-26474 (USAF)	MIL-R-26484 (USAF)	MIL-R-27173 (USAF)	MIL-R-26667 (USAF)	MIL-R-26674 (USAF)	MIL-STD-441 (DOD)	MIL-R-22256 (AER)	AFBM 58-10 (USAF)
I. Equipment									
A. Weapon systems						x			x
B. Electronic equipment	x	x	x	x	x		x	x	
1. General equipment	x		x		x		x	x	
2. Ground equipment		x		x					
C. Phase									
1. Design			x	x		x	x	x	x
2. Development	x	x	x	x	x	x	x	x	x
3. Production	x	x		x	x	x			x
II. Scope									
A. Requirements (reliability)									
1. General	x	x	x	x		x	x	x	x
2. Program		*a*							
a. Organization—operational procedure and guide	x	x	x	x		x		x	x
b. Demonstration					x				
c. Design and reliability analysis							x	x	
III. Requirements									
A. Program									
1. Organization	x		x	x		x		x	x
2. Indoctrination	x	x	x	x		x		x	x
3. Subcontractor monitor and control	x	x	x			x			x
4. Quality control	x	x	x	x		x		x	x
B. MTBF									
1. Propose	x							x	x
2. Specify in detail specification	x	x	x	x	x	x		x	
3. Minimum (if not specified)									
a. Given			x	x				x	
b. Computation equation or curve given		x						x	
C. Operating life			x	x					
D. Prediction									
1. Preliminary	x	x	x	x	x	x	x	x	x
2. Interim	x	x	x	x	x	x		x	x
3. Final		x	x	x	x		x	x	
E. Reliability analysis									
1. Preliminary analysis	x	x	x	x	x	x	x	x	*b*
2. Trade-off	x	x	x	x		x	x	x	
3. Part selection and application		x	x	x		x	x	x	
4. Design consideration	x	x	x	x		x	x	x	
5. Interim analysis	x	x				x			
6. Final analysis		x	x	x		x			
F. Failure reports									
1. Maintain records			x	x	x			x	x
2. Reporting program	x		x	x	x	x		x	x
3. Corrective action	x		x	x		x		x	x
4. Failure analysis	x		x	x	x	x		x	x
G. Redundancy		x	x	x	x	x	x	x	
1. Procuring activity review required		x	x	x				x	

TABLE 14-2. TABULAR COMPARISON OF RELIABILITY SPECIFICATIONS (*Continued*)

Subject	Specification								
	MIL-R-25717C (USAF)	MIL-R-26474 (USAF)	MIL-R-26484 (USAF)	MIL-R-27173 (USAF)	MIL-R-26667 (USAF)	MIL-R-26674 (USAF)	MIL-STD-441 (DOD)	MIL-R-22256 (AER)	AFBM 58-10 (USAF)
2. Procuring activity applicable required			x					x	
H. Maintenance						x	x	x	x
1. Periods			x	x	x				
2. Replacement lists			x	x	x				
3. Analysis	x				x				
I. Operating time									
1. Elapsed time indicators		x	x	x	x			x	
2. Elapsed time record	x					x			x
J. Reports						x			
1. Initial	x	x	x	x	x	[c]	x	x	x
2. Special	x	x			x				
3. Interim			x	x	x			x	x
4. Quarterly	x	x							
5. Final		x	x	x			x	x	
6. Other	x		x	x	x				x
IV. Test program									
A. Types									
1. General (nonspecific)							x		
2. Qualification						x	x	x	x
3. Preproduction	x	x			x	x		x	x
4. Production	x	x			x	x			x
5. Debugging			x	x	x				
6. Demonstration	x	x	x	x	x	x			x
B. Conditions									
1. Accept-reject criteria		x			x				
2. Test limitation provision		x				x			x
3. Ambient		x			x				
4. Environmental			x		x			[d]	
5. Specified in contract	x		x	x	x	x	x	x	
6. Quantity to be tested			x	x	x				
C. Reports									
1. Results	x		x	x	x			x	

[a] References MIL-STD-441.
[b] Accord with MIL-D-9310 and MIL-W-9411.
[c] Accord with MIL-D-9310.
[d] References MIL-T-5422.

Table 14-2 gives a tabular comparison of those specifications in Table 14-1 which are most frequently used.[1]

As can be gleaned from Table 14-2, nine basic reliability specifications are compared in four sections entitled equipment, scope, requirements, and test program. The nine specifications are as follows:

[1] Tables 14-1 and 14-2 are based on the work of Christian M. Kahl and T. L. Commons, "Contract Technical Requirements," The Martin Co., Baltimore, Md., March, 1960.

1. MIL-R-25717C (USAF) Reliability Assurance Program for Electronic Equipment, March 9, 1959.
2. MIL-R-26474 (USAF) Reliability Requirements for Production Ground Electronic Equipment, June 10, 1959.
3. MIL-R-26484A (USAF) Reliability Requirements for Development of Electronic Subsystems or Equipment, April 18, 1960.
4. MIL-R-27173 (USAF) Reliability Requirements for Electronic Ground Checkout Equipment, July 6, 1959.
5. MIL-R-26667A (USAF) Reliability and Longevity Requirements Electronic Equipment, General Specification for, June 2, 1959.
6. MIL-R-26674 (USAF) Reliability Requirements for Weapon Systems, June 18, 1959.
7. MIL-STD-441 Reliability of Military Electronic Equipment, June 20, 1958.
8. MIL-R-22256 (AER) Reliability Requirements for Design of Electronic Equipment or Systems, November 20, 1959.
9. AFBM Exhibit 58-10 Reliability Program for Ballistic Missile and Space Systems, June 1, 1959.

Table 14-2 provides the following information:

Section I indicates the type of equipment and phase of the program which each specification covers.

Section II indicates the main general scope, purpose, and intent of each specification. The specification checked under the appropriate heading indicates the following:

(*a*) Provides the requirements of an organization, operational procedure and guides for a reliability program. (*b*) Provides the requirements for conducting a reliability test demonstration program. (*c*) Provides the procedure and requirements for conducting design and reliability analysis during a design and development program.

Section III outlines the various detail requirements provided for in the specifications listed.

(*A*) Indicates the type of program effort required by each specification. (*B*) Indicates the specifications requiring that a mean time between failures (mtbf) be specified and when or how it shall be computed. (*C*) Indicates the specifications which require a minimum equipment operating life. (*D*) Indicates the specifications that require prediction studies and when they shall be performed. (*E*) Indicates the specifications which require reliability analyses and when the analyses shall be conducted. Some of the specific analyses called out by some specifications are included. In some instances, these analyses are nothing more than prediction studies. (*F*) Indicates the specifications which require a failure reporting effort and the extent of this effort. (*G*) Indicates the specifications which mention the use of redundancy and also those specifications which call out procuring activity review or approval of the redundant techniques proposed. (*H*) Indicates the specifications which mention a maintenance requirement and the detail to which this requirement is specified. (*I*) Indicates the specifications requiring the recording of the elapsed operating time of the equipment. Some specifications call for elapsed time meters. (*J*) Indicates the reports that are to be submitted to the procuring activity.

Section IV indicates the test effort required in each of the specifications. This includes the type of tests, conditions of the tests, and reporting of the test results.

It is obvious from Tables 14-1 and 14-2 that many of the provisions of the listed specifications overlap. For example, MIL-STD-441, MIL-R-26484, and MIL-R-22256 are similar and are oriented toward the design and development phase. Other specifications, such as MIL-R-26474 and MIL-R-25717, are oriented with respect to the production phase. The remaining specifications are general in nature and not oriented to any specific phase of a program.

An examination of individual specifications will reveal that an appreciable amount of repetition occurs in these documents. In many instances, entire paragraphs are repeated in different specifications. In other cases, we find the content of some specifications, such as MIL-R-26484 and MIL-R-27173, to be almost identical.

In addition to actual duplications in requirements, most of the specifications are similar in several respects despite the fact that the wording of the narrative might differ. For example, with regard to mean time between failures, as can be observed from Table 14-2, seven of the nine specifications listed state that the mtbf is to be specified in the detail specification. Three of the specifications specify values ranging from 300 to 4,000 hr. Also, three of the specifications include a formula for computing the value of mtbf based on the number of parts of a particular type which are used.

With respect to reliability prediction, Table 14-2 shows that all the specifications listed specify preliminary predictions, eight specifications require interim reliability predictions, and six require final reliability predictions.

From the above illustrations, it is obvious that a high degree of redundancy exists in the listed specifications. It should, therefore, be possible to reduce their number and probably devise one general specification applicable to research and development and another applicable to production equipment. This will probably be achieved at some later date when the various government agencies realize that reliability programs can be administered more economically and effectively by coordinating the various requirements in a consolidated general reliability specification. In this manner a procurement contract can refer to applicable portions of the general specification as well as list other desired requirements.

An interesting observation is that all of the specifications are deficient with respect to a measure of maintainability. Although it is mentioned in some instances, it is not quantitatively specified, nor are provisions made for its measurement. Actually, this is a serious shortcoming of current reliability specifications, because the maintainability aspect, as we have seen in Chaps. 9 to 12, is a vital component of availability. Perhaps maintainability has been neglected because, until very recently, its impact on availability was not too clearly established.

The measure of the maintenance-action rate in terms of mean time to repair (mttr) is most important. However, just as in the case of reliability measurement, applicable specifications should clearly delineate the maintenance discipline which is applicable to a specific mttr. To state, for example, that the mttr is one half hour is meaningless unless we further indicate the conditions under which the repair is being made. The specification of the maintenance discipline is a necessity because the repair rate is a function of such parameters as the skill of the repairmen, available tools and test equipment, and the surrounding physical environment. A mature design attempts to minimize the effects of human variability on maintainability by the use of automatic fault-location and isolation devices and by the incorporation of replaceable modules as an equipment characteristic. In this manner, in the event that a failure occurs, it can be eliminated very rapidly by means of a simple primary maintenance action. The primary maintenance action in this case would be the withdrawal of the defective module and replacement with a good unit. In those instances where automatic fault location and isolation is employed by means of a red light or similar device indicating that the trouble is in a specific module, the primary maintenance action is relatively simple, because all that is required is the replacement of a bad unit by a good one. In this case, the maintenance-action rate is relatively constant because the effect of the skills of maintenance personnel is not a primary factor. However, in those instances where it is necessary to trouble-shoot to isolate the failure, the skill of the maintenance man is most important. In these cases, it becomes relatively difficult to specify a maintenance-action rate, particularly in those instances where the variability in personal skills is great. It is for this reason that primary maintenance should be automated to the highest possible degree, particularly when the end item will be put to critical use and must either be continuously operable or have a high degree of availability. A good reliability specification will specify the conditions applicable to the maintenance-action rate referred to therein.

In the case of secondary maintenance, the variability of the secondary maintenance-action rate is not as critical as that of the primary maintenance-action rate. The reason is that if an adequate supply of spares is on hand, the time to repair on the bench could vary appreciably without affecting the availability of the equipment, because there would always be spare modules on hand to replace defective units. However, in the interests of economy, it is also recommended that the secondary maintenance-action rate be kept within controlled limits in order to reduce the required spares which would be necessary to sustain a specified availability.

Section 11-6 gives a good discussion of the importance of plotting the maintenance-action rate to assure that it stays within statistical control limits. The material in this section has been adapted by the author in a production reliability specification which plots the number of failures and the maintenance-action rate on a subsample f chart. This particular specification, which has been used at International Electric Corporation, Paramus, New Jersey, serves a dual purpose. It is used as an acceptance medium as well as a control chart to highlight problem areas in both the reliability and maintainability area. By its use, corrective action can be assured on a timely basis. Abstracts of this specification are discussed in App. 3 of this book.

14-2. Typical Provisions of Current Reliability Specifications. MIL-R-26474 entitled "Reliability Requirements for Production Ground Electronic Equipment" includes a formula for calculating the minimum mean time between failures. For example, in Paragraph 3.2 of this specification, it states:

> The mean operating time between independent failures shall be not less than that specified in the detailed equipment specification. When not specified therein the mean operating time between independent failures shall be not less than that given by the following formula:
>
> $$\text{mtbf (in hours)} = 1/F_r$$
>
> where $F_r = 30 \times 10^{-6} \times N_t + 15 \times 10^{-6} \times N_m + 2 \times 10^{-6} \times N_s + 0.5 \times 10^{-6} \times N_c$
>
> N_t = Total number of tubes (envelopes) included in the equipment parts complement
>
> N_m = Total number of motors and relays included in the equipment parts complement
>
> N_s = Total number of semi-conductors (transistors, diodes, et cetera) included in the equipment parts complement
>
> N_c = Total number of remaining electrical and electrical mechanical parts included in the equipment parts complement

In this case, the attempt has been made to use a mathematical model applicable to ground electronic equipment for calculating the minimum permissible mean time between failures. An analysis of the model will show that it is nothing more than the application of arbitrary failure rates to tubes, motors, relays, semiconductors, and other residual parts. Therefore, the values of mtbf computed from the mathematical model represent a conservative reliability prediction which it is felt can be met by electronic equipment of good design and workmanship. It should be emphasized, however, that when the reliability is specified in the detailed equipment specification, the mathematical model does not govern.

Another highlight of MIL-R-26474, which is comparable to that of other specifications, is referred to in its Paragraph 3.3, entitled "Reliability Program." It states that "the contractor shall establish a reliability program, the scope of which shall be consistent with the requirements of MIL-STD-441, the requirements of this specification, and the results of the reliability analysis. Where the requirements of MIL-STD-441 and this specification conflict the requirements of this specification shall govern. The reliability program shall include, but not necessarily be limited to, the following requirements or applicable portions thereof."

It should be noted that the reliability program called for references MIL-STD-441. This latter specification covers the design and development of electronic equipment to assure a certain required inherent reliability. It is intended to be applied to the development and design of all electronic equipment whether used in aircraft, shipboard, ground, or other categories of use and expendability.

Under a general heading called "Development Procedure" MIL-STD-441 covers two phases: Phase 1 is entitled "Study and Planning" and phase 2 is called "Design and Construction of Prototype." In a sense, MIL-R-26474 should not refer to all the applicable requirements of MIL-STD-441, since the former specification is intended to be applicable to production equipment, whereas the latter is applicable to design and development. Therefore, MIL-R-26474 should not refer to MIL-STD-441 except for those instances when major design changes are made in the production equipment. This is due to the fact that for production equipment of mature design, all that is necessary is to assess the reliability of the production devices by means of subsample f charts or effective time sampling techniques, since the reliability of design would have been previously demonstrated by the use of MIL-STD-441 techniques. Therefore, to avoid confusion, MIL-R-26474 should state clearly that if the reliability of a production device is unknown, then an analysis should be made in accordance with the provisions of MIL-STD-441 before any hardware is produced. Paragraph 3.4 of MIL-R-26474 implies that this is the case, because it states that "the contractor shall analyze existing design to determine compatibility with the requirements of this specification."

From the foregoing discussion it should be obvious that when the reliability and maintainability characteristics of production equipment have been previously verified, subsequent effort should confine itself to the continual assessment of production reliability by means of statistical time sampling methods. MIL-R-26474 specifies a method of achieving this objective for preproduction units by specifying a sequential time sampling plan. This is shown in Table 14-3 which is taken from MIL-R-26474, Table 1.

TABLE 14-3. FAILURE RATE TESTING (ACCEPT-REJECT CRITERIA)

Multiples of MTBF[a]	Accumulated number of failures[b]		
	Rejection[c]	Continuation of testing[d]	Acceptance[e]
3.00	8	2–7	1
3.32	8	2–7	1
3.58	9	3–8	2
4.01	9	3–8	2
4.27	10	4–9	3
4.70	10	4–9	3
4.96	11	5–10	4
5.39	11	5–10	4
5.65	12	6–11	5
6.08	12	6–11	5
6.34	13	7–12	6
6.77	13	7–12	6
7.03	14	8–13	7
7.46	14	8–13	7
7.72	15	9–14	8
8.15	15	9–14	8
8.41	15	10–14	9
9.10	15	11–14	10
9.79	15	12–14	11
10.10	15	.....	14

[a] Column 1 indicates the ratio of total operating time (accumulated by all equipments under test) divided by MTBF. Required MTBF shall be as specified in 3.2.

[b] Failures listed are total independent failures experienced by all equipments under test.

[c] Rejection shall occur if the number of failures indicated occur on or before the time indicated in Column 1.

[d] Testing will be continued if the number of failures falls within the range indicated at the time indicated in Column 1. (See 4.3.3 for exception.)

[e] Acceptance shall occur if no more than the number of failures indicated occur by the time indicated in Column 1.

An examination of the table indicates that the testing time is specified in multiples of the mean time between failures. The acceptance and rejection numbers are also specified and the continued test region is shown. This is a very ingenious method of preparing a sequential sampling table because it makes its use universal in application. It is based on the fact that the reciprocal of the mean time between failures is the failure rate; therefore, when the test time equals the mtbf, the expectation, or in other words, the average number of failures, must equal unity. Thus

$$rT = \frac{1}{m} \times m = 1$$

where m is another symbol used for mtbf.

Therefore, for multiples of the mtbf, the expectation is equal to the value of the multiple. When the multiple of the mean time between failures is shown as 3.32, this means that 3.32 failures is the average number of failures expected in 3.32 hr of testing.

The difficulty with this type of sequential table is that when a specified mean time between failures is large, a long test time is required before an accept-reject decision is made. As a matter of fact, if the results of the testing fall in the continued-test column, the testing could go on for a relatively long time unless some provision were made to terminate the test. A method of test truncation is specified in Paragraph 4.3.3, entitled "Termination of Test," which is quoted as follows:

> In those cases where the allowed test time is limited by the equipment specification or where the combination of high required mtbf and limited sample size will result in excessive test time, the accept-reject criteria shall conform to the following.
>
> 4.3.3.1 "Condition No. 1"—When any one of the equipments under test has accumulated 1500 hours of operation, the sum of operating time for all equipments is no less than three times the required mtbf and the number of failures falls into the "continue test" category of Table 1, the equipment will be accepted and the test terminated provided the final reliability estimate is clearly greater than the required mtbf.

With respect to reliability testing of production reliability articles, MIL-R-26474 attempts to specify these requirements in Paragraph 4.4, which reads as follows:

> "Production Reliability Test"—The production reliability test shall be performed by the contractor on specified numbers of production equipment, chosen at random from those which have passed all the acceptance tests specified in the detailed equipment specification. The test shall demonstrate the continued compliance of parts, quality and workmanship within the required reliability levels, and performed as specified in the detailed equipment specification. When not specified therein, the contractor shall, upon approval of pre-production tests, propose sample sizes and selection methods which will most effectively utilize his routine quality control and inspection procedures in providing a high confidence, low cost type of production reliability test procedure. These procedures shall be subject to the approval of the procuring activity.

It should be noted that this paragraph grants wide latitude for the contractor to propose his own sampling test methods to assure that the production reliability specification has been achieved. In a sense, although it is desirable to grant a contractor the prerogative of specifying his own test methods, subject to the approval of the procuring agency, controls should be exercised by the latter. This is done in MIL-R-26474 by the requirement that the program which is devised should achieve the specified goals of the procuring activity at a minimum cost and test time. Such a program has been devised by the author and has been used at the International Electric Corporation. Its basic advantage is that it minimizes the consumer's reliability risk (i.e., the procuring activity) at the expense of the producer. Although this might appear to be unfair to the producer, the technique is used only when the producer feels that his product is capable of meeting the required reliability and therefore the calculated risk of rejection is justified, or when the government requests minimum test time in the interests of expediting shipment. The procedure used at IEC is included in its production reliability specification and labeled Appendix A, abstracts of which are quoted in App. 3 of this book.

14-3. Reliability Requirements. One of the tendencies in current specifications is the overspecifying of reliability requirements. It appears that some design engineers, in their desire to achieve a high degree of reliability, oftentimes specify extremely high figures for mean time between failures and then stipulate that these be verified in accordance with the provisions of a production reliability specification such as MIL-R-26474. As an example, in one instance, the specified reliability for an item was an mtbf of 10^7 hr. This is equivalent to 10,000,000 hr and, therefore, in order to verify reliability in accordance with Table 1 of MIL-R-26474, it would require a minimum test time of 30,000,000 hr of test on a single unit, which is equal to about 3,100 years in time. The only way that the test time per unit could be reduced is by testing a larger number of units, each for proportionately lesser time. However, when the quantity on order is small, the test time per unit, although less than what would be required on a single unit, would still be extremely high, and, therefore, it would be impractical and unduly expensive to do this type of testing.

The reliability engineer could have achieved the same high order of reliability either by specifying known parts or components of known reliability or by resorting to redundancy or other techniques; or he could have specified the reliability of the next higher assembly or assemblies, of which the component was but a part, and in this way reduce the test time by testing at that level.

Oftentimes it is apparent that the high order of reliability which has been specified is not at all necessary. In an analogous sense this is comparable to the mechanical designer who places extra-tight tolerances on drawings when looser ones would have done just as well. The end result in either case is increased costs of production and test.

Of course, when items are relatively low in cost and produced in large number, it is relatively simple to test to high orders of reliability, but when the contrary is true, it is impractical to do so. Therefore, in those instances judgment, experience, and grouping of subassemblies or assemblies to higher orders of complexity should be resorted to and the tests conducted at these levels.

In any case, the conditions of test should clearly specify such requirements as the environmental conditions, the order and level of reliability testing, and the time sampling plan which will apply. The sampling plan should also specify the allowable number of failures and the definition of successful performance or failure.

14-4. Availability Requirements. For a required usage, it would probably be more advantageous to specify availability rather than reliability. As we have seen, availability is made up of both reliability and maintainability. Hence, reliability is a part of availability. The two can, therefore, be considered synonymous only when no maintainability is possible or desired, i.e., when the maintenance-action rate μ equals zero.

In the early days, when the science of reliability was relatively new, it appeared logical to talk in terms of reliability rather than availability because the then current techniques were being applied to devices which were intended for short tactical missions such as aircraft radio or vehicular communications equipment. In those instances, since maintenance of the device while in use was relatively impractical, reliability was the natural ingredient which was specified, and the basis for its prediction was the exponential failure law. To a degree, this is still the case for these devices and for such others as missiles which have a short mission time. Since it is impractical to repair these devices while in use or flight, their availability is naturally a function of the inherent reliability which they possess. Unfortunately, as a result of precedent, the practice of specifying reliability instead of availability became so commonplace that it is still carried on today, without regard to the type of device to which it is applicable.

In ground-based continuously operable and maintainable equipment, as we have seen from Fig. 9-5, it is possible to achieve a specified availability at minimum cost by selecting the optimum combination of reliability and maintainability. It therefore appears logical that availability should be prescribed for these devices instead of reliability. The reason is that availability can be universally applied to continuously operating maintainable equipment as well as to equipment of finite mission time. This is because the maintenance-action rate for the latter can be considered to be zero, and, therefore, the only availability component remaining is the reliability of the device.

An effective specification should not only specify availability but it should also indicate the type desired, i.e., either mission or equipment availability. It should also specify the operating conditions as well as the specific methods, procedures, tools, and equipment which should be employed in the maintenance activity. The method of recording and measuring the time of maintenance action should also be specified as well as the type of maintenance activity, since the former affects the availability assessment. This requires that the preventive or scheduled maintenance cycle be clearly defined to distinguish it from corrective maintenance.

The specification should also clearly provide for a classification of failures by type and causative agent or any other classification deemed necessary. The conditions constituting failure should also be clearly defined. This should include a description of the fall-back devices which are permissible for a given time constraint, the combination of components or minimum number of components which must be operable, or any other criteria which differentiate between successful and unsatisfactory performance.

The conditions of test are also important, and, therefore, it should be made clear whether the test is to be conducted under ambient or environmental conditions, and the exact parameters to be measured as well as the required performance limits should be defined. The time sample size and the method of recording the time and number of failures should also be specified.

It is conceded that during the preliminary design phase, before the actual hardware is built, it is difficult to specify all these requirements. However, it should be possible to determine the required characteristics if the mission requirements are known. These could then be included in the general availability specification. The details for implementing the general requirements can then be embodied in the specific equipment specification as the necessary information becomes available.

A good reliability test specification will also highlight the methods for demonstrating or verifying that the required availability has been achieved. The procedure should be based on considerations of economy and efficiency without the sacrifice of confidence in the results obtained. This can be achieved by specifying the subassembly or complete assembly level which should be tested to verify reliability. When possible, testing the completed device as a complete system will result in the greatest economy. The reason is that the mean time between failures is usually less for the complete system than for its individual components. If this should be impractical because of the large size or complexity of the device, then the next lower natural level should be tested. The availability of the complete system can then be verified by synthesizing all the effects by means of the mathematical models which have been previously discussed.

An advantage of testing completed devices as a whole is that this method constitutes a measure of the performance of each part and component in its own environment under the conditions of actual use. Thus, we see that not only is this a more economical method, but the data are gathered under more realistic conditions than are possible when testing is under assumed or operating conditions and environments. When this is done, any substandard components can be readily evaluated and corrective action instituted. This type of test makes possible the constant evaluation of product improvement due to a combination of aging effects due to testing and the corrective design actions which are the result of studies based on failure data. A good reliability specification will require that continual assessment of availability be made and that a measure of its growth be presented by periodic curves and reports. In any case, the methods, procedures, type, format, and frequency of required reports, and mathematical models and required data should always be specified.

In conclusion, the above outline only briefly discusses some of the highlights, and these are certainly not all-inclusive. As long as the engineer remembers that a good specification should have the basic elements outlined in Sec. 14-1 and that these are specified qualitatively and, when possible, quantitatively, he will find that he should produce an effective reliability specification.

CHAPTER 15

UNITED STATES GOVERNMENT REQUIREMENTS

15-1. General. In this chapter we shall attempt to outline those government requirements which are generally applicable in any government contract in which a definite level of quality and reliability has been specified for the products on order. In order to do this logically, it appears that a relatively brief outline of the evaluation of present-day practices would be timely and apropos.

For many years the government followed the principle of *Caveat emptor*, meaning "let the buyer beware." In other words, the government, as the buyer, put little faith in the vendor as a supplier. The theory was that the only way to be assured of receiving articles of high quality was to have them appraised by government inspectors. This attitude was responsible for the development of the inspection departments of the various services. These departments were vested with authority guaranteed by statutory contractual clauses which ensured that the authority to inspect became an inherent right guaranteed by the contract. Typical of such clauses are:

INSPECTION (*a*) All materials and workmanship shall be subject to inspection and test at all times and places and, when practicable, during manufacture. In case any articles are found to be defective in material or workmanship or otherwise not in conformity with the specification requirements, the government shall have the right to reject such articles or require their correction. Rejected articles and/or articles requiring correction shall be removed by and at the expense of the contractor promptly after notice so to do. If the contractor fails to promptly remove such articles and to proceed promptly with the replacement and/or correction thereof, the government may by contract or otherwise replace and/or correct such articles and charge to the contractor the excess cost occasioned the government thereby, or the government may terminate the right of the contractor to proceed as provided in paragraph __ of this contract, the contractor and surety being liable for any damage to the same extent as provided in said paragraph __ for terminations thereunder.

(*b*) If inspection and test, whether preliminary or final, are made on the premises of the contractor or subcontractor, *the contractor shall furnish, without addi-*

tional charge, all reasonable facilities and assistance for the safe and convenient inspections and tests required by the inspectors in the performance of their duty. All inspections and tests by the government shall be performed in such a manner as not to unduly delay the work. Special and performance tests shall be as described in the specifications. The government reserves the right to charge to the contractor any additional cost of inspection and test when articles are not ready at the time inspection is requested by the contractor.

(*c*) Final inspection and acceptance of materials and finished articles will be made after delivery, unless otherwise stated. If final inspection is made at a point other than the premises of the contractor or subcontractor, it shall be at the expense of the government except for the value of samples used in case of rejection. Final inspection shall be conclusive except as regards latent defects, fraud, or such gross mistakes as amount to fraud. Final inspection and acceptance or rejection of the materials or supplies shall be made as promptly as practicable, but failure to inspect and accept or reject materials or supplies shall not impose liability on the government for such materials or supplies as are not in accordance with the specifications. In the event public necessity requires the use of materials or supplies not conforming to the specifications, payment therefor shall be made at a proper reduction in price.

An examination of the inspection clause quoted above indicates that the government regarded the inspection of its purchases as a dire necessity. This is still true today but the techniques of inspection have been changed from minute detailed inspection procedures on the part of government inspectors to a surveillance of contractor performance.

15-2. The Necessity for Inspection. The government inspects its purchased items because it must have assurance that material of specified type and quality is supplied to the armed forces. Therefore, as we have seen, it protects its right to *be assured* by inserting the appropriate inspection clauses in all purchase contracts; but, even were this not the case, it is the basic inalienable right of any customer to examine the article he is about to buy to make sure that he gets what he wants with respect to description, quality, reliability, and quantity. If he has the misfortune to buy the wrong type of material or something that is inferior or useless, or if he is given a short count, he frequently has little or no recourse against the supplier. This is true despite efforts in recent years to give legal protection to the buyer. This has resulted in limited protection which is restricted to requirements for correct labeling of articles of food and medicine and to the maintenance of certain standards of sanitation in their manufacture. The Food and Drug Act is practically the only extensive legislation for the protection of the customer.

The needs of the armed forces for supplies and equipment, both for subsistence and for tactical operations, have always been borne almost completely by private industry. The manufacturing facilities of the armed forces are quite limited and can be depended upon to supply only a very small fraction of the requirements. This is especially true in time of war, when almost all the manufacturing potential of the nation is devoted to the manufacture of war matériel.

Ordinarily, even in peacetime operation, manufacturers perform inspection of their products in one form or another in order to produce items of a certain level of quality and performance. In order to avoid return of matériel for replacement, they attempt to provide only articles of proper quality. Inspection by the manufacturer is then primarily an economic consideration, because if he ships satisfactory products, he will save the cost of replacement and repair of defective items, and the resulting by-product is a good reputation and "good will."

Since inspection is an economic factor, its costs as related to the over-all costs of manufacture should be normally considered in the price. Thus, the more expensive or precise the item, the greater should be the extent of its inspection. However, the extent of government inspection cannot be based solely on economic considerations. For example, a small, low-cost item furnished to the troops may be more vital to the successful completion of their mission than a large, highly expensive piece of equipment. The government must therefore use whatever techniques it considers essential in order to assure itself of the quality of items it purchases.

The following items are typical of the elements which are of concern to a government inspector:

1. *The requirements and procedures governing source inspection.*

2. *Proper packaging, packing, and preservation.* It is necessary that the contractor package and pack the product so that it will be properly protected against damage in transit, rough handling, or deterioration due to prolonged storage or stacking in depots. In some cases, items must be individually packaged so that the external container may be opened and part of the contents removed without exposing the remainder to spoilage. Moreover, there is little use in testing, packing, and shipping an item if it arrives at its destination in an inoperable condition or if the item becomes damaged in storage. It is because of these important considerations that the services insist on inspecting for packaging, packing, and preservation.

3. *Proper marking.* It is of vital concern that the manufacturer mark the shipping container so that the contents are correctly identified and addressed to assure their reaching the intended destination with a minimum of error or delay. It is also very vital to comply with shipping instructions and proper execution of shipping documents. Wartime experience demonstrated that one of the serious weaknesses of the supply system was the failure of manufacturers to identify adequately packages and shipping containers to indicate their contents without the necessity of opening them. This led to much waste, because it became necessary to open packages in order to identify the contents and the opened pack-

ages containing items not desired at the time often deteriorated in storage, since the act of unpackaging also destroyed the preservatives which had been originally used.

4. *Assurance of proper count.* Before shipments can be made, proper shipping documents to accompany the shipments must be approved and signed by the inspector. The shipping document used for this purpose is known as the Material Inspection and Receiving Report (MIRR). After assuring himself that the alleged quantities noted on the shipping document are actually being shipped, the inspector signs the certificate on the MIRR to that effect, creating a record which the contractor uses in obtaining payment. The inspector is therefore in a position to ensure proper count and agreement of shipped items with contractual requirements.

5. *Assurance of adequate reliability, quality, and performance.* The quality and reliability of an item are assured by applicable specifications and drawings as specified in the procurement contract. If it passes the quality check, the item must then be subjected to reliability testing to make certain it will function as intended under specified environmental conditions for the period of time required, prior to release for shipment.

6. *Records of government property.* The cognizant inspection agency maintains records of quantities of government property furnished the contractor, either for test or sample purposes, or for incorporation into the end item. The inspector or quality assurance representative, as he is also called, maintains adequate surveillance over such material at the contractor's plant and reports instances of negligence by the contractor. He also checks and verifies all shortages, damages, or discrepancies claimed by the contractor on incoming shipments of the government-furnished property. In general, the inspector performs a valuable function in the eventual clearance of property accounts.

7. *Protection of the government financial interests.* The inspector is the only government representative who is constantly stationed at a contractor's plant and, because of this, is in an excellent position to protect the government's financial interests. In this capacity he not only inspects material for compliance with quality standards, but also executes a number of other duties as a representative of the contracting officer.

For example, in the case of a contract termination, the inspector is responsible for taking an inventory of stock, completed assemblies and subassemblies, as well as of the work in process, and for certifying the accuracy of the count. He also makes production estimates for the contracting officer as required. Another of his functions is to sign all shipping documents for accepted material, thereby authorizing the finance officer to pay the contractor. The contractor's future business relations with the government depend upon the inspector to a major degree, since the periodic reports furnished the government on contractor performance determine whether the contractor remains on the approved list of bidders for future government business.

15-3. The Effect of Absence of Government Inspection. In order to evaluate government inspection fully, it is well to consider the possible effects of its absence or elimination. Some manufacturers, not necessarily of the greatest size, have the integrity and ability to produce a product adequate to meet all of the specification requirements uniformly and fully, to pack the product in an acceptable manner, to mark containers as required, to maintain proper records of incoming and outgoing government furnished equipment, and to be most cooperative in furnishing reports and forecasts of production to the government. Years of experience have indicated, however, that dependence upon all industrial concerns in general to conduct themselves in this manner is not practical, because economic considerations are of prime importance to some manufacturers. In watching their costs, they tend to achieve the bare minimum of the requirements, and, because of this, it is most essential that surveillance be maintained by government personnel to ensure that all terms of the contract are complied with. In time of war there is even more reason for such surveillance. At such times, even companies which have had excellent reputations and years of experience in the manufacture of certain items may prove unreliable. One of the reasons is the considerable influx of new personnel due to increased production demands and losses to the draft. These new employees, in most cases, have little or no manufacturing experience so that, in spite of the very best of intentions, an inferior or a defective product may result. Moreover, the usual commercial practices which may be sound for the ordinary civilian supply of goods cannot be fully applied to defense procurement. In civilian procurement, time is not necessarily a vital factor, and means are available to the customer for correcting deficiencies in items he procures. In fact, in some cases the manufacturer saves the cost of inspection completely and offers a certificate or guarantee of repair or replacement to the purchaser, thus making the purchaser the inspector. However, time is of the essence for the services, and it is impossible, or far less practical, to repair or replace equipment arriving at destination in a defective condition. Not only may such failures result in the loss of military objectives, but repeated occurrences may cause the troops to lose faith in the equipment and in the defense supply program, with a consequent serious lowering of their morale. In the operations of the armed forces, dependence upon equipment must be based not merely on faith, but on definite assurances of quality and reliability.

15-4. Statistical Quality Control. Statistical quality control was developed by industry as a mathematical method of controlling quality throughout production. This system has been adapted by the technical services to determine whether a given lot of material is acceptable. Through the use of a scientific plan of inspection procedures and the efficient use of recorded inspection results, much has been accomplished toward regulating inspection and limiting it to the degree and amount necessary for adequate assurance that accepted material will be of satisfactory quality. Under statistical quality control, detailed procedures are specified covering the number of samples to be taken as representative of a definite unit of production and the manner in which these samples will be inspected and the data recorded. Through the analysis of the recorded data, the amount of inspection may be reduced if the quality of material produced by the contractor is consistently maintained at an acceptable quality level. On the other hand, if the quality level declines, the amount of inspection is increased and the acceptance criteria are made more strict in order to induce the contractor to improve his production processes.

15-5. The Evolution of the Quality Assurance Technique. The development of statistical quality control methods resulted in the realization that the 100 per cent inspection that had been practiced by the services for many years did not necessarily guarantee perfect material. This led to the development and implementation of quality assurance methods. This meant that the specific job of inspection and controlling production was left to the vendor, while the services adopted techniques of quality assurance. As a result of this new attitude, government departments changed the titles of their quality departments from "Inspection" to "Quality Assurance Representatives." As the name implies, quality assurance is a method of satisfying oneself that the quality of the product is in accordance with prescribed standards. The question was how was this to be done? Each of the services had its own ideas on this subject and therefore its own quality control specifications. For example, the Air Force had been governed by MIL-Q-5923C, while the Signal Corps had been using RIQAP. The provisions of each will be explained in due course. It is interesting to note that although these specifications appear to differ, their basic objectives are similar. These are summarized as follows:

1. The job of process control belongs to the contractor and not the government.
2. The inspection function, by contract and by basic concept, belongs to the contractor.
3. The amount and degree of government inspection is reduced or completely eliminated in direct proportion to the quality performances of the contractor.
4. The role of the government is to act as an auditor, i.e., to assess

quality in accordance with prescribed sampling or surveillance techniques and reject or accept products on that basis.

As a result of these common objectives, in April, 1959, the Department of Defense developed a new specification, MIL-Q-9858, entitled "Quality Control System Requirements." This specification has been approved by the Department of Defense and is mandatory for use by the Department of the Army, the Navy, and the Air Force. The most important requirements of this specification are abstracted in App. 4 by paragraph number as used therein in order to give the reader firsthand knowledge of its requirements.

15-6. The Evolution of Reliability Assurance. We have seen how, for many years, the government has been concerned with quality. Moreover, we have studied the evolution of quality methods from 100 per cent inspection to sampling to present-day quality assurance techniques.

It is only very recently that the government has become interested in reliability. This is because of the increased complexity of present-day military equipment, which requires a high order of reliability. However, in the past the government did not have a reliability design specification for assuring itself that the equipment which would be produced exhibited the required degree of reliability. The reason was that there were no known dependable reliability parameters then in existence which could be used as a basis for reliability assessment. Because of this, the contractor had no idea of how well his equipment would perform. Therefore he consoled himself by believing he could attain the desired reliability objective by prescribing the best in parts and components. He reasoned that as long as he had done this and used good quality control practices during manufacture, his responsibility with respect to reliability had been fulfilled. This was far from the truth, but the state of the art had not yet advanced to a stage which would permit disputing such an attitude. As a matter of fact, many government engineers condoned and supported the contractor's attitude, since they had great confidence in the MIL specifications which were applicable to most parts commonly used in military equipment. Moreover, for many years, government inspectors had been accustomed to strict adherence to military specifications. Thus, it was only natural to consider a manufacturer who complied with their requirements as one who produced exceptionally fine equipment. The concept of reliability, which in a sense is a measure of quality with respect to time, had not yet permeated the inspection atmosphere. The inspector reasoned that as long as the equipment passed the specified inspections and tests at the time they were performed, the equipment should be considered satisfactory. Not one thought of whether or not it would function after delivery ever entered the inspector's mind. How wrong this attitude was became evident in World War II, when it was found that a high percentage of equipments removed from their shipping cartons would not function initially. This situation caused much concern to the military and was the subject of much discussion, but there was not much that could be done, because the true cause of the trouble was not known. The simplest explanation always appeared to be related to quality. Hence, whenever explanations were required from higher authority, somebody was invariably accused of being a poor quality producer, and the trouble was usually placed at the doorstep of a parts manufacturer. The greatest offender was always the tube manufacturer, since vacuum tubes caused most of the trouble. It never entered anybody's mind that perhaps the application of the tube in the circuit might be wrong or that the tube might be working at the limits of its capacity or that some transient effect might be responsible for its short life.

As we have seen, the stress on quality was responsible for the development of such quality assurance procedures as the Air Force MIL-Q-5923C, the Signal Corps RIQAP, and finally the Department of Defense Specification MIL-Q-9858. These specifications were very effective in improving quality, but they did not do enough to improve reliability appreciably. It was therefore apparent that similar procedures or requirements were needed to do for reliability what the quality assurance specifications had been able to do for quality. The specifications quoted in Chap. 14 are a mixture of technical requirements and procedures. However, they fall short of specifying exactly what reliability monitoring points the government considers necessary to control reliability, nor do they specify the details of an effective reliability organization.

Two USAF Specification bulletins were developed to cover these two areas. These are USAF Specification Bulletin 506 dated May 11, 1959, entitled "Reliability Monitoring Program for Use in the Design, Development, and Production of Air Weapon Systems and Support Systems" and USAF Specification Bulletin 510 dated June 30, 1959, entitled "Guides for Reliability Organization."

Appendixes 5 and 6, respectively, are made up of abstracts and quotes taken directly from these USAF Specification bulletins. Number references are those specified in the bulletins.

CHAPTER 16

MANAGEMENT OF A RELIABILITY PROGRAM

16-1. Product Assurance. There is no known standard method of managing a reliability assurance program. There are probably as many ideas on this subject as there are individuals concerned with it. The subject of management in itself is still considered an art by many, although there are those who believe that sufficient progress has been made in recent years to warrant its being called a science. Management becomes a still more controversial matter when it is applied to the relatively new science of reliability. Hence, the management aspect of reliability will be discussed strictly from the author's own background and experience, with the hope that it will provide sufficient background and motivation for the reader to stand him in good stead in his daily work.

In general, it is agreed that reliability begins with design, but this is not the only means to the desired end. It is but one of the factors which comprise reliability. If design were the only consideration, our job would be simple, since all that should be necessary is to prescribe good design, which would presumably solve the problem. However, the solution to the problem is not so simple, firstly, because reliable design is not something which we can arbitrarily legislate and, secondly, because design, as was previously stated, is only one of the many factors which affect reliability. However, if we were to consider the design factor alone, we would still require a more positive approach than the simple expedient of legislating that engineers design reliable equipment. The reason is simple and basic. Nowadays engineers specialize in particular fields in the same manner as is prevalent in other professions, such as law and medicine. Therefore, just as some doctors specialize in the eye, heart, brain, or other organ, so will an engineer specialize in particular branches of his field. For example, an electrical engineer is wont to specialize in specific fields such as transmission, microwave, audio, or data processing. The reason is that complex systems require such extensive knowledge that it is practically impossible for an individual to be a specialist in everything. Moreover, even if one were not concerned with complex systems but considered himself an expert in a particular endeavor, it is most unlikely that this same individual would be cognizant of all of the aspects of the science of reliability, since it is a specialty all its own. Therefore, to assure reliable designs, it is necessary to have specialists in reliability, just as there are specialists in other fields. This situation has led to the evolution of the reliability engineer. It is the reliability engineer who constantly studies all the latest techniques of the science so that he can act as a consultant to the design engineer as well as assess the effectiveness of a design from the reliability point of view. We shall see how these engineers can work together as a team as we develop the management techniques. We have also seen from Chap. 15 that the quality engineer also plays an important role in reliability, and therefore we must have a means of welding the achievements of both the reliability and quality engineers together. We achieve this marriage of the two through the means of product assurance.

The product-assurance department, when properly organized, should include all those functions of reliability and quality which will assure a high degree of reliability in the end item. Therefore, product assurance should report to top management and be vested with the responsibility and authority to exercise surveillance in all those related operations which are responsible for assuring reliability, typical of which are design, purchasing, production, and field installation and maintenance.

There are various possible ways of organizing a product-assurance division. The author will therefore suggest and recommend what he considers to be the most effective system.

The product-assurance division should be composed of two major departments whose functions, duties, and responsibilities include all the necessary elements to assure a reliable product. These two departments are quality assurance and reliability assurance. Under one division director, they should encompass all the necessary duties and responsibilities which are necessary to the shipment of a reliable product. In general, the basic elements of concern should be the same for each of these departments whether they are a part of a systems management company or directly integrated into a manufacturing organization. However, the methods of accomplishing specific tasks might differ.

Figure 16-1 shows a typical factory product-assurance organization. Figures 16-2 and 16-3 show typical system management product-assurance organizations. A weapon system management company does comparatively little manufacturing of its own and for the most part subcontracts the major portion of the work to others. Its functions are therefore mostly of a management nature. In our discussions of management of a product-assurance organization, perhaps the best way to proceed is to discuss the major tasks which must be accomplished and later to discuss the organizations we are suggesting to implement them.

16-2. Major Tasks of a Product-assurance Division. In this section we shall discuss first the tasks of a reliability assurance organization and then the tasks of a quality assurance organization.

Tasks of a Reliability Assurance Department. Before proceeding with an outline of the tasks of a reliability assurance organization, we shall briefly discuss its relationships with other departments and some basic philosophy.

The purpose of reliability assurance is, as its name implies, to assure that the product it is interested in is reliable. This responsibility is all-inclusive and extends from the very inception of design to ultimate use in the field. This does not imply that reliability assurance should usurp the design responsibility of engineering, but it does mean that engineering must fulfill its reliability design obligations. Some argue that if this is so, then reliability assurance should be an integral part of the engineering department. A little thought should convince the reader that it would be a mistake to do this, because the responsibilities of reliability assurance extend far beyond the requirements of design alone. Moreover, even if this were not the case, it is never wise to make a department subservient to the group it is supposed to evaluate. In summary, a person cannot successfully evaluate himself, nor can a department. Another consideration is that the interests of reliability assurance continue long after the design engineer considers his task complete. This means that this department is vitally interested in reliability throughout the manufacturing and operating period and is constantly making measurements, gathering data, and evaluating reliability long after other departments considered their tasks complete. It also frequently prepares reports and recommends improvements. We shall see how this is done as we study the following tasks of this department.

1. Contractual Requirements. One of the first tasks of the reliability assurance group is to study the reliability contractual requirements and all subsidiary specifications to make certain that they are clearly understood. If there are any problem areas they should be resolved with the customer, through channels, at the earliest possible practical date.

2. Preparation of Reliability Specifications. On the basis of the study of customer reliability requirements, the reliability assurance group prepares a general company reliability specification which embodies all the necessary procedures, measurements, and reports which the company must follow in order to assure the fulfillment of the contractual reliability requirement. Provision is also made for the revision of this specification, on a timely basis, to reflect changes in customer requirements as described in modifications to the contract. Changes are also made to effect an improvement in reliability or to comply with any reevaluation following prototype testing.

3. Provide Consulting Services. The reliability assurance department should provide consulting services as requested by other company departments and subcontractors if these requests are necessary and justified. Such assistance may involve the actual assignment of specialists on a full-time basis for an extended period or for limited periods in accordance with the situation. The type of service may vary from the performance of statistical studies for the systems engineering group to the study of reliability allocations for parts and components. In the case of subcontracts, reliability engineers may be requested by the contracts group to interpret subcontractor reliability proposals and associated costs and to render an opinion on their justification. Or the systems engineering group might request the assignment of a statistician to assist them in the preparation of systems reliability requirements and specifications. This is not in conflict with any general specification prepared by reliability assurance, because the system engineering specification is detailed.

4. Preliminary Plant or Subcontractor Survey and Orientation. Since reliability requirements are generally not fully appreciated by other than technical personnel, it is very important that orientation be provided to all who will participate in the contract to a marked degree. Personnel of reliability assurance will therefore participate in meetings either with other company departments or with subcontractors to brief them on the contractual reliability requirements and to interpret and clarify any questions as necessary. In the case of subcontractors, such orientation will precede any formal survey of the subcontractor's plant.

5. Plant Evaluation and Surveillance. In order to be certain that each plant involved is cognizant of and is following all applicable requirements and specifications, it becomes necessary to survey the plant. This may be the prime contractor's plant or one of his plants, or that of a subcontractor; however, the basic elements of the survey will be the same. For example, one of the first elements of the survey is a study of the plant organization, responsibilities of key personnel, manpower, and the facilities necessary to achieve reliability. A study is also made of the plant product-assurance manual to assure that it covers all the necessary programs, plans, and procedures which are considered a must in order to achieve the required quality and reliability. Another very important area is to assure that the proper system reliability approach presents an accurate picture of all the parameters that may affect reliability. This means that the proposed designs should have been evaluated for reliability of operation and that the optimum combination of parts and components needed to achieve simplification of equipment, controls, and maintenance was considered. Of equal importance is the plant's reliability efforts with respect to the selection and application of circuits and

the selection and application of parts. The foregoing are some of the factors which will determine whether the design of the equipment reflects an understanding of the methods of achieving reliability and of implementing these methods in the design.

Another important consideration during the plant surveillance is to analyze the estimates of equipment reliability, including inherent, use, and operational reliability. This analysis should be based on a study of reports of the study and planning phase to assure that all factors are considered in ensuring that the equipment is suitable for operational use and that the provisions for maintenance are adequate and consistent with the estimated reliability required of the system. Again, with respect to design, it is imperative to evaluate the selection of assembly design techniques in terms of such factors as ease of maintenance and compatibility with environmental considerations such as temperature, humidity, shock, and vibration. Moreover, it is important that the design provide for equipment's having the lowest probability of human error. Techniques for assuring this include printed circuits, mechanical standardization, and incorporation of factors requiring a minimum of hand assembly.

Upon completion of the design and after the required prototypes have been built and tested, it is necessary to analyze test data, corrective action reports, equipment evaluation studies, and recommended modifications. This action should assure that production units following the prototypes are of the highest order of reliability.

6. Analysis, Correlation, Interpretation, and Feedback of Data. Another task of reliability assurance is to analyze, correlate, and interpret data received from plants and the field in order to assess the reliability of the equipment. In each case, it is most important that these data be fed back to the source which can make the best use of them to improve the reliability. Or the analysis of the findings based on available data may be summarized to determine spare parts and maintainability requirements. These data may also be used for periodic evaluation of efforts of engineers, subcontractors, and their suppliers, as required.

7. Design of Experiments. In those instances where the cause of unreliability cannot be readily determined, it may be necessary to design statistical experiments in order to determine the prevailing cause-effect relationships involved. This is a specialty that requires much mature statistical judgment and experience and one for which a good reliability group is eminently qualified.

8. Recommendations. Since the reliability assurance department is constantly reassessing the over-all system reliability on a periodic basis, it is in an excellent position to make recommendations for reliability improvement. In most instances, these recommendations are directed to the engineering department, but they are by no means restricted to

one area of interest, because they do not always involve design. There are many times when workmanship could be improved, when the use of a better manufacturing process would result in better reliability, or when better methods of packaging and packing for shipment would improve reliability because it would eliminate many potential hazards which are a function of shipment.

9. Reports. The reliability assurance department, because of the investigative nature of its work, must issue reports periodically and as required. This is an important part of its function. Some of these reports are presented as required by the department, while others are in accordance with customer requirements. Various types of periodic reports which are issued include weekly, monthly, and quarterly progress reports; status reports as required by the customer; special reports summarizing the findings of a plant surveillance; or reports of reliability achievement.

10. Plan of Operation. The reliability assurance department must have a plan with sufficient details to indicate the procedure for accomplishing its tasks. This plan should be included together with the quality assurance department plan as a part of the over-all product-assurance planning effort. In the next section we give suggested methods for accomplishing the various tasks, but these are subject to modification in accordance with the specific requirements of the product on order and the requirements of the contractor. Therefore, the tasks which we are outlining in this chapter should not be construed as totally encompassing all that is required. They are merely given as major elements in an illustrative sense and as a guide for the more detailed analysis that a particular contract might require. A similar outline to the one presented for reliability assurance will therefore be presented for quality assurance, and the two, when tied together, will constitute the total requirement for product assurance.

Suggested Methods of Accomplishing the Reliability Tasks. Regardless of whether the company is a weapon system contractor or an integrated manufacturing company with a limited number of subcontractors, the method of accomplishing tasks 1 and 2 is similar. The first step is to prepare a list of all reliability contractual requirements including all applicable government specifications. The provisions of these specifications are then incorporated in an over-all company specification which is all-inclusive and also lists all government specifications as subsidiary to it. The reason for this latter step is to assure everyone concerned that there is a definite basis for the company specification and also to guard against some inadvertent omissions. At this point, a word of caution is necessary regarding the acceptance of all provisions in government specifications. Some companies, in their desire to be awarded a contract, accept it without fully realizing the implications of the clauses in subsidiary specifications. Some of these are so formidable that they probably spell the difference between a profit or loss on a contract. For example, a contract which requires a higher degree of reliability than the contractor is capable of realizing with his present product involves much additional engineering work with the consequent expenditure of funds, which the contractor might not have anticipated. Consequently, it is the function of the reliability group to point these things out and to ask for clarification from the government through the proper company contractual channels. It is conceded that the proper time to do this is during negotiation, but some companies do not always consult with all their technical groups prior to signing a contract and leave the negotiations and acceptance of the contract to administrative or financial officers who have limited technical ability. When such a situation occurs, it is up to the reliability group to point out any technical difficulties regarding reliability requirements as early as possible.

The reliability assurance group should also prepare the applicable paragraphs of the statement of work for any subcontracts which the company intends to award to make certain that the reliability requirements of the prime contract are made mandatory on the subcontractor.

The method of accomplishing task 3 varies with the company structure. Fortunately, in general, the basic elements of organization for each company are relatively the same, and, because of this, it is possible to suggest a method of operation. Consultation services should be provided the engineering department through the assignment of reliability engineers to the project-design engineer. These engineers should be considered as being on detached service, working with the project-design engineer and advising him on the reliability practices which are necessary in order to fulfill the requirements of the contract. In this capacity, the reliability engineer will receive administrative direction from the project engineer but technical direction from the manager of reliability assurance. The outstanding advantage of this procedure is that the reliability engineer becomes intimately familiar with the design of the product, and, therefore, after his services are no longer necessary as a consultant to the design engineers, his familiarity with the product makes it easier for him to carry out his duties during the manufacturing and installation phase.

The reliability engineer also provides technical consultation to the subcontracting group or to subcontractors, when so requested. He also works closely with government technical representatives and consults with them or advises them on steps he considers necessary to achieve the contractual reliability.

Another medium that is used by some companies to provide a mutual understanding of reliability problems is the reliability control group

(RCG). A typical composition of such an RCG is shown in Fig. 16-1. As a member of the RCG, the product-reliability engineer provides consulting services or advice as requested or required.

Tasks 4 and 5 are accomplished through a series of steps which are interrelated, and therefore these two tasks will be considered jointly. When a potential subcontractor is under consideration for an award, the subcontracting department asks the product-assurance division to conduct a preliminary survey. The reliability department conducts the reliability phase of the survey, and the quality control department the quality phase. So that all subcontractors may be surveyed impartially, it is imperative that the same standards be used as a basis of judgment. Therefore, a survey questionnaire is prepared for this purpose. It includes pertinent questions about the subcontractor's reliability organization and the responsibilities of the various individuals involved; it covers the adequacy of facilities and inquires about the reliability plans and programs of the company. Other questions relate to techniques for implementing reliability in design, preparation of reliability manuals, failure reporting systems, and the system of data feedback. This latter item is concerned with the system used for feeding information back to design engineers on all malfunctions and failures starting from the initial test phase and continuing through the completion of the contract. The purpose of the latter questions is to assure that the feedback system is organized in such a manner as to provide for automatically feeding back essential data to all cognizant groups and to follow up to assure that all required changes are made. The subcontractor's design techniques to assure compatibility with environmental conditions and ease of maintenance are also made a matter of record on the survey questionnaire. He is also queried on his prediction techniques. This involves the prediction of inherent, use, and operational reliability. The methods the potential subcontractor uses to control his vendors to assure that they use approved reliability methods and procedures are also studied.

The results of the reliability assurance survey constitute one of the factors upon which approval or disapproval of a potential subcontractor is based.

After the award of a contract to a subcontractor, task 5 begins. This constitutes a periodic plant evaluation and a continuous surveillance of the plant for its performance with respect to the particular contract awarded. This differs from the preliminary survey, which was concerned solely with evaluating a potential subcontractor's ability to perform. After the award, the stress is shifted to actual performance with respect to criteria furnished by the prime or weapon system contractor, as the case may be. In any event, one of the first items is to assure that the subcontractor is taking the proper reliability approach and has an accurate picture of all the parameters which will affect performance. This is accomplished by the assignment of a reliability engineer to the subcontractor's plant whose function it is to check all major factors involved.

In order to make certain that the reliability engineer does not neglect important factors, he should be provided with a practices-and-procedures manual. This manual covers such items as a statement of the reliability engineer's authority and responsibility; abstracts of statements of work taken from the applicable contract; abstracts of important standard practices of the prime or weapons system contractor by whom he is employed, whichever is applicable; and a listing of all applicable specifications. Typical of such specifications is MIL-STD-441, which has been referred to in Chap. 14. Additional references will be made to company specifications which cover such subjects as reliability measurement by time sampling techniques, human factors, maintenance concepts, and the general design specification. This manual also contains special instructions governing such items as visits to second-tier suppliers, approval of nonstandard parts, waivers and deviations, and the qualification of parts suppliers based on approved sources of reliability data for these parts.

The reliability practices-and-procedures manual also contains many data and reliability procedures. Typical data are failure rate indices and derating factors for standard military parts and for commonly used, and therefore usually approved, commercial parts. The manual also contains methods of reliability prediction, ARL sampling tables, a procedure for failure reporting, and the criteria for evaluating subsystem reliability. Several forms are also included, such as requests for corrective action, requests for expediting corrective action, and a weekly summary report procedure.

The items listed above are only typical of what a good procedures manual should contain. For example, there are many more items which could be included, such as security measures, which cover transmission and maintenance of classified data and the procedures applicable for security clearance. Other procedures included are of an administrative nature and include such routine matters as project work orders, petty cash, advances, expense accounts, budgetary reports, facilities reports, protocol of relationship with subcontractor and customer, personal conduct, etc.

Task 6, the analysis and correlation of data, requires much training and experience on the part of the reliability engineer, because he must be capable of evaluating data which are representative of many different phases of the total reliability area. As an illustrative example, he must be capable of analyzing reliability-evaluation test data for completeness, adequacy, and accuracy and refining these data to enable rapid interpretation. He must also be able to evaluate statistically or categorize these data in order to highlight critical reliability problems. Moreover, the reliability engineer must evaluate any modifications made by the subcontractor to correct deficiencies by assessing the adequacy of the modification, the practicability of the change, the effect on standardization and maintenance, and the ease of manufacture. He must also devise methods for testing and evaluating the change and feed back the results of his analysis to all cognizant and interested groups. In those cases where the corrective action is not obvious, the reliability engineer assumes the responsibilities of what we formerly described as task 7, design of experiments. A statistical experiment is one in which the several suspected factors or influences causing the trouble are caused to influence each other by means of planned combinations; in this way it is possible to determine which combination is most likely to be the cause of trouble so that the actual problem can be isolated. The setting up of such experiments is not a one-man determination, since they usually are expensive because of the additional facilities, test equipment, and manpower required. Therefore, the reliability engineer must work cooperatively with the subcontractor's reliability engineers in setting up such experiments. In the case of involved projects, the home office statistical group usually participates, and the additional costs and delays must also be approved through the subcontracting organization.

The results of these experiments are one of the contributory elements to the execution of task 8 involving recommendations, since on the basis of the results obtained, recommendations are usually made to the engineering department outlining the suggested changes or summarizing the cause of the trouble. However, recommendations for improvements do not come only from statistical experiments. They emanate from any source: from observation, past experience, subcontractor recommendations, or from data analysis.

Task 9, reports, varies with the type of order and organization, but there are certain basic reports which we shall outline here. The administrative report covers elements of expense, travel itinerary, conferences, and activity and status. The technical reports may be on failure rates, trouble areas and recommendations for corrective action, plant performance logs, performance against schedule and requirements, reliability predictions and evaluations, engineering requirements, and test and evaluation studies. The frequency and format of the reports are a function of the type of contract and customer requirements. Usually, weekly, monthly, and quarterly reports are required as well as final reports upon the completion of a particular phase of the order.

Task 10, plan of operation, is of the utmost importance, since it is the major administrative vehicle which provides for efficient operation. In the past paragraphs, we have been discussing some of the methods of accomplishing some of the typical tasks which were listed, but this was not meant to be a complete plan. A complete departmental plan must not only list the tasks and the proposed methods of accomplishing these tasks, but must also schedule each major task and each of its subelements in a master phasing schedule showing the time it is expected that the task will begin and end. It must also include manpower requirements and show the build-up required to accomplish the tasks. In addition, budgetary and travel requirements must also be listed. In this manner, it is always possible for the manager of the department to measure his performance against his plan. This plan need not be considered static. As a matter of fact, it should be amended periodically to reflect changing concepts, i.e., other methods of accomplishing the tasks, or to represent the changed contractual requirements.

The outstanding advantage of such a plan is that it facilitates a periodic assessment of performance against the plan. This makes possible the comparison of the progress of each task as it relates to others in the same department or comparison between departments. It also provides a medium by which progress can be measured. Therefore, in periodic reports the per cent completion in comparison with the schedule shown on the phasing charts can be readily and simply shown.

Tasks of a Quality Assurance Department. The general outline of the tasks of a quality assurance department is similar in scope to that of reliability assurance, but not so in terms of objectives or detail. In this discussion, we shall outline the major tasks and then, in a manner similar to our reliability discussion, outline the details necessary to accomplish these tasks.

1. Analysis of Contractual Requirements. The quality assurance department studies, analyzes, and interprets all quality contractual requirements and applicable subsidiary specifications to assure that all factors are taken into consideration.

2. Preparation of Quality Specifications. The quality assurance department prepares company specifications for the purpose of supplementing and implementing contractual requirements.

3. Preliminary Plant Survey and Orientation. Prior to the award of a contract to any subcontractor, the quality assurance department will conduct a preliminary survey of the prospective subcontractor and orient him on the quality requirements of the job. The purpose is to determine the feasibility of considering him as a possible subcontractor.

4. Drawing and Specification Review. The purpose of this task is to assure that the subcontractor's quality control department reviews all engineering drawings for all quality requirements. The quality assurance department performs the same task for the prime or weapons system contractor of which it is a part.

5. Subcontract and Purchase Order Review. The quality assurance department reviews all subcontracts to assure that the quality provisions of the prime contract are imposed on the subcontractor. It also checks the effectiveness with which the subcontractor's quality control department executes a similar function at that level.

6. Plant Surveillance and Evaluation. After the award of a subcontract, the quality assurance department conducts a constant surveillance of plant quality operations to assure that they comply with contractual requirements and result in the production of a high-quality product.

7. Consultation Services. The quality assurance department provides consulting services when they are requested by any department or subcontractor engaged in the contract.

8. Analysis, Correlation, Interpretation, and Feedback of Data. After collecting data from responsible services, the quality assurance representative analyzes, correlates, and interprets them and feeds back the information to all interested parties either verbally or in the form of reports, depending on the degree of importance or required speed of transmittal. In cases of extreme emergency, the data are transmitted verbally and followed by formal reports.

9. Recommendations. On the basis of observations which are part of the surveillance function or as the result of data analysis, recommendations for quality improvement are made to the cognizant party.

10. Reports. Periodic reports to management are a regular operation of a quality assurance organization. The type and frequency of the report are a function of the requirements of the contract.

11. Plan of Operation. This is the entire plan of operation which describes each of the tasks, their method of accomplishment, and the schedule for each and constitutes part of the product-assurance plan. (See this same item under task 10 of the reliability assurance section.)

Suggested Methods of Accomplishing the Quality Tasks. The analysis of contractual requirements, task 1, is accomplished by reviewing the contract for all quality items and recording them. All impractical or questionable items should be questioned by the quality assurance group through proper company channels. Particular attention should be devoted to prescribed acceptable quality levels and sampling plans. Likewise, the standards of workmanship and classification of defects should be analyzed and either accepted as is or commented on. Moreover, the general acceptance test requirements should be studied to determine if they are clearly understood. When all the outstanding questions have been clarified, the quality assurance group is ready to interpret the quality requirements for any interested group and to prepare company quality specifications as required. In those instances where the company intends to subcontract part of the work, the quality assurance department will prepare the applicable quality subcontract clauses to assure that the subcontractor meets his quality obligations. These clauses should also cover the rights and facilities which the subcontractor is expected to furnish liaison engineers of the quality assurance department who are assigned to his plant.

The other details of the quality requirements are covered in general and detailed specifications prepared and issued by the quality assurance department. These specifications, as part of the subcontract, protect the rights of the liaison engineers and clearly define the tasks of the subcontractor. This is what we have listed as task 2. The following are illustrative of the type of specifications usually issued: (1) quality control requirements, (2) workmanship standards, (3) interchangeability standards, (4) test and inspection standards, (5) tool and gauge calibration, standardization, and cataloguing, and other procedures as required.

Task 3 is another important function of the quality assurance group, because it prevents the award of subcontracts to organizations who are not quality producers. This is a service which is rendered the subcontracting department to assist it in performing its function effectively. In this instance, the quality assurance department visits the potential subcontractor's plant to study the quality control practices he is currently using. This preliminary survey is conducted in accordance with a survey questionnaire prepared specifically for this purpose. This questionnaire generally covers the following areas: incoming inspection procedures, method of material review, surveillance of incoming raw materials and methods of storing accepted materials, methods of reviewing purchase orders, calibration procedures for tools, gauge and test equipment, methods of quality reporting and corrective action, etc. On the basis of this survey, a report is written either recommending or disapproving the potential subcontractor. This report is then consolidated with the reliability assurance report and forms a part of the over-all product-assurance recommendation. A good recommendation should not be without a soul. This means that the impressions of the potential subcontractor gained during orientation discussions should constitute part of the record. For example, the survey may show that he is lacking in some important attribute; but if he has an earnest desire to participate in the program, and if finances, facilities, and the will to overcome the deficiency are noted, then due consideration should be given him because of his favorable attitude. After award of the contract, the continual surveillance which follows will assure that he meets his promises, or else the contract may be terminated.

Task 4, drawing and specification review, although very important, is many times neglected. Much of the quality problems encountered in production are due to omissions or tolerance clashes in drawings. For example, specifically on government contracts, the types of fastening used for terminals and the nomenclature of parts are particularly important. Such items could be omitted from drawings simply because design engineers are not familiar with government standards. Drawings should also be reviewed for adequate specifications of plating, painting, or general finishing requirements. Moreover, the compatibility of what appears on the drawing and applicable specifications should also be investigated. It is the function of the quality assurance engineer to satisfy himself that the subcontractor's quality organization has procedures and methods for performing a drawing review and to evaluate the effectiveness of its performance.

Another important precaution is that of task 5, subcontract and purchase order review. Immediately after a subcontract is awarded, or preferably, prior to award, the responsible cognizant quality assurance liaison engineer should compare all the provisions of the subcontract with the original prime contract to assure that there are no omissions or conflicts between the two. He should also insist that the subcontractors' quality organization perform the same type of review and report any apparent discrepancies or disagreements to him.

When the subcontractor begins his purchasing cycle, the liaison quality engineer should assure himself that his quality organization is reviewing each purchase order thoroughly to determine that it contains all the necessary clauses, provisions for source inspection, or other special statements of work which must be mandatory on the vendor if the terms of the prime contract are to be satisfied. The liaison engineer satisfies himself that this action is thorough and complete by selecting random samples of approved purchase orders and reviewing them to determine if all is in order.

After the manufacturing cycle begins the liaison engineer starts on task 6, the plant surveillance and evaluation phase. This work is conducted in accordance with procedures outlined in a surveillance manual prepared for this purpose. In general, the manual contains applicable contracts and statements of work. If the contract is classified, then proper precautions are taken. The manual also covers applicable procedures and standard practices of the prime or weapons system contractor who employs the liaison engineer. Moreover, it contains typical government specifications, such as MIL-Q-9858, which is the general government quality control standard, and MIL-STD-105A, which is composed of standard military sampling tables, etc. The types of prime contractor specifications which should be included in the manual are quality control requirements, standards of workmanship, plant survey and evaluation, preparation for shipment, test requirements, interchangeability, etc.

Other factors which should be evaluated and the procedures for which should be part of the manual are incoming inspection procedures for both purchased parts and raw materials, procedures for stock control and fabrication as well as for assembly, inspection, test, packaging, packing, and shipping.

The manual should also contain instructions for verifying a number of other quality operations. Among these are the methods used in purchase order review, material review-board operations, verification of certification of special processes such as plating, painting, annealing, etc., and specification and drawing revisions.

Tasks 7, 8, and 9 are interrelated and will therefore be discussed jointly. As a result of his analysis of data, the quality liaison engineer is in a good position to make recommendations for product improvement. In this respect, he acts as a consultant to the engineering department or to the subcontracting department when questions involving the necessity for an engineering change arise. The liaison engineer's actions need not be based on statistical evaluations but may be the result of past experience, observation, or the subcontractor's recommendations, with which he may or may not concur. In any event, because of his intimate association with the project, his close relationship with government quality assurance representatives, and his tie-in with the reliability assurance engineers, who are also stationed at the subcontractor's plant, the quality liaison engineer is an excellent source for information and recommendations. Moreover, through the periodic technical reports which he issues, much improvement in the quality of the product is possible.

Task 10 involves the issuing of reports by quality assurance. Many types of reports are required, depending on the requirements of the contract. However, some reports are of an administrative nature and are required under any type of contract, while others are of a special nature as required by the contract. In general, reports may be subdivided into a number of categories such as administrative and personal, technical, and special, involving among other things the transmission of information and data.

Personal reports involve such items as expenses, hours, activities, plans, and problem areas. Other periodic reports are prepared as may be required in the contract, such as weekly, monthly, or quarterly status or progress reports. Although these reports may emanate from the liaison engineer, they usually constitute a home office requirement, since the information provided therein is the source from which the product-assurance department prepares its reports.

Technical reports cover such details as material review-board actions, amount of product accepted or rejected, plant quality ratings, deviations, waivers, engineering change proposals, etc.

Reports involving the transmission of information or data are reports to other product-assurance departments or subdivisions, management, the government, or other interested departments. The information transmitted usually involves such items as the quality performance of the subcontractor as compared with the established standards, the forecast of the subcontractor's performance against his established schedule, or any other similar report or request by the prime contractor's interested departments.

In some instances, the nature of the requested information may not strictly involve quality alone; but usually a liaison engineer furnishes it anyway, because he is resident at the subcontractor's plant and the information is available to him.

The discussion of task 11, the preparation of a plan of operation, is similar to that of task 10 for reliability assurance and therefore the reader is referred to this section. Only the tasks of quality assurance are different; the basic elements of the plan itself are the same.

16-3. Scope of Product-assurance Operations. The previous discussion stressed the role of the two major departments of product assurance by emphasizing specific tasks in the manufacturing or subcontractor's plant. However, the scope of product assurance as an integrated division extends far beyond this area. As a matter of fact, a well-organized product-assurance division is involved in practically all phases of operations from the initial design until final installation and deals in one way or another with all company departments.

Since the concept of product assurance is relatively new and is practiced only by relatively few companies in the United States, it is almost a necessity that the author continue with his explanations of how the division should operate and hope that, in this way, a standard will be established for the use of industry.

In the discussions of Sec. 16-2, we constantly referred to the tasks of reliability assurance and quality assurance, which probably created the impression that these two areas are separate and distinct. The real reason for the distinction is that as of the present writing there is an actual shortage of engineers who are qualified to perform the full product-assurance function. Hence the division into the two basic elements of reliability and quality assurance. In his present position of director of product assurance of the International Electric Corporation, Paramus, New Jersey, the author follows the two-department principle in his product-assurance organization but is fast attempting, by means of training, to develop a group of product-assurance engineers. When this is ultimately achieved, the field liaison engineer will perform both the quality and reliability function at any location, i.e., either at a plant or at the installation location. Accordingly, additional areas, duties, and responsibilities of a fully qualified and properly implemented product-assurance organization will now be discussed.

One of the principal activities of product-assurance engineers in the field is related to maintainability. As we have seen in Chap. 9, maintainability is the probability that after maintenance action a device will be restored to its operational effectiveness within a given period of time. During the theoretical analysis phase of maintainability, which is conducted in accordance with techniques discussed in Chaps. 9 and 12, it is possible to make an estimate of the maintainability that is expected in the field. However, one is never certain that theory and practice correlate until sufficient field data are accumulated to verify whatever maintainability prediction might have been made. The gathering of such field data is one of the jobs of the product-assurance engineer. This task requires much experience, skill, and maturity on the part of those making a maintainability study in order to assure that only bona fide data are recorded. This means that the product-assurance engineer must be closely associated with service personnel and engineers who are actually engaged in operating the equipment under actual field conditions.

When a failure occurs, it is most important that a determination first be made about whether or not it is actually a failure and, if so, whether it can be charged to the equipment or to associated equipment. Moreover, it is important to assess whether the failure is due to the fact that inexperienced personnel were operating the equipment or whether it is attributable to other reasons, such as the environment or usage. The time, means, and methods of restoring the equipment to operating condition should also be noted and recorded. These data are very necessary in order to determine the maintenance-action rate, which would be very small if a standby unit were available which could be immediately switched into service in case of failure. Therefore, when reporting on this phase it is important that the dual installation be mentioned. On the other hand, if there were no redundant unit available at the time of failure and operation had to be restored by repair of the defective unit, the resulting μ for such an installation would be very different from that of the dual, redundant installation. Hence, the product-assurance engineer in the field must be very accurate and detailed about the circumstances involved in maintenance of field equipment. Lack of detail is one of the shortcomings of data-processing methods, which cannot always provide the intimate information which is necessary to effect permanent product improvement.

As a by-product of his maintainability studies, the product-assurance field engineer is in an excellent position to provide data for determining spare-parts requirements based on estimates of part failure rates calculated under actual conditions of use. This work is most important and results in much economy to the user, because the number of spare parts stocked is in proportion to the quantity in use and their probability of failure. Without this type of study, parts would be ordered and stocked in an unscientific manner which would result in a needless waste of money and materials. Another responsibility of major importance of the field liaison product-assurance engineer is to act as a representative of his company in the field. In this capacity, he is helpful to the customer and creates good will for his own company by providing immediate and efficient service in all instances. Moreover, through his liaison activities, he learns firsthand what the customer's requirements are and what they might be in the future. This information, when fed to the home office, may result in additional business for the company as well as provide a much-needed service to the customer.

Moreover, by virtue of these data, comments, and statistical analysis received from the field, engineering changes are constantly made to provide for better operational reliability by improving equipment inherent reliability and maintainability.

The product-assurance engineer also assists the customer in training his personnel on the proper use and maintenance of the equipment and, in cases of emergency, at the request of the customer, even participates in the actual work of restoring inoperative equipment to service. The good will so created may result in the customer's contracting for his service work with the company on a sustained basis. Many companies have secured millions of dollars of service work in the field as a result of the good will created by their field representatives.

However, the product-assurance engineer in the field at installation sites is not alone in his work. Through his home office, his work is coordinated with product-assurance liaison engineers at the plant or at the subcontractor's plants. These latter engineers are engaged in work similar to that which is going on at the installation sites, except that the environments and conditions of use vary. Information from the installation site is fed to the factory, and vice versa, for the purpose of comparing experiences and evaluating the effects more thoroughly. For example, if the installation site reports that maintenance manuals and tools are inadequate to do the job effectively, the factory product-assurance engineer's responsibility is to follow through and insist, through proper contractual channels, that the required corrective action is taken. On the other hand, in the case of an engineering change in the factory, product assurance might ask that its effectiveness be evaluated in a particular manner at the installation. The plant and installation site are therefore closely knit together through the medium of the product-assurance division, resulting in benefits to both company and customer.

16-4. Liaison with the Government Quality Assurance Representative. As was previously explained, the government now refers to its inspectors in contractors' plants as government quality assurance representatives, which term is usually abbreviated as QArep. However, the two terms are still used interchangeably by many and therefore are oftentimes intermingled; but in each case their meaning is synonymous. Since the product-assurance manager's duties are so closely interwoven with the interests of the government, it is obvious that a close relationship must exist between the product-assurance manager and the QArep. It is for this very reason that the product-assurance manager is usually designated as the liaison between the company and the QArep. In the preceding discussions we have outlined the duties of the product-assurance manager. In order to understand how his liaison activity with the QArep can be most effective, we should discuss some of the basic government philosophy underlying the behavior of the QArep.

The government QArep occupies a position of trust and responsibility. Generally speaking, his duties and responsibilities are fairly well defined. He is usually conscientious in his approach and attempts to do a good job. Oftentimes friction occurs between the QArep and contractor because each does not appreciate the other's job. The QArep feels that the contractor's sole motivation is to produce and ship products regardless of whether quality and reliability are up to standard, while the contractor feels the QArep is lacking in judgment and common sense, suffering from "deskitis" and engulfed by rules and regulations which he does not care to interpret logically. When this kind of situation exists, it is obvious that friction will occur and that both parties will suffer.

Actually most problems would disappear if each party would attempt to work harmoniously with the other by trying to see the opposite viewpoint objectively. Usually, a product-assurance man who is skilled in his profession can get along very well with government personnel. This is only natural, since his objective is the same as that of the government, i.e., the production of goods of high quality and reliability. Therefore, a good plan of action is to designate the product-assurance engineer as liaison between the company and the government QArep in all matters pertaining to quality. Usually, when this is done, good relations are established.

For a long time the government also has recognized the need for cooperation between its QArep and the contractor. The following quote is taken directly from the Munitions Board publication dated April, 1951, and called "Introduction to Procurement Inspection"; the same rules are still in effect today.

THE INSPECTOR'S PLACE IN THE SUPPLY PROGRAM[1]

Section 400

401 Duties of Inspectors

401.1 It should not be concluded, from the text of the preceding sections, that the inspector's job is exclusively technical. In fact, the inspector's adminis-

[1] The author was a member of the committee responsible for drawing up this code. It was one of the last duties he performed as a production inspection officer in World War II. Subsequent to this assignment, he received his honorable discharge.

trative functions as a representative of the United States Government are equal in importance to his technical functions. A competent technical man may fail completely as an inspector if he is unable to deal competently with the contractor and government agencies.

401.2 In the administration of the contract at the contractor's plant, the inspector should be viewed as an ambassador to assist in carrying out the general government supply program. There is a definite code of ethics and procedure that the inspector should follow in his relations with the contractor and with other government agencies. These relations are discussed in the two succeeding sections.

402 The Inspector-Contractor Relation

402.1 The standards and relation that the inspector establishes with the contractor and his employees are of vital importance to the effectiveness of his work in the plant. Every contractor is interested in having his plant operate smoothly, at peak load, and without interruptions. The inspector is likewise interested in quantity production, but he is also charged with the responsibility of seeing that set standards of quality are maintained. That is his special mission. The inspector should conduct himself in such a manner within the plant that he can best accomplish his important mission. Much that the inspector may accomplish depends on the attitude with which he regards the contractor and the plant employees and the respect which he in turn receives from them. His greatest assets are the sincerity, purposefulness, and effectiveness with which he does his work; the exercise of calm judgment and tact at all times. He can best convince the contractor of his ability and his integrity by knowing his inspection job thoroughly and by doing it in a businesslike manner.

402.2 One of the inspector's first interests, on taking over a new assignment, is in being shown through the plant. Such trips are never for the purpose of criticism but to enable the inspector to study and analyze the contractor's layout and methods of operation. It should always be remembered that the inspector has no right whatever to tell a contractor how to run his plant. It is only after the inspector has carefully studied every phase of the contractor's fabrication program that he is qualified to exercise maximum protection of the government's interest and to make helpful suggestions, if and when asked. The contractor has the right to demand that the inspector, in his zeal to protect the interests of the government and to obtain high grade products, do nothing to injure him in his business by making unreasonable demands.

402.3 The inspector should be courteous in all of his contacts with the contractor, his executives, and his employees. He should realize that a man who has successfully manufactured a certain product for a number of years will naturally feel that he possesses the general and technical information required to manufacture the product that he has contracted to furnish the government. If the inspector, speaking without thorough knowledge of the point involved, were to make an ill-founded suggestion, the fact would be immediately apparent to the manufacturer. This would create the impression that the inspector did not understand his work. It might also produce a feeling on the part of the manufacturer that the inspector could be readily deceived. When a question arises which the inspector is unable to solve, he should admit the fact but without fail seek the answer from his superior.

402.4 Inspectors must not interfere with the duties of the employees of the factory. They must not, other than in exceptional circumstances, order the cessation of any manufacturing operation. If it is noted that any of the material employed or its manipulation is at variance with the specification requirements, such information must be reported at once to the appropriate company official. If the violation is serious or flagrant and is not immediately corrected, it must be reported to the central inspection office. The report must be in detail and must thoroughly explain the difficulty; that is, whether inferior materials are being used or some non-specification operation in manufacture is being performed. It must also tell when and to whom the inspector reported the matter in the plant, and the reaction his report evoked.

402.5 The inspector should avoid giving the impression that the honorable intentions of the contractor or his personnel are in question and should not hastily conclude that every violation of the specification is deliberate. Even with the most earnest cooperation on the part of the manufacturer and his employees, it may still be difficult to maintain at all times the highest standard of workmanship and materials which specifications require. While the inspector should ever be watchful and alert for poor workmanship and non-specification material, he must not, without positive proof, consider such deficiencies as a judgment against the contractor's integrity. He should impress upon the minds of those with whom he comes in contact that he considers the interests of the manufacturer as well as the government. The best results are obtained when a minimum of friction exists between the inspector and the factory personnel.

402.6 The inspector is often the sole government official in personal contact with the contractor and therefore fills a position of great trust and responsibility. He is expected to conduct himself, both on and off duty, as a person of dignity, good character, ability, and sound judgment, and as one who at all times is conscious of his responsibility as a representative of the Government of the United States.

403 Inspector's Rules of Conduct

403.1 The inspector shall not assume authority not clearly given him by his supervisor, nor shall he accept orders or be influenced in the performance of his duties, except through his superior inspection authority.

403.2 He shall be tactful and courteous and shall cooperate with the contractor within the limits of his assigned duties and responsibilities.

403.3 He shall be impartial in action and judgment, render prompt decisions, and give a tolerant hearing to the opinions of others.

403.4 He shall not accept gratuities, nor place himself under obligation to the contractor, and he shall avoid undue familiarity with officials and employees of the contractor.

403.5 He shall observe the rules and regulations pertaining to personal conduct, safety, and security established in the contractor's plant where he is stationed.

403.6 He shall observe the working hours established for the inspection unit to which he is assigned and shall comply with work schedules set up by his supervisor.

403.7 He shall not discuss his own or the contractor's affairs or plant operations or techniques except with his supervisor or other properly authorized persons, and shall confine his activities to the portions of manufacturing plants necessary to the performance of his duties.

403.8 He shall carry his official identification pass on his person at all times while in the plant. If the contractor's regulations so require he shall wear, while in the plant, a badge furnished by the contractor.

From the foregoing it is obvious that the government has established a code of ethics for its inspection personnel; isn't it only logical that industry should do the same and establish a code for industry personnel? The suggested code for industry indicates desirable attitudes and actions for the product-assurance representative.

SUGGESTED CODE FOR PRODUCT-ASSURANCE REPRESENTATIVE

1. Be courteous and tactful at all times. Sell ideas; do not bludgeon them through. Use of threats or appeals to higher authority or reference to influence should be strictly avoided.
2. Speak with authority based on facts and not on conjecture or rumor.
3. Never make a promise for expediency's sake which there is no intention of fulfilling at a later date.
4. Take the government QArep into your confidence regarding product-assurance planning and procedures so that they don't come as a shock. Many a good plan has been defeated because people associated with it were not consulted in advance.
5. Do not be deceitful and attempt to bypass government regulations. This constitutes a breach of contract. If a short cut is necessary, discuss it with the government QArep and elicit his cooperation.
6. Build mutual trust and confidence by following a spirit of forthrightness and honesty.
7. Review contractual product-assurance requirements at the beginning of the contract and any change-orders throughout the contract and discuss any special problems with the QArep frankly and freely.
8. Develop a spirit of give and take. If the QArep makes a mistake, don't try to hang him. He may hang you next. Give, even if at times it may not be a contractual requirement (unless much cost is involved), and you will in turn receive when most in need.
9. Be prepared to take an adverse decision unemotionally and without resorting to vituperation.
10. Do not go over the QArep's head to complain to superior authority unless the QArep is first courteously informed of this intention and is invited to participate.
11. Make the QArep part of the product-assurance team.

16-5. Typical Product-assurance Organizational Structures. Figure 16-1 shows a typical factory product-assurance organization. Reporting directly to the director are the manager of the environmental test laboratories, the manager of incoming inspection, and the managers of product assurance for as many projects as are in existence. Also reporting to the director are two staff engineers, one for quality and the other for reliability assurance. The supporting structure for each of these first-line groups is a function of their requirements. A typical substructure for a manager of product assurance (project A) is shown in Fig. 16-1. Reporting to him are an engineer in charge of reliability-testing procedures, an engineer in charge of the statistical analysis and evaluation section, a project-reliability engineer, and a quality engineer. The manager of product assurance for a specific project also acts as chairman of the reliability control group (RCG) which is comprised of personnel shown on the organization chart.

Since we discussed the functions of quality assurance and reliability assurance in the previous sections, we shall only briefly highlight the functions of each group in Fig. 16-1 to outline its major responsibilities.

The environmental test laboratories are used to test parts and equipments under the environmental conditions specified by the appropriate specifications. The members are a service group and are available to all departments of the product-assurance director's organization. They record the data and furnish them to the party for whom the tests were conducted.

The incoming-inspection department inspects and tests all incoming parts and materials in accordance with applicable requirements. If special tests are required, the environmental test laboratory is available for this purpose.

The manager of product assurance for a specific job bears the over-all responsibility for assuring that the quality and reliability for his product are in conformance with applicable requirements. In this capacity, he has direct-line control over the departments which report to him and also the services of the staff groups reporting to the director. He also presides over the RCG and coordinates the recommendations made with the rest of his organization in order to provide for their early implementation and/or required action.

The staff reliability engineer acts as an adviser to the director and also establishes standards, criteria, and general operating procedures for the guidance of all reliability personnel.

The staff quality assurance engineer is in a similar capacity to the staff reliability engineer, except that the standards, criteria, and general operating procedures are established for the guidance of all quality control personnel.

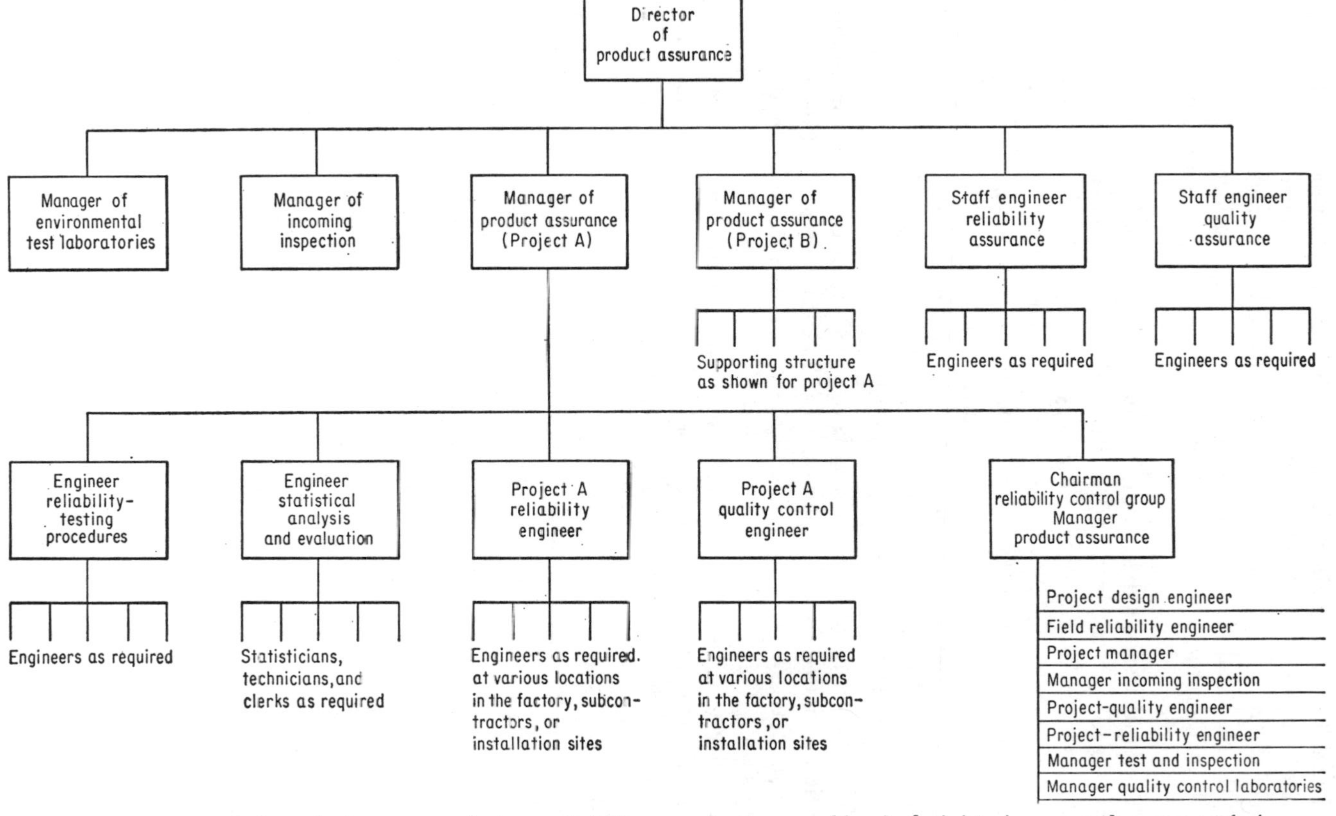

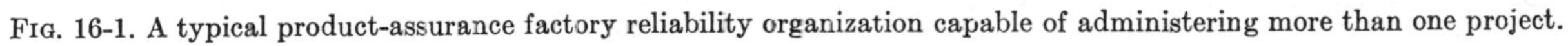

FIG. 16-1. A typical product-assurance factory reliability organization capable of administering more than one project.

The duty of the engineer in charge of reliability-testing procedures is to devise test procedures for parts and equipment in order to verify that they are of the required reliability within specified confidence levels. He also has the responsibility of planning special tests to be utilized by the quality control laboratory for either debugging or screening purposes for the parts on his project.

The statistical analysis and evaluation section collates data received from the factory and installation sites and analyzes them to determine such items as mean time between failures, equipment failure rates, part failure rates, mission availability, time availability, probability of survival, maintenance-action rate, and equipment availability. Each of the above factors provides a means of accurately measuring the individual equipment's performance and capabilities. They also provide a means of comparing the predicted reliability and maintainability design goals with those obtained under actual operating conditions.

The project-reliability engineer is responsible for coordinating all information reviewed from all sources. In order to keep abreast of the latest developments, he stations reliability engineers in the factory and installation sites. Using his staff as a back-up, he is responsible for publishing reliability assurance procedures, establishing requirements for the procurement of test equipment for the reliability testing of parts and components, and for the recording of all failures. He also routes rejected or defective parts from line and field to the environmental laboratory for verification of failures and determination of the assignable causes and refers those parts exhibiting a failure trend to the RCG for action. In addition, he issues reliability reports, writes special instructions for the incoming-inspection group or the laboratory to perform special tests on parts when data indicate they are necessary, and arranges for withdrawing these parts from stock for the special tests. At the conclusion of the special tests, he reports the results to the RCG or other interested groups. His department also maintains a record of the failure rate of parts and components tested under particular conditions.

In addition to the afore-mentioned, the project-reliability engineer makes requests of his field engineers to furnish specific data such as the number of accumulative trouble-free hours of operation; the number of accumulative hours out of service; the number of part, equipment, or subsystem failures within a specified period; unavailability of replacement parts; normal mission duration; and scheduled and nonscheduled maintenance times.

With respect to improvements, the project-reliability engineer makes recommendations regarding siting, adequacy of operational and instruction handbooks, spare parts, and engineering changes.

The project quality control engineer is responsible for the many and varied activities described in the previous sections. However, in addition he cooperates very closely with the reliability engineers by screening failures on the factory floor to make certain that they truly are the cause of trouble. If he so determines, he makes certain that the tag used to convey the defective parts to the laboratories has all required information such as:

1. Nomenclature and manufacturer of parts
2. Part number
3. Symptom
4. Circuit symbol or equivalent designation
5. Date and time of failure
6. Time in service before failure
7. Secondary failures, if any
8. Nature of failure if known, e.g., wear-out, catastrophic, etc.
9. Area in which component failed
10. Any pertinent remarks which will assist investigation

As we mentioned previously, the manager of product assurance also acts as the chairman of the RCG. The RCG, under the direction of its chairman, makes recommendations for actions to be taken based upon the data it receives from the project-reliability engineer. These recommendations encompass such items as:

1. Further research and design
2. Additional tests to be made on components or parts
3. Changes in specifications or procedures
4. Substitution of parts
5. Disposition of rejected items or materials
6. Request for studies, reports, or actions

Figure 16-2 shows the organization chart of a typical systems management company in which the reliability assurance and quality assurance groups are completely separate and are integrated only through the director of product assurance, since they both report to him. Each of the two departments also has its own supervisor of field liaison engineers. The disadvantage to this type of organization is that a subcontractor must deal with two different representatives of the same division unless one is arbitrarily designated as a common representative or spokesman. This type of organization is usually employed when product-assurance engineers who can perform a dual function are not available.

Figure 16-3 is similar to Fig. 16-2 except that there is a common manager of field operations who supervises all field liaison engineers charged with assignments from both the reliability and quality assurance departments. This type of organization depends upon the availability of quali-

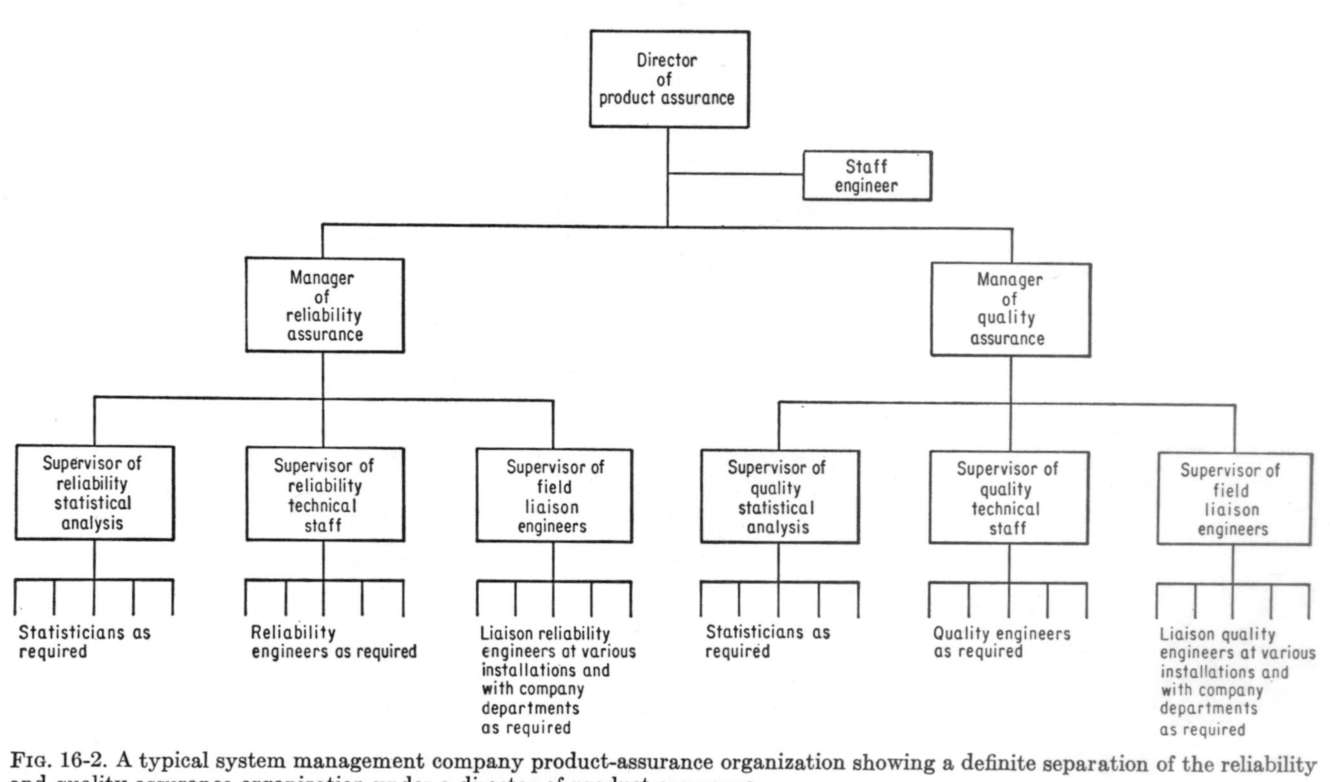

FIG. 16-2. A typical system management company product-assurance organization showing a definite separation of the reliability and quality assurance organization under a director of product assurance.

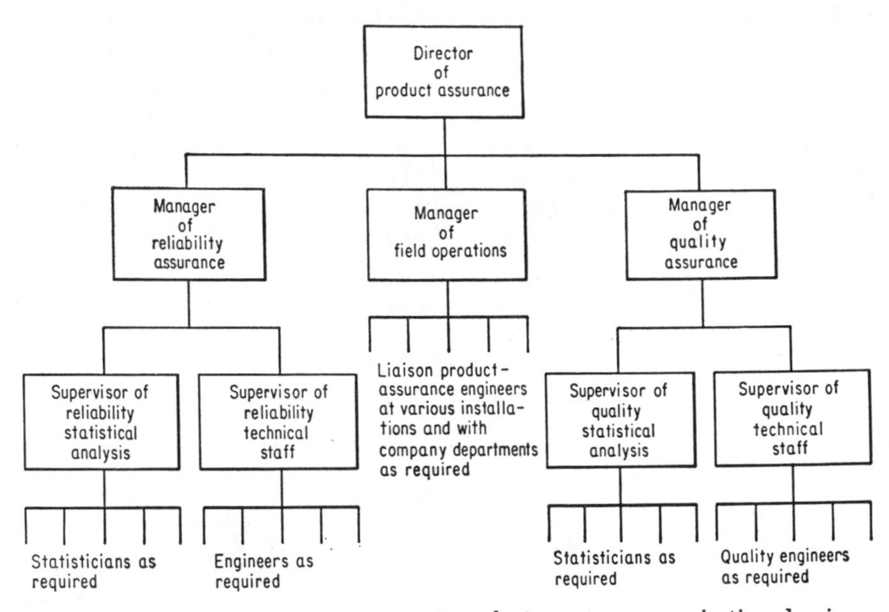

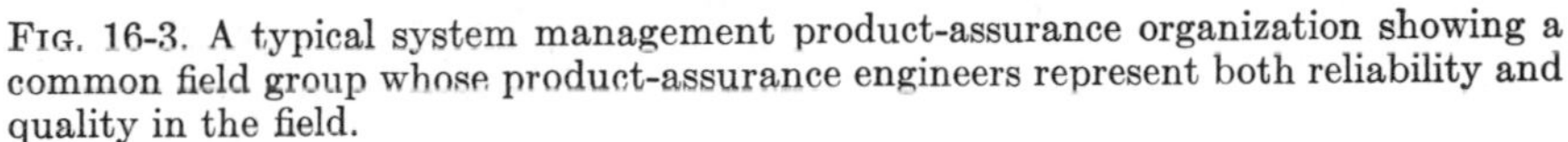
FIG. 16-3. A typical system management product-assurance organization showing a common field group whose product-assurance engineers represent both reliability and quality in the field.

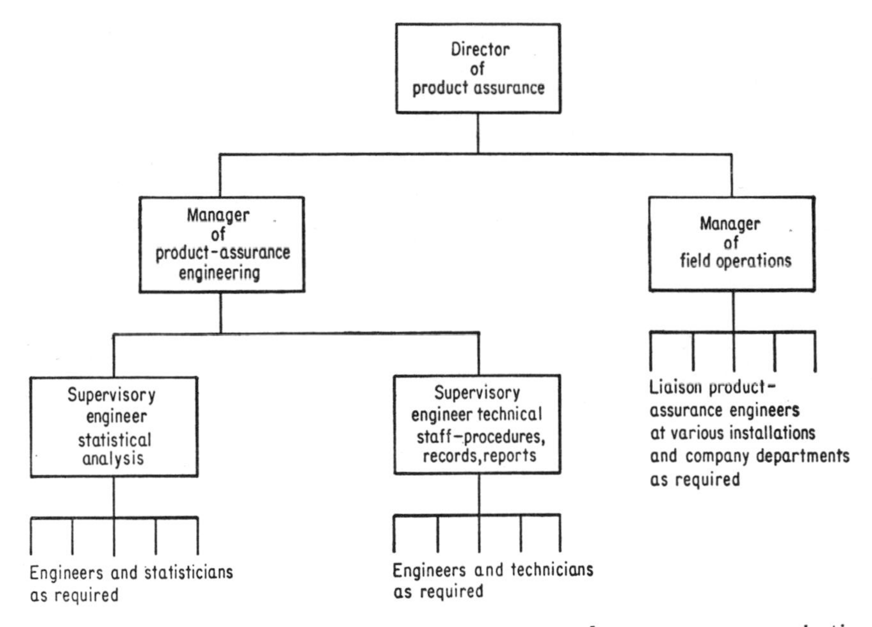

FIG. 16-4. A well-integrated system management product-assurance organization which is possible only when highly qualified product-assurance engineers are available.

fied product-assurance engineers, without whom it cannot be properly implemented. It also has the advantage of providing one common division representative in the field and at various subcontractor plants.

Figure 16-4 shows the final transition from the previous organizations. In this type of operation, all product-assurance engineering activity is centered under one manager and all field operations under another. This is a very efficient organization but requires a very high-caliber experienced staff of product-assurance engineers for its implementation.

In all these various organizations, the basic duties and responsibilities of the various groups are similar, so it is unnecessary to repeat them for every different organization configuration.

APPENDIX 1

DEFINITIONS FREQUENTLY USED

Acceptance. The act of assuming ownership, by the buyer's representative, of supplies or services which are adjudged by him to comply with all the terms and conditions of the applicable contract.

Acceptance, Probability of. The probability that a lot of a particular quality or reliability will be accepted when measured in accordance with the applicable criteria of the prescribed sampling plan.

Action, Maintenance. The act of maintaining a device in its operating condition. It may consist of a series of events such as repair, test, or preventive maintenance, as required.

Action, Mean Time to Maintenance. The average time required to perform a maintenance action on a device.

Analysis, Statistical. The mathematical treatment and evaluation of data for the purpose of evaluating various effects and interrelationships or formulating a conclusion.

Assembly. An article which consists of parts and/or subassemblies or a combination thereof and as such comprises an element of a component and performs functions necessary to the operation of the component as a whole.

Assurance, Product. A technical management operation reporting to top management and devoted to the study, planning, and implementation of the required design, controls, methods, and techniques to assure a reliable product in conformity with its applicable specification.

Assurance, Quality. A technique or science which uses all the known methods of quality control and quality engineering to assure the manufacture of a product of acceptable quality standards.

Assurance, Reliability. A technique or science which assesses the reliability of a product by means of a surveillance and measure of those factors of design and production which affect it.

Attribute. A characteristic or property which is evaluted on a go–no-go basis, i.e., in terms of whether it is, or is not, present.

Attributes, Method of. An evaluation of quality by the means of recording the presence or absence of an attribute in each item of a sample or lot undergoing inspection.

Availability. The probability that a device will operate at any specific instant of time. This is also referred to as up-time ratio (UTR) or time availability. The equipment or mission availability is, respectively, the probability that a stated per cent of equipment or missions will provide adequate performance in time T with no down-time interval exceeding the maintenance time constraint.

Average. The arithmetic mean, calculated as the quotient of the sum of a series of values to the number of such values.

Average, Process. The average quality being produced. It is computed as the ratio of the total number of defectives or defects in a sample to the total number of units in the sample. It is usually expressed in terms of per cent defective or defects per hundred units.

Bit, Failure. One-thousandth of the per cent failure per thousand hours.

Burn-in (Debugging). A reliability conditioning procedure which is a method of aging by operating the equipment under specified environmental test conditions in accordance with an established test conditioning specification. The purpose is to eliminate early failures and thus age or stabilize the operation of the equipment.

Cause, Assignable. An identifiable factor which is responsible for, or contributes to, the variation in reliability or quality.

Characteristic. A trait, property, or quality of a specified item, type of item, or group of items.

Characteristic, Operating. The curve which describes the probability of acceptance of a lot for various values of process average.

Characteristic, Reliability Operating. A curve which describes the probability of acceptance of a reliability time sample for various values of mean reliability level.

Chi Square (Distribution). The sum of squares of ν independent normal deviates with zero mean and unit standard deviation.

Component. An article which is normally a combination of parts, subassemblies, or assemblies and is a self-contained element of a complete operating equipment and performs a function necessary to the operation of that equipment.

Confidence, Degree of. See *Level, Confidence.*

Control, Quality. A scientific system involving the application of all known industrial and statistical techniques to control the quality of the manufactured article throughout its various production phases.

Data, Censored. Data whose exact values are not known but must be greater or less than some number.

Defect. A characteristic which does not conform to applicable specification requirements and which adversely affects or potentially affects the quality of a device.

Defective, Per Cent. The number of defective units in a lot divided by the total number of units times 100.

Defects, Classification of. A method of categorizing defects into certain classes in accordance with their importance or severity. The most common classes used are critical, major, minor, and control.

Degradation. A gradual deterioration in performance as a function of time.

Derating. The practice of subjecting parts or components to lesser electrical or mechanical stresses than they can withstand in order to increase the life expectancy of the part or component.

Deviation, Fractional Reliability. See *Index, Reliability.* The ratio of the values of the mean reliability level and the acceptable reliability level.

Deviation, Standard. The root-mean-square deviation of the observed values from their average.

Distribution, Binomial. A distribution which is applicable when the items in an infinite population can assume only two values, such as 0 or 1, or good or bad. The distribution therefore describes the probability of occurrence of various combinations of these two values in samples of size n selected at random from the parent population.

Distribution, Exponential. The exponential variation of the probability of occurrence of a variable y with respect to a change in time t.

Distribution, Frequency. A distribution whose outline describes the relative frequency of occurrence of the variables which make up the distribution.

Distribution, Normal. The distribution of a continuous variable about a mean value in accordance with a frequency of occurrence as defined by the outline of the distribution curve. This outline is similar in shape to a bell; hence this distribution is often referred to as the bell-shaped curve.

Distribution, Parent. See *Population.*

Distribution, Poisson. The probability of occurrence of specific values in a continuum of time, area, or volume for a population of known expectation.

Distribution, Probability. A distribution which describes the probability of occurrence of the chance or event under consideration.

Effectiveness. The capability of the weapon system or device to destroy all or a portion of the target and to perform other facets of its mission.

Element. A term used generally to refer to a part, assembly, component, equipment, complete operating equipment, and/or system.

Engineering, Human. The science of studying the man-machine relationships in order to minimize the effects of human error and fatigue and thereby provide a more reliable operating system.

Engineering, Quality. The science of establishing quality acceptance and evaluation criteria such as acceptance sampling plans, control charts, classification of defects, and tests.

Engineering, Reliability. The science of including those factors in the basic design which will assure the required degree of reliability.

Environment. The aggregate of all the conditions and influences which affect the operation of a device.

Environment, Immediate Part. The environmental conditions local and immediate to each individual part when the equipment is utilized under the conditions of specified end use.

Equipment, Complete Operating. An equipment together with the necessary parts, accessories, and components, or any combination thereof, required for the performance of a specified operational function.

Expectation. The mean or expected value of a population.

f Chart, Sequential. A series of samples taken in steps after each of which the number of failures, if any, is plotted on the chart and a decision is made to accept, continue the test, or reject on the basis of the number of failures found at any particular time with relation to the governing chart limits at that time.

f Chart, Subsample. A time sample control chart which plots the number of failures at periodic intervals during the reliability test for the purpose of studying the reliability trend as well as making a decision at a point in time regarding acceptance or rejection of the test.

Factor, Safety. The degree of conservativeness exercised in the application of component parts to ensure their not being operated at their limiting stress point or at their maximum energy dissipation limit, whichever is the controlling parameter.

Failure. A malfunction which cannot be corrected by the operator by means of controls normally accessible to him during routine operation of the device and which results in inoperativeness or substandard performance.

Failure, Catastrophic. A failure which occurs suddenly without warning and which results in a complete failure of the mission.

Failure, Independent. A failure which occurs in a random manner and which has no relationship or association with any other failure.

Failure, Manifestation. The apparent evidence which occurs to the observer that a failure has occurred.

Failure, Mean Time to. The measured operating time of a single device divided by the total number of failures during this period. This calculation is accomplished by repairing the device after each failure and continuing the test.

Failure, Mean Time to First. The average time to first failure of several equipments. It is used to determine the apparent approach of the equipment life characteristic to its random constant failure rate.

Failure, Part. A failure which usually involves an unrepairable breakdown and immediate end of life for a part which is subsequently permanently replaced.

Failure, Secondary (Dependent). A failure which occurs as a by-product of an independent failure.

Failure, Test to. The practice of inducing increased electrical and mechanical stresses in order to determine the maximum capability of a device so that conservative usage in subsequent applications will thereby increase its life through the derating determined by these tests.

Failure, Wear-out. Those failures which can be predicted on the basis of a known wear-out characteristic and which can therefore be prevented by appropriate preventive maintenance.

Failure Degradation (Creeping). A failure due to a latent fault which occurs as the result of the change of a parameter or attribute with time, environment, or usage.

Failure Law, Exponential. The exponential failure law states that the probability of survival P_s of an equipment operating for a time T is a function of the mean life m or of failure rate r as expressed by the following formulas:

$$P_s = e^{-T/m}$$
$$P_s = e^{-rT}$$

Failure Rate, Ratio. The ratio of the number of failures which occur during a unit interval of time to the total number of items at the start of the reliability test.

Failure Rate or Hazard r. The failure rate is expressed in terms of failures per unit of time, such as failures per hour or failures per hundred or thousand hours. It is computed as a simple ratio of the number of failures during a test interval to the total or aggregate test time of all units for that interval. During the operating period this failure rate is essentially constant from interval to interval.

Failures, Mean Time Between m. The average time of satisfactory operation of a population of equipments. It is calculated by dividing the sum of the total operating time by the total number of failures.

Fault. An attribute which adversely affects the reliability of a device.

Freedom, Degrees of. The number of observations that are free to vary at random regardless of the restrictions imposed by the statistics describing the distribution.

Frequency, Relative. The frequency of occurrence of an observation divided by the total frequency of occurrence of all observations.

Function, Distribution. The relative frequency of occurrence of specific observations comprising the distribution. In a cumulative distribution the sum of these frequencies represents the probability of the total's being equal to or less than some value X.

Hazard. See *Failure Rate.*

Index, Reliability. The ratio of the mean reliability level to the acceptable reliability level.

Interval, Confidence. The maximum and minimum limits which define the range within which it is expected that an observation will occur in accordance with a degree of certainty as dictated by the confidence level.

Interval, Test. A specific period of time during which one or more units may be undergoing test simultaneously.

Kappa Square. A statistic providing a two-tailed confidence level which can be used for various statistical determinations as explained in mathematical techniques developed by the author.

Level, Acceptable Quality. The quality standard associated with a given producer's risk which is prescribed by the customer or quality engineer for the products on order. It is usually expressed in terms of per cent defective or defects per hundred units.

Level, Acceptable Reliability. A nominal value expressed as a failure rate or mean time between failures applicable to a specific product undergoing reliability testing under specified conditions.

Level, Confidence. The degree of certainty, expressed as a percentage, that a given hypothesis is true within a specified limit or limits.

Level, Confidence (One Tail). The degree of certainty, expressed as a percentage, that a given hypothesis is true within a specified limit and that the probability of its being outside this limit is proportional to the area of the tail of the probability distribution which exceeds the limit.

Level, Confidence (Two Tail). The degree of certainty, expressed as a percentage, that a given hypothesis is true within specified limits and that the probability of its being outside these limits is proportional to the area in the left and right tails of the probability distribution.

Level of Significance. A significance level associated with the test of a hypothesis which defines the expected number of times the hypothesis will be rejected even though it might be true.

Life, Shelf. The length of time an item can be stored under specified conditions and still meet specification or operational requirements.

Life, Useful. The total operating time between debugging and wear-out.

Life Characteristic of Equipment. The relationship which exists between the failure rate of the equipment and operating or test time.

Limit, Average Outgoing-quality. The limiting value or poorest quality that it is possible will ever be shipped to the customer as the result of the use of an average outgoing-quality sampling plan.

Limits, Probability. The extreme or lower and upper limits which define the range within which it is likely that an event will occur.

Limits, Sigma. The interval about the mean expressed in units of standard deviation. For example, in the case of a normal distribution, 1-sigma limits on each side of the mean would include about 68 per cent of the population.

Lot. A collection or group of units of product from which a sample is drawn and inspected to determine compliance with applicable criteria or specifications and drawings.

Lot Tolerance Fraction Reliability Deviation. That extreme value of fractional reliability deviation which is tolerable within the significance level specified by the consumer's reliability risk for the applicable time sampling plan.

Maintainability. The probability that a device will be restored to operational effectiveness within a given period of time, when the maintenance action is performed in accordance with prescribed procedures.

Maintenance, Emergency. A maintenance procedure required as the result of a failure.

Maintenance, Precautionary. A procedure of periodically reconditioning a product, but not in accordance with specific instructions or scheduling, to prevent or reduce the probability of failure or deterioration while the product is in service.

Maintenance, Preventive. A procedure of periodically reconditioning a product in accordance with specific instructions and scheduling to prevent or reduce the probability of failure or deterioration while the product is in service. This type of maintenance may be either scheduled or nonscheduled in nature.

Maintenance Ratio. The number of maintenance man-hours of down time required to support each hour of operation.

Malfunction. A malfunction denotes unsatisfactory operation of a device because of one or more faults.

Mean, Arithmetic. See *Average.*

Measure, Statistical. The representation or estimator of the value of a parameter in a probability distribution.

Median. The point of a continuous random variable which divides the distribution into two equal halves such that one-half the values are greater and one-half smaller than the value of the subdividing point.

Mode. The maximum point of the frequency distribution of a continuous random variable.

Mortality, Infant. The premature catastrophic-type failures occurring at a rate substantially greater than that observed during the operating period prior to wear-out.

Multiple Censorship. Multiple censorship, in a systems-reliability evaluation, is a series of incomplete observations of time between failures which are incomplete for reasons entirely disassociated from system failure.

Number, Acceptance. The largest number of defects or failures which are permissible as a condition of acceptance in an attributes or reliability time sampling plan.

Number, Rejection. The smallest number of defects or failures which will result in rejection in an attributes or reliability time sampling plan, respectively.

Number, Working Acceptance. An acceptance number which may be a decimal fraction and is used for the purpose of facilitating interim calculations by the use of special techniques developed by the author.

Part. An article which is an element of a subassembly or an assembly and is of such construction that it is not practically or economically amenable to further disassembly for maintenance purposes.

Per Cent Defective, Lot Tolerance. The extreme value of fraction per cent defective which is tolerable within the significance level specified by the consumer's risk for the applicable attribute sampling plan.

Plan, Acceptance Sampling. A plan used to determine the acceptability of an inspection lot. It usually consists of criteria such as a classification of defects and a statement of the sample size and acceptance number to be used.

Plan, Average Outgoing-quality Sampling. A plan which specifies values of sample size, acceptance numbers, and process averages necessary to assure an average outgoing quality.

Plan, Double Sampling. A sampling plan which permits the use of a first and, if necessary, second sample before a final decision is made to reject the lot.

Plan, Multiple Sampling. A sampling plan which permits the use of several samples up to a specific point before a final decision to reject the lot is made.

Plan, Single Sampling. A sampling plan which calls for a decision to accept or reject on the basis of the results of a single random sample selected from a lot.

Plan, Time Sampling. A sampling plan which uses time as a sample and hence which defines the number of hours of testing and the number of failures permitted (acceptance number) before a decision to accept or reject the reliability is made.

Population. A statistical term which is used to define an infinite number of similar or like units.

Probability. A measure of the likelihood of occurrence of a chance or event varying in degree from 0 to 1, the former number being indicative of no likelihood and the latter of 100 per cent likelihood.

Production, Pilot. The initial production of a device utilizing the basic tools, methods, and procedures which will be used in production in order to prove the effectiveness of the device as well as the production system used.

Quality. A measure of the degree to which a device conforms to applicable specification and workmanship standards.

Quality, Average Outgoing. The ultimate average quality of products shipped to the customer which are the result of the composite techniques of sampling and screening.

Quality, Lot. The ratio of the number of units defective to the total number of units comprising the lot.

Randomness. The occurrence of an event in accordance with the laws of chance.

Range. The difference between the greatest and least value of a set of numbers.

Range, Acceptable Environmental Test. A test conducted within the range of environmental conditions in which an equipment can be expected to function properly.

Range, Environmental. The range of environment throughout which a system or portion thereof is capable of operation at not less than the specified level of reliability.

Range, Moving. The difference between the greatest and least values in a number of consecutive measurements, recalculated after each group of measurements.

Rate, In-commission. The percentage of the total operational time (24 hr a day, 7 days a week) during which the equipment is entirely ready for operation or is in operation.

Rate, Maintenance-action. The number of maintenance actions per hour which can be completed when performed in accordance with a prescribed maintenance procedure.

Rate, Mission Success. That percentage of the total missions uninterrupted by failure of the equipment.

Ratio, Cost. The ratio of the cost of maintenance per year for a given equipment or system to the initial cost.

Ratio, Down-time. The ratio of down time to total operating time.

Readiness, Operational. The probability that a device will perform satisfactorily at any point in calendar time. This is analogous to up-time ratio.

Redundancy. The existence of more than one means for accomplishing a given task, where all means must fail before there is an over-all failure to the system. Parallel redundancy applies to systems where both means are working at the same time to accomplish the task, and either of the systems is capable of handling the job itself in case of failure of the other system. Stand-by redundancy applies to a system where there is an alternate means of accomplishing the task that is switched in by a malfunction-sensing device when the primary system fails.

Reliability. The probability of a device's performing its purpose adequately for the period of time intended under the operating conditions encountered.

Reliability, Inherent. The probability that a device will perform in accordance with applicable specifications for a specified period of time under prescribed factory or laboratory test conditions.

Reliability, Intrinsic. The probability that a device will perform its specified function, determined on the basis of a statistical analysis of the failure rates and other characteristics of the parts and components which comprise the device.

Reliability, One-hundred-hour. The probability of a device or devices surviving a reliability test of 100-hr duration.

Reliability, Operational. The probability that a device of known inherent and use reliability will perform its mission for the required time when utilized in the manner and for the purpose intended.

Reliability, Parallel. The over-all reliability of a device whose parts are connected functionally in parallel.

Reliability, Series. The over-all reliability of a device whose parts are connected functionally in series.

Reliability, Use. The probability that a device of known inherent reliability will perform its prescribed mission for a specified time when subjected to field human engineering factors and maintenance practices.

Repair, Mean Time to. The average time required to repair a failure under the operating conditions encountered.

Replaceability. The ease with which a part of a component can be replaced without the necessity for extensive disassembly of adjacent hardware.

Risk. The probability of rendering the wrong decision based on pessimistic data or analysis.

Risk, Consumer's. The probability of acceptance of lots whose quality is of lot tolerance per cent defective.

Risk, Consumer's Reliability. The probability of acceptance of a sample of lot tolerance fractional reliability deviation as a result of an optimistic time sample.

Risk, Producer's. The risk in a sampling plan that a lot of acceptable quality will be rejected because of sampling variations which occasionally result in a pessimistic-looking sample's being drawn. This risk is usually of the order of magnitude of 5 to 10 per cent.

Risk, Producer's Reliability. The risk that a lot of an acceptable reliability level will be rejected by a reliability time sampling plan as a result of a pessimistic-looking grouping of random failures.

Rule, Product. The product rule states that if a number of parts or components are connected *functionally* in series, the reliability of the complex is the product of the probabilities of the respective parts.

Sample. A random selection of units from a lot for the purpose of evaluating the characteristics or acceptability of the lot. The sample may be either in terms of units or in terms of time.

Sample, Time. A number of hours of reliability testing conducted on one or more units for the purpose of evaluating the reliability of the lot within a certain confidence level.

Sampling, Acceptance. The art or science that describes procedures which outline the criteria and method of examining samples for the purpose of determining whether to accept or reject.

Screening. The process of performing 100 per cent inspection on rejected lots and removing the defective units therefrom.

Serviceability. Those properties of an equipment design that make it easy to service and repair while the equipment is in operation.

Significance Level. See *Level of Significance.*

Simulation. A set of test conditions designed to duplicate operating and usage environment and conditions as closely as possible.

Stress, Component. The stresses on component parts during testing or usage which affect the failure rate and hence the reliability of the parts. Voltage, power, temperature, and thermal environmental stress are included.

Subassembly. A subdivision of a component which consists of an assemblage of parts which may or may not be removable as a unit from the component.

Subgroup. One of a series of groups of observations obtained by subdividing a larger group of observations.

Subsample. One of a series of time samples which provide an interim measure of reliability during a reliability testing program but with a lesser degree of confidence than is possible for a total time sample.

Survival, Probability of. A numerical expression of reliability with the accepted nomenclature of P_s and a range of 0 to 1.0 indicating the likelihood that a device will still be functioning at some point in time.

Survivors. Those units which have not failed during a particular time interval of a reliability test.

Switchover. The technique of transferring a function from an element which has failed to a stand-by or redundant element for the purpose of improving reliability.

System. Any combination of complete operating equipment, components, accessories, or parts, interconnected or interrelated in such a manner as to perform a specific operational function or functions.

Test, Acceptance. Tests performed on all equipment prior to shipment to a customer in order to ensure a satisfactory product.

Test, Destructive. The intentional operation of a device to ultimate failure in order to discover design weaknesses which might adversely affect reliability.

Test, Individual Operational. A test performed on all equipment 100 per cent to ensure operation for major characteristics only. This type of test usually precedes an individual reliability test.

Test, Individual Reliability. This is usually a burn-in of 100 per cent of all equipment for a specified time. During the course of the test, changes in major characteristics as established in these individual tests are studied, and corrective action is taken.

Test, Material. A test of the basic materials from which parts and components are made.

Test, Sampling Reliability. The same type of test as the individual reliability test, except that it is longer in duration and is performed on samples selected at random from equipment successfully passing the requirements of the individual reliability tests.

Test, Sequential. A series of samples taken in steps, after each of which a decision to accept, select additional samples, or reject the lot or hypothesis is made, depending on the number of defects found with relation to the chart control limits.

Test, Special Reliability. A test usually run on samples of products passing all other types of testing. It is customarily of the life-test variety and involves testing of samples over long periods of time under particular environmental conditions.

Test, Statistical. A procedure which compares the observations of a test with a standard statistical hypothesis in order to make a decision about conformance or nonconformance.

Test, Type. A test which generally determines the quality of design of typical equipment. These tests are frequently destructive or time-consuming and involve the application of all the specified extremes of environment to a few equipments.

Testing, Attribute. A test procedure in which the items are classified according to qualitative rather than quantitative characteristics.

Testing, Marginal. A form of prediction testing, since it tests the system under conditions which are more severe than those encountered under normal usage. The purpose is to reveal an intermittent malfunction or potential failure by causing it to become a steady failure.

Testing, Nondestructive. Testing which in no manner affects the function or the life expectancy of the device undergoing test.

Testing, Variables. A test procedure in which the items are classified according to quantitative rather than qualitative characteristics.

Tests, Design Type (Qualification). Type tests performed on engineering samples.

Tests, Periodic Type (Verification). Type tests performed on samples of production equipment at periodic intervals to ensure that they comply at all times with standards previously established.

Tests, Preproduction. Type tests performed on first production equipment, using factory-type test equipment and methods to ensure that processes and methods are capable of producing a satisfactory product.

Tests, Reliability. Tests and analysis, in addition to other type tests, which are designed to evaluate the level and uniformity of reliability in a product or in parts or systems and the dependability or stability of this level with time and use under various environmental conditions.

Tests, Screening. Tests which are especially tailored for the purpose of isolating particular characteristics by testing them on a 100 per cent basis and rejecting or correcting any deficiency, when possible, by reworking and retesting.

Tests, Service. Tests which are conducted by the government in the field under actual service conditions in order to evaluate the suitability of the device for the purpose intended.

Time, Available. Usable time minus preventive maintenance time.

Time, Down. The time during which equipments are not capable of doing useful work because of malfunction. This does not include preventive maintenance time.

Time, Mission. The period of time in which a device must perform a specified mission under the operating conditions required.

Time, Operating. The time during which a device is energized by the application of power and is functioning in a satisfactory manner.

Time, Readiness. The time required to place an operable device in service so that it can perform its mission under required operating conditions.

Time, Repair. The time actually required to replace defective items and ensure satisfactory operation. This time does not include delays due to shortages of parts, transportation, or other nonrelated factors.

Time, Stand-by. Time in which the system is operable but has a partial application of power and is ready for full instantaneous use upon demand.

Time, Trouble-shooting. The time required to determine or isolate the cause of a system malfunction. It does not include the time required to replace or repair the units in which the fault occurred.

Time, Usable. The time during which equipments are capable of doing useful work. Part of this time may be used for preventive maintenance.

Time, Warm-up. The time from the application of power to an operable device to the instant when it is capable of functioning in its intended fashion.

Tolerance, Environmental. The capability of a system or portion thereof to operate within a specified environmental range.

Usage. A function of various factors, such as skill of maintenance personnel, operating and maintenance procedures, maintainability, storage, effects of shipping and handling, environment, etc.

Universe. See *Population.*

Unreliability. The probability of a device's failing in a given period of time. It is calculated as 1 minus the reliability of the device.

Use, End. The ultimate use for which the consumer intends to utilize the device.

Variance. The square of the standard deviation.

APPENDIX 2

SPECIAL TECHNIQUES

1. Concept of Working Acceptance Number. An approximate equation for determining the working acceptance number A_w for any reliability sampling plan for a stipulated failure rate r and confidence level Z is

$$A_w = rT + Z\sqrt{rT} \tag{1}$$

or, since rT equals d, we can also write

$$A_w = d + Z\sqrt{d} \tag{1a}$$

where A_w = acceptance number (working)
T = sample size in hours
Z = a variable measured from the universe mean in standard deviations which determines area under normal curve from $-\infty$ to Z (use Z_α and Z_β as in Sec. 2)
r = failure rate in failures per hour
d = expected number of failures

and the equation for finding the acceptance number A is

$$A = A_w - 0.5 \tag{2}$$

The above findings were derived by assuming a normal distribution as shown in Fig. A-1, despite the fact that it is known that the distribution is

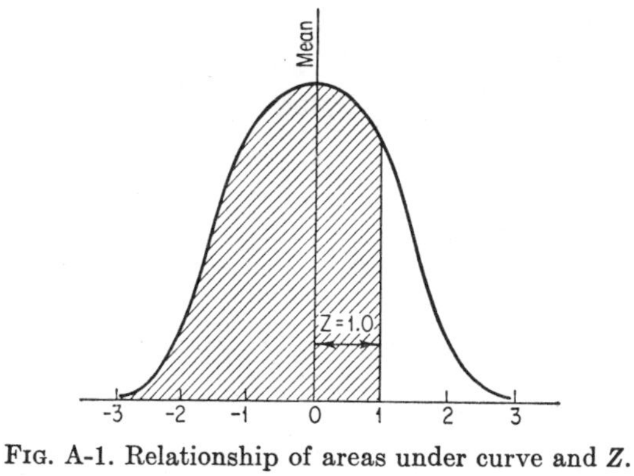

FIG. A-1. Relationship of areas under curve and Z.

not always exactly normal. However, by use of the concept of the working acceptance number A_w and subtraction of 0.5 from it, a good approximation of the actual acceptance number A is obtained.

At this point, it is important to realize the reason for using A and A_w. The acceptance number A must always be a whole number, since it represents the number of failures associated with a specified confidence level; however A_w can be a decimal fraction, since it is used only in interim calculations to arrive at more accurate conclusions than are possible if the whole number A is used. This is why we refer to A_w as the working acceptance number.

2. Derivation of Value of rT, Z, r, and T. If Eq. (1) is solved for rT, the result is

$$rT = \frac{2A_w + Z^2 \pm \sqrt{4A_wZ^2 + Z^4}}{2} \tag{3}$$

The derivation of this equation is as follows:

From Eq. (1) $\quad A_w = rT + Z\sqrt{rT}$

Transposing $\quad A_w - rT = Z\sqrt{rT}$

Squaring $\quad (A_w - rT)^2 = Z^2(rT) = A_w{}^2 - 2A_wrT + (rT)^2$

Simplifying $\quad (rT)^2 - (Z^2 + 2A_w)rT + A_w{}^2 = 0$

Let

$$x = rT$$
$$b = -(Z^2 + 2A_w)$$
$$c = A_w{}^2$$

Rewriting we get

$$x^2 + bx + c = 0$$

But the above is a quadratic equation whose general solution is given in any elementary book on algebra as

$$x = \frac{-b \pm \sqrt{b^2 - 4ac}}{2a}$$

in which case c is merely an algebraic constant used to facilitate the solution of the equation and does not mean number of defects.

Substituting values for x, b, and c ($a = 1$),

$$rT = \frac{Z^2 + 2A_w \pm \sqrt{(Z^2 + 2A_w)^2 - 4(1)A_w{}^2}}{2}$$

Simplifying $\quad rT = \dfrac{2A_w + Z^2 \pm \sqrt{4A_wZ^2 + Z^4}}{2} \qquad$ QED

For Z_α, which applies to R_α, the equation is written as

$$rT = \frac{2A_w + Z_\alpha{}^2 + \sqrt{4A_wZ_\alpha{}^2 + Z_\alpha{}^4}}{2}$$

For Z_β, which applies to R_β, the equation is

$$rT = \frac{2A_w + Z_\beta^2 - \sqrt{4A_wZ_\beta^2 + Z_\beta^4}}{2}$$

Other formulas can be derived from Eqs. (1) and (3), such as

$$Z = \frac{A_w - rT}{\sqrt{rT}} \tag{4}$$

$$r = \frac{2A_w + Z_\alpha^2 + \sqrt{4A_wZ_\alpha^2 + Z_\alpha^4}}{2T} \tag{5}$$

$$T = \frac{2A_w + Z_\alpha^2 - \sqrt{4A_wZ_\alpha^2 + Z_\alpha^4}}{2r} \tag{6}$$

Equations similar to (5) and (6) may be written for Z_β.

3. Derivation of Value of T for Specified R_α and R_β. When a sampling plan is desired that stipulates a particular producer's reliability risk R_α and consumer's reliability risk R_β, it may be derived by the use of Eq. (7).

$$T = \left[\frac{Z_\alpha\sqrt{r_1} - Z_\beta\sqrt{r_2}}{r_2 - r_1}\right]^2 \tag{7}$$

where Z_α = value of Z for specified R_α
Z_β = value of Z for specified R_β
r_1 = acceptable reliability level (ARL)
r_2 = lot tolerance fraction reliability deviation (LTFRD)

Equation (7) is derived by equating the formulas for determining the working acceptance numbers for R_α and R_β (since they must be equal) and solving for T. Thus

$$r_1T + Z_\alpha\sqrt{r_1T} = r_2T + Z_\beta\sqrt{r_2T}$$

$$r_2T - r_1T = Z_\alpha\sqrt{r_1T} - Z_\beta\sqrt{r_2T}$$

$$T(r_2 - r_1) = \sqrt{T}\,(Z_\alpha\sqrt{r_1} - Z_\beta\sqrt{r_2})$$

$$T = \frac{\sqrt{T}\,(Z_\alpha\sqrt{r_1} - Z_\beta\sqrt{r_2})}{r_2 - r_1}$$

$$\frac{T}{\sqrt{T}} = \frac{Z_\alpha\sqrt{r_1} - Z_\beta\sqrt{r_2}}{r_2 - r_1}$$

$$\sqrt{T} = \frac{Z_\alpha\sqrt{r_1} - Z_\beta\sqrt{r_2}}{r_2 - r_1}$$

$$T^* = \left[\frac{Z_\alpha\sqrt{r_1} - Z_\beta\sqrt{r_2}}{r_2 - r_1}\right]^2 \qquad \text{QED}$$

* When substituting in this equation, it should be remembered that the sign for Z_α is plus and for Z_β is minus.

Equation (7) can be rewritten as shown in Eq. (8) by introducing k, which is the ratio of r_2 and r_1. The advantage of using Eq. (8) is that it gives general solutions for any ratio of r_2 and r_1. The derivation of Eq. (8) follows:

Rewrite Eq. (7),

$$T = \left[\frac{Z_\alpha \sqrt{r_1} - Z_\beta \sqrt{r_2}}{r_2 - r_1}\right]^2$$

Let $r_2 = kr_1$ and substitute

$$T = \left[\frac{Z_\alpha \sqrt{r_1} - Z_\beta \sqrt{kr_1}}{kr_1 - r_1}\right]^2$$

$$T = \left[\frac{\sqrt{r_1}\,(Z_\alpha - Z_\beta \sqrt{k})}{r_1(k - 1)}\right]^2$$

$$T = \left[\frac{Z_\alpha - Z_\beta \sqrt{k}}{\sqrt{r_1}\,(k - 1)}\right]^2 \tag{8}$$

If it is desired to obtain an expression for r_1T in terms of Z_α, Z_β, and k, multiply each side of Eq. (8) by r_1, and Eq. (9) is the result.

$$r_1T = \left[\frac{Z_\alpha - Z_\beta \sqrt{k}}{k - 1}\right]^2 \tag{9}$$

Equations (8) and (9) have a decided advantage over Eq. (7) because calculations are more rapid, since less extracting of square roots is required and in all instances extraction of decimal square roots is seldom required.

4. Derivation of Kappa Square K². Since $r = 1/m$, Eq. (8) may be rewritten as

$$T = m\left[\frac{Z_\alpha - Z_\beta \sqrt{k}}{k - 1}\right]^2 \tag{10}$$

If the statistic in brackets is called K, then the squared bracketed expression is K^2, and the equation may be written as

$$T = mK^2 \tag{11}$$

Note 1. Whenever any of the preceding equations which have been described in this appendix are used, care should be exercised to use the proper plus or minus sign for Z. All values of Z to the right of the mean are plus and labeled Z_α, while all values to the left of the mean are minus and called Z_β.

Note 2. Relationship of Failure Rate and Per Cent Defective. The same equations presented in the preceding part of this appendix can be used to determine AQL attributes single sampling plans by substitution of p for r and n for T. In this case, p represents defects per unit, and n the number of units in a sample. The designation of subscripts and primes or bars is a function of whether standard per cent defective p' or process average $\bar{p}$ is intended.

5. Derivation of Poisson Probability Ratio Equations for the Sequential *f* Chart. The RO-C curve of Fig. A-2 indicates the relationship of the various reliability risks, which are defined as follows:

R_{α_1} = probability of rejecting lot of r_1 reliability
R_{α_2} = probability of rejecting lot of r_2 reliability
R_{β_1} = probability of accepting lot of r_1 reliability
R_{β_2} = probability of accepting lot of r_2 reliability

Moreover, the following relationships govern:

$$\text{ARL} = r_1 \qquad R_{\beta_1} = 1 - R_{\alpha_1}$$
$$\text{LTFRD} = r_2 \qquad R_{\beta_2} = 1 - R_{\alpha_2}$$

The probability ratio (PR) is defined as

$$\text{PR} = \frac{\text{probability of } f \text{ failures when sample is of } r_2 \text{ reliability}}{\text{probability of } f \text{ failures when sample is of } r_1 \text{ reliability}}$$

The Poisson probability of f failures when the sample is of r_2 reliability is $(r_2T)^f e^{-r_2T}/f!$ The Poisson probability of f failures when the lot is of r_1

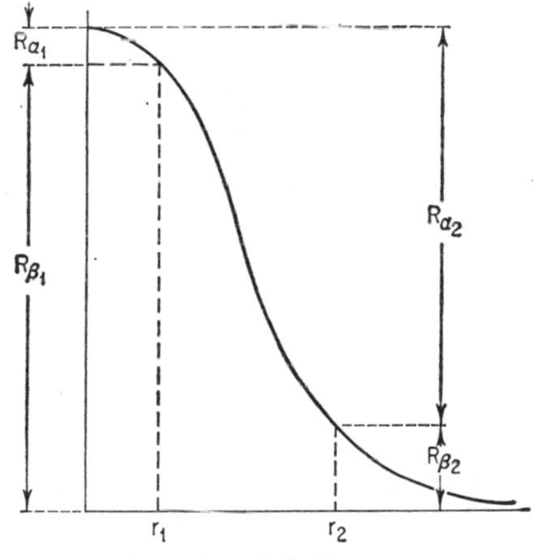

FIG. A-2. RO-C curve.

reliability is $(r_1T)^f e^{-r_1T}/f!$ Therefore, by substitution, the probability ratio is

$$\text{PR} = \frac{(r_2T)^f e^{-r_2T}}{(r_1T)^f e^{-r_1T}} = \left(\frac{r_2}{r_1}\right)^f e^{-T(r_2-r_1)}$$

Let $k = r_2/r_1$ and substitute in above, and we get

$$\text{PR} = k^f e^{-r_1T(k-1)}$$

To determine a point on the acceptance line of a sequential f chart for any sample T, make PR $= R_{\beta_2}/R_{\beta_1}$ and substitute A_T for f in Eq. (12). This substitution is done to show that A_T is the acceptance number applicable to time T. Equation (12) then becomes the equation of the acceptance line and is written as

$$\frac{R_{\beta_2}}{R_{\beta_1}} = k^{A_T}e^{-r_1T(k-1)} \qquad \text{(acceptance-line equation)} \tag{12}$$

If we make $r_1T = d$, the final form of the equation is

$$\frac{R_{\beta_2}}{R_{\beta_1}} = k^{A_T}e^{-d(k-1)} \tag{13}$$

To determine a point on the rejection line for any time T, we make PR $= \dfrac{R_{\alpha_2}}{R_{\alpha_1}}$ and substitute the rejection number R_T for f in Eq. (12). The rejection-line equation then becomes

$$\frac{R_{\alpha_2}}{R_{\alpha_1}} = k^{R_T}e^{-d(k-1)} \tag{14}$$

6. Derivation of Equations for Maintainability and Availability.

a. Derivation of Maintainability M. The average number of maintenance actions which can be completed in time t is

$$b = \mu t \tag{15}$$

or

$$b = t/\phi \tag{15a}$$

Since the probability of performing 0, 1, 2, etc., maintenance actions in time t is described by the Poisson distribution, then the probability of completing zero maintenance actions in an increment of time t is $e^{-\mu t}$, and the probability of completing one or more maintenance actions is $1 - e^{-\mu t}$. This is the expression for maintainability M.

$$M = 1 - e^{-\mu t} \tag{16}$$

which also may be written as

$$M = 1 - e^{-t/\phi} \tag{16a}$$

b. Derivation of Maintainability Increment M_Δ. The maintainability increment is the product of the probability of one or more failures P_f in time T and the maintainability M for a time constraint t.

$$\begin{aligned} M_\Delta &= P_fM \\ &= (1 - e^{-rT})(1 - e^{-\mu t}) \\ &= 1 - e^{-rT} - e^{-\mu t} + e^{-rT}e^{-\mu t} \\ &= 1 - e^{-rT} - e^{-\mu t}(1 - e^{-rT}) \end{aligned} \tag{17}$$

Thus we see that the maintainability increment is the percentage of failed units which can be restored to operation within time t.

c. Equipment Availability. This approximate derivation is based on the following assumptions:

1. There is a maintenance crew available to repair each failure or perform a maintenance action when required.
2. The maintenance time is small and therefore is ignored when T is measured.
3. Time to failure and repair are not mutually exclusive.
4. The device is not altered after maintenance, and therefore its failure rate is the same as before.
5. The distributions of failure and repair are exponential. Hence the probability of making zero repairs is the first term of the Poisson and equals $G = e^{-\mu t}$. Likewise, the probability of zero failures is also the first term of the Poisson and equals $P_0 = e^{-rT}$. Since this is equivalent to survival, it is also usually labeled P_s. Let $P_1 + P_2 + P_3 + \cdots + P_f$ represent the sum of the probabilities of exactly $1, 2, 3, + \cdots + f$ failures in time T which are subjected to a maintenance action as soon as they occur.

The probability of not completing the maintenance action in time t is

$$GP_1 + GP_2 + GP_3 + \cdots + GP_f = G(P_1 + P_2 + P_3 + \cdots + P_f)$$

But $P_1 + P_2 + P_3 + \cdots + P_f$ is the sum of the probability of one or more failures in time T, which can be expressed by summing the terms of the Poisson distribution as follows:

$$\sum_{i=1}^{f} \frac{(rT)^i e^{-rT}}{i!} = \sum (P_1 + P_2 + P_3 + \cdots + P_f)$$

where the product rT equals the expectation, or average number of failures, and i is the number of consecutive failures.

Therefore it follows that the probability of not being able to effect the maintenance action is the unavailability U.

$$U = G \sum_{i=1}^{f} \frac{(rT)^i}{i!} e^{-rT}$$

But since the probability of more than one failure in time T equals the total probability minus the probability of exactly zero failures e^{-rT}, we have

$$U = G \sum_{i=1}^{f} \frac{(rT)^i}{i!} e^{-rT} = G \left[\sum_{i=0}^{f} \frac{(rT)^i}{i!} e^{-rT} - e^{-rT} \right]$$

and unavailability is

$$U = Ge^{-rT}\left[\sum_{i=0}^{f} \frac{(rT)^i}{i!} - 1\right]$$
$$= Ge^{-rT}(e^{rT} - 1)$$

And since $G = e^{-\mu t}$,

$$U = e^{-\mu t}(1 - e^{-rT})$$

But $\Lambda_E = 1 - U$, and therefore the availability is

$$\Lambda_E = 1 - e^{-\mu t}(1 - e^{-rT}) \tag{18}$$

With some thought we could have written Eq. (18) intuitively, since the sum of the probabilities of f consecutive failures from zero to infinity must equal unity. Therefore, if we subtract the probability of zero failures e^{-rT} from the total, the remainder must be the probability of one or more failures, which is $1 - e^{-rT}$. Multiplying $1 - e^{-rT}$ by the probability of zero repair $e^{-\mu t}$, we get the unavailability $U = e^{-\mu t}(1 - e^{-rT})$. Since $1 - U = \Lambda$, the result is the same as Eq. (18).

d. Mission Availability. Let us use the same assumptions as were made for equipment availability; remember, however, that we are now interested in finding the probability of success of the mission, which means the probability of repairing all of n consecutive failures within the allowable time constraint t.

The probability of i consecutive failures in time T is given by the Poisson distribution, viz.:

$$\frac{(rT)^i e^{-rT}}{i!}$$

Therefore the probability of repairing all consecutive failures is

$$\sum_{i=1}^{\infty} \frac{(rT)^i e^{-rT}(1 - G)^i}{i!} = e^{-rT} \sum_{i=1}^{\infty} \left[\frac{rT(1 - G)}{i!}\right]^i$$
$$= e^{-rT}e^{+rT(1-G)}$$
$$= e^{-rTG}$$

But $$G = e^{-\mu t}$$

Therefore $$\Lambda_M = \exp(-rTe^{-\mu t}) \tag{19}$$

e. Derivation of the General Redundancy Equation. An exact classical derivation of the general redundancy equation is beyond the scope of this text. Therefore a practical, simple empirical method is presented which is accurate to better than the fourth decimal place. (For the classical derivation for a duplex system see article by B. Epstein, "Reliability of Some Two Unit Redundant Systems.")

The effective mean time between failures of a system of n redundant elements can be thought of as consisting of two components: the reliability component P_c and the maintainability component M_c. The sum of the two components is the effective mean time between failures.

We call the first component the reliability component because it represents the effective mean time between failures of a system of n redundant elements when no maintenance is practiced. We label the second component the maintainability component because it depends upon repair or maintenance.

Thus, the effective mean time between failures m_n for n elements in redundance is

$$m_n = P_c + M_c$$

The reliability component is the product of $1/r$ and the sum of the harmonic series, viz.:

$$P_c = \frac{1}{r}\left[1 + \frac{1}{2} + \frac{1}{3} + \cdots + \frac{1}{n}\right]$$
$$= \frac{1}{r}\sum_{i=1}^{n}\frac{1}{i}$$

which indicates the summation of terms.

The maintainability component is found as follows:

$$r\phi = \text{average number of failures in time } \phi$$

But $$\phi = \text{mean time to repair}$$

and $$\mu = \frac{1}{\phi} = \text{maintenance-action rate}$$

$$\frac{r}{\mu} = r\phi$$

Assume a system of n redundant elements, in which ϕ is relatively small as compared with T. In this case, the average number of failures per element for a small time ϕ is $r\phi$. This is also the probability of failure. Therefore, the probability of all n units' failing in the same ϕ interval of time (overlapping failures) is $(r\phi)^n = (r/\mu)^n$. This constitutes a failure of a redundant system of n elements. But since r/μ is approximately equal to the down-time ratio (DTR), then the nth power of DTR is the down-time ratio of a system of n redundant elements.

It can be shown that the mean time to repair a redundant system of n elements ϕ_n is equal to $1/n$ times the mean time to repair each redundant element of the system ϕ, provided that all elements are similar. Thus

$$\phi_n = \frac{\phi}{n}$$

Therefore, the maintenance-action rate of a system of n redundant elements μ_n (since $\mu = 1/\phi$) is

$$\frac{1}{\mu_n} = \frac{1}{n\mu}$$

or

$$\mu_n = n\mu$$

In other words the maintenance-action rate of a system of n redundant elements equals n times the maintenance-action rate of the individual elements.

It can also be shown that $(\mathrm{DTR})/\phi = (\mathrm{DTR})\mu = r$. Therefore, since the DTR of the system is approximately $(r/\mu)^n$, and $\mu_n = n\mu$, this product is the contribution of the maintenance component to the equivalent failure rate, viz.:

$$\left(\frac{r}{\mu}\right)^n n\mu = \frac{nr^n}{\mu^{n-1}}$$

and its reciprocal represents the equivalent mean time between failures, or $M_c = \mu^{n-1}/nr^n$.

Adding, we get

$$m_n = \frac{1}{r}\left(\sum_{i=1}^{n} \frac{1}{i}\right) + \frac{\mu^{n-1}}{nr^n}$$

and since the equivalent failure rate for n redundant elements is $r_n = 1/m_n$, we get

$$r_n = \frac{nr^n}{n\left(\sum_{i=1}^{n} \frac{1}{i}\right) r^{n-1} + \mu^{n-1}}$$

But the mission availability is $\Lambda_M = e^{-r_nT}$. Substituting,

$$\Lambda_M = \exp\left[\frac{-nr^nT}{n\left(\sum_{i=1}^{n} \frac{1}{i}\right) r^{n-1} + \mu^{n-1}}\right] \tag{20}$$

If we institute a maintenance time constraint t, the resulting mission availability equation is

$$\Lambda_M = \exp\left[\frac{-nr^nTe^{-n\mu t}}{n\left(\sum_{i=1}^{n} \frac{1}{i}\right) r^{n-1} + \mu^{n-1}}\right] \tag{21}$$

For a simplex system the maintainability component μ^{n-1}/nr^n becomes zero (since if repairs are impossible, $\mu = 0$), and therefore only the reliability component $1/r\left(\sum_{i=1}^{n} 1/i\right)$ exists. Therefore, when $n = 1$ (simplex), $m = 1/r$, and Eq. (20) reduces to $P_s = e^{-rT}$, which is the customary form.

The equipment availability equation is

$$\Lambda_E = 1 - e^{-n\mu t}(1 - e^{-r_n T})$$

Substituting the value for r_n we get

$$\Lambda_E = 1 - e^{-n\mu t}\left\{1 - \exp\left[\frac{-nr^nT}{n\left(\sum_{i=1}^{n}\frac{1}{i}\right)r^{n-1} + \mu^{n-1}}\right]\right\} \tag{22}$$

APPENDIX 3

RELIABILITY CONTROL AND ACCEPTANCE PROCEDURE[1]

10. Scope

This procedure covers the methods, techniques, and criteria which will be employed to demonstrate and assure that the contractually specified reliability requirements have been achieved. Its provisions are applicable to maintainable ground equipment as well as to nonmaintainable devices.

When approved by the procuring activity, this procedure may be used to implement the requirements of Paragraph 4.4 of MIL-R-26474 (USAF).

20. Reliability Sampling Methods

Two types of time samples are permissible in the reliability testing provision of this procedure, the time sample and subsample.

20.1 The Time Sample. The purpose of the time sample is to ensure that the contractually specified reliability has been achieved within the required confidence level. The method of obtaining this level of confidence is to test selected units during the production cycle for a period of time as determined in Paragraph 20.1.1. The time sample so computed is considered the minimum reliability test time permissible to provide the specified degree of confidence in the results obtained. The subsample provides a lesser degree of confidence.

20.1.1 Determination of the Time Sample. Figure A-1[2] of this specification entitled "Values of Kappa Square for Various Confidence Levels" shall be used to determine size of the time sample in accordance with the following procedure:

On the abscissa, locate the value of k which is specified by the procuring activity. This k value simply represents the ratio of the Lot Tolerance Fractional Reliability Deviation (LTFRD), r_2, to the Acceptable Reliability Level (ARL), r_1. The value of Kappa Square (K^2) can then be determined for any specified confidence level. This is done by locating the point of intersection of the required confidence level curve and the vertical line representing k. By following the point in a horizontal line parallel to the abscissa until it intersects the ordinate we read the value of K^2.

The size of the time sample to provide the required confidence level may then

[1] Abstract of Appendix A of International Electric Corp. Production Reliability Specification IEC-60020B. See Chap. 14 for discussion of this abstract.

[2] See Fig. 11-1 of this text in lieu of Fig. A-1, which has been omitted from this abstract to avoid repetition.

be determined as the ratio of K^2 and the ARL. (Thus: T = time sample, i.e., number of hours of reliability testing)

$$T = \frac{K^2}{ARL} \tag{1}$$

or since $1/ARL = m$, we can express T as

$$T = mK^2 \tag{1a}$$

This is the time sample required to assure a specified confidence level γ for a given k. Unless otherwise specified, a two-tailed confidence level, $\gamma = 90\%$, will be used. This will provide for a producer's reliability risk R_α of 5% and a consumer's reliability risk R_β equal to 5%.

During production the aggregate number of reliability test hours shall not be less than one time sample as calculated from Equation (1). Unless otherwise specified, at least one time sample of reliability testing will be provided per month.

20.1.2 Acceptance and Rejection Numbers. The acceptance number A for the time sample as calculated from Equation (1) is found by using Equation (2), and the rejection number R is found by the use of Equation (3).

$$A = K^2 + Z\sqrt{K^2} - 0.5 \tag{2}$$
$$R = K^2 + Z\sqrt{K^2} + 0.5 \tag{3}$$

The values of Z for various confidence levels are shown in Table A-1. If the result obtained for A and R is a decimal, the nearest whole number will be used.

TABLE A-1. VALUES OF Z FOR VARIOUS CONFIDENCE LEVELS γ EXPRESSED IN PER CENT

Z	γ	Z	γ
0.68	50	1.64	90
0.84	60	1.96	95
1.04	70	2.24	97.5
1.28	80	3.0	99.73
1.44	85	3.54	99.96

20.2 The Sub-time Sample. A sub-sample is a fraction of the time sample and may be used to reduce the required number of test hours when maturity of design and effective quality control has been demonstrated. This is a form of reduced sampling which is instituted to reduce costs, and under specific conditions permits shipment of individual devices prior to the actual completion of the required number of hours constituting the time sample.

Since the use of a sub-sample increases the producer's and consumer's reliability risks, it should only be used in accordance with the provisions specified in Sections 20.4 and 20.5 of this procedure.

20.2.1 Determination of the Sub-time Sample. The size of the sub-sample and acceptance and rejection numbers for specific values of k, Z and γ may be computed as explained in Paragraph 20.1.1 and 20.1.2 of this procedure. Thus the size of the sub-sample will vary for differing values of γ. However, a simpler method may be used which uses m hours as the sub-sample size. This is the method we will describe here. The simplified method of calculating a sub-sample is to use the specified value of m as the sub-sample size and a rejection number of either 1 or 2. A rejection number of 1 is used when prior reliability test data indicates marginal reliability compliance, and a rejection number of 2 is used when prior data indicates a high degree of reliability achievement which justifies this reduced sampling plan. The specific procedures to be used will be in accordance with Paragraph 20.4 of this procedure.

20.3 Selection of Units to Be Tested. During quantity production, the physical number of units to be tested per time sample should never be less than the value of K^2 as determined in Paragraph 20.1.1. Thus, if $K^2 = 5$ for a value of $\gamma = 90\%$ and $k = 3$, a minimum of 5 units will be tested for an aggregate test time equal to the number of hours required for a time sample. However, this does not preclude the testing of more than K^2 units provided that the aggregate test time is no less than the number of hours necessary to constitute a time sample.

In selecting a sample, the procuring activity shall select units at random in such a manner as to assure a true cross-section of the reliability of the product that is undergoing test. No unit will be selected for test which has not been previously accepted on the basis of normal production acceptance testing. In addition, prior to qualifying for reliability test, the minimum test time on each unit must be such as to assure that it has been tested to a point exceeding its infant mortality period. All units on order, whether or not they are selected as part of the time sample, shall have been tested for at least this minimum time prior to shipment.

When it is apparent that the provisions of this procedure will result in requiring an exceedingly large time sample which will delay shipment because of the effects of low production and a high degree of specified reliability, the matter will be referred to the procuring activity for resolution.

20.4 Preparation of the Sub-sample f Chart. The Sub-sample f Chart will be used to record all failures which occur during the reliability test.

20.4.1 Plot of Failures. Figure A-2[1] demonstrates a typical Sub-sample f Chart for an ARL of 0.005 failures per hour, which corresponds to a value of $m = 1/\text{ARL} = 200$ hours. Assuming a k value of 3, the LTFRD is 0.015 f/hour.

Using the simplified approach, a sub-sample equal to $m = 200$ hours was selected. The rejection number for the sub-sample is $R = 2$, corresponding to an acceptance number $A = 1$, which provides a value of $R_\beta = 20\%$. This is indicated on the chart as $R_{200(0.20)}$,* showing that for a sub-sample of 200 hours there is a 20% risk of accepting a product equal to or exceeding the value of the LTFRD of 0.015 f/hr.

For a cumulative sample of $2m$ or $2(200) = 400$ hours, since $K^2 = T/m$, the corresponding value of $K^2 = 400/200 = 2$ and the intersection of the lines for $K^2 = 2$ and $k = 3$ (see Figure A-1) shows a two-tailed confidence level of 70%; therefore, one tail is 15%, which corresponds to an $R_\beta = 15\%$. The rejection number for this cumulative sample for a two-tailed confidence level, $\gamma = 70\%$ ($Z = 1.04$ from Table A-1), is

$$R = K^2 + Z\sqrt{K^2} + 0.5 = 2 + 1.04\sqrt{2} + 0.5 = 3$$

[1] See Fig. 11-3 in text in lieu of Fig. A-2, which is sufficiently similar to illustrate methods of plotting the chart.

* In this case the consumer's risk β is shown as a subscript of R instead of the confidence level, as shown in Fig. 11-3. Either method is permissible.

The rejection numbers for the other multiples of m which are shown on the chart are calculated in a similar manner.

The number of failures per sub-sample are shown on the chart. The cumulative rejection number corresponding to the cumulative sample is also shown. The cumulative sample is the sum of the sub-samples. The actual number of cumulative failures are plotted on the lower part of the box under the cumulative rejection number.

The body of the chart also lists the reason for failure and the failure report number which summarizes all details with respect to the failure in question. For example, Failure Code No. 4 is shown as reported on Failure Report No. 4 which appears in the body of the chart directly under the plotted point indicating the failure. An advantage of this type of presentation is that the frequency of any particular type of failure is obvious from an examination of the chart.

20.4.2 Plot of Mean-Time-to-Repair. The lower part of the Sub-sample f Chart is used to plot the mean-time-to-repair for sub-groups of four failures. For example, the time to repair the first four failures is respectively shown as 24, 20, 16 and 8 minutes. This gives a mean-time-to-repair equal to 17 minutes for the first sub-group of four. This is indicated with a triangular point following the 4 first plotted failures. Subsequent groups of 4 are also shown by succeeding triangular points which were determined in a similar manner.

20.4.3 Control Limits for Mean-Time-to-Repair. The upper and lower control limits for the mean-time-to-repair ϕ can be calculated by the use of the chi-square statistic χ^2 in accordance with accepted statistical practices. However, a simpler and less involved procedure which gives comparable results is to use the Poisson curves of Figure 3.[1] The procedure is to find the curve on Figure 3[1] corresponding to one less than the number of repairs. Thus if we were interested in finding the upper and lower control limits for sub-groups of four failures, we would locate the curve $f = 3$. Having located the curve we would then find the value of rT corresponding to each limit of the specified confidence level. If we divided the total repair time by each of these values of rT respectively, we would get the upper and lower confidence level.

For example, suppose we wanted to find the upper and lower control limits for sub-groups of 4 repairs, of the specified value of $\phi = 10$ minutes, and the confidence level required was 90%; we would proceed as follows:

On Figure 3 locate the line $f = 3$. Find the point of intersection of $f = 3$ and the horizontal line for probability of occurrence of 95%. This gives a value of $rT = 1.35$. Similarly the intersection of $f = 3$ with the horizontal line for 5% gives a value of rT equal to 7.75.

The expected repair time is the number of failures multiplied by the specified mean-time-to-repair (i.e., $4 \times 10 = 40$). If we divided 40 by the respective values of 1.35 and 7.75, the result is the upper and lower control limits respectively.

Thus:

$$\text{UCL} = \frac{4 \times 10}{1.35} = 29.6 \text{ minutes}$$

$$\text{LCL} = \frac{4 \times 10}{7.75} = 5.16 \text{ minutes}$$

[1] See Fig. 6-1 of this text in lieu of Fig. 3, which has been omitted from this abstract to avoid repetition.

If the mean-time-to-repair for any sub-groups of 4 failures exceeds the UCL, the reasons for this occurrence should be thoroughly investigated since the probability of this occurring is only 5% or one time out of twenty.

20.5 Longevity Testing

20.5.1 In those instances where no prior knowledge exists about the length of time of the infant mortality period or of the behavior of the failure rate as a function of time, this will be determined by longevity testing of the first article or articles as required. The longevity test time will not be less than one time sample as required to assure a confidence level of 90% as described in Paragraph 20.1.1 but when practical should be of sufficient duration to provide an estimate of the life characteristic of the equipment. As a minimum requirement the infant and operating periods should be clearly demonstrated if it is not feasible because of time constraints to continue testing until the wear-out stage becomes evident.

20.5.2 If the number of failures during longevity testing does not exceed the rejection number calculated in accordance with Paragraph 20.1.2, the unit under test will be adjudged as conforming and may be shipped after specified refurbishing, or held for further testing if additional data is considered a necessity.

20.5.3 Upon the satisfactory conclusion of longevity testing, the reliability test time may be reduced to the appropriate sub-sample as specified in Paragraph 20.2.1, and the accept-reject criteria will be in accordance with the provisions of Paragraph 20.6.

20.6 Accept-Reject Criteria

20.6.1 If the number of failures equals or exceeds the rejection number for an individual sub-sample or cumulative sample, the device or devices undergoing test will be considered to be non-conforming. When this occurs the contractor shall conduct an investigation and analysis to determine the causes of failure and take whatever appropriate action is necessary to assure the elimination of the causes of failure. The contractor shall also furnish the IEC Product Assurance Engineer whatever objective evidence is necessary to satisfy IEC that the action he has taken will result in the elimination of failure without deleterious effects upon other parameters and that the end result will be an improved product of greater reliability.

20.6.2 When the IEC Product Assurance Engineer is satisfied that the causes of failure have been determined and that the proposed corrective action is satisfactory, he will eliminate those failures from consideration and accept or reject on the basis of the remaining chargeable failures.

20.6.3 Any costs which may accrue as a result of the contractor's investigation of failures per 20.6.1 shall not be chargeable to the reliability test program. However, upon request, all data, results and conclusions resulting from the contractor's activity as required by 20.6.1 shall be made available to the IEC Product Assurance Engineer for his information, guidance and necessary action.

20.6.4 After satisfactory evidence that maturity of design has been achieved and that a satisfactory level of reliability has been achieved, shipments may be made on the basis of sub-sample testing. However, if an excessive number of failures for a series of sub-samples, although less than the rejection number, are in evidence, shipments may be stopped at the discretion of the IEC Product Assurance Engineer until the cumulative sample equals the time sample, after which a decision to accept or reject will be made on the basis of criteria applicable to the time sample.

20.6.5 When, in the best interests of the procuring activity, time does not permit longevity testing and a demonstration that a maturity of design exists, shipments may be made on the basis of sub-sample testing on a tightened basis. In this case, for a sub-sample of m hours the acceptance number A will equal zero, i.e., rejection number $R = 1$. This will provide a value of $R_\beta = 5\%$ and $R_\alpha = 63\%$ when $k = 3$.

NOTICE: When Government drawings, specifications, or other data are used for any purpose other than in connection with a definitely related Government procurement operation, the United States Government thereby incurs no responsibility nor any obligation whatsoever; and the fact that the Government may have formulated, furnished, or in any way supplied the said drawings, specifications, or other data is not to be regarded by implication or otherwise as in any manner licensing the holder or any other person or corporation, or conveying any rights or permission to manufacture, use, or sell any patented invention that may in any way be related thereto.

APPENDIX 4

QUALITY CONTROL SYSTEM REQUIREMENTS[1]

1. Scope

1.1 Scope. This specification requires the establishment of a quality control system by the contractor to assure that supplies or services meet the quality standards established by the contract. This system, including procedures, is subject to surveillance by the Government representative. The procedures shall be designed by the contractor. The contractor's procedures used to implement the requirements of this specification shall be subject to the disapproval of the Government representative.

3. Requirements

3.1 Outline. The contractor shall maintain an effective and economical quality control system planned and developed in conjunction with other planning functions. The system, including procedures, shall be adjusted to suit the type and phase (research, development, production) of procurement. The system shall be based upon consideration of the complexity of product design, quantity under procurement, interchangeability and reliability requirements, and manufacturing techniques. The system shall assure that adequate control of quality is maintained throughout all areas of contract performance, including, as applicable, the receipt, identification, stocking and issue of material, and the entire process of manufacture, packaging, shipping, storage and maintenance. All supplies or services under the contract, whether manufactured or performed within the contractor's plant or at any other source, shall be subject to control at such points as necessary to assure conformance to contractual requirements. The system shall provide for the prevention and ready detection of discrepancies and for timely and positive corrective action. The contractor shall make objective evidence of quality performance readily available to the Government representative.

3.2 Description of Procedures. The contractor shall provide and maintain a description of procedures for control of quality. To the extent necessary, written inspection and test procedures shall be prepared to supplement the applicable drawings and specifications, and shall make clear the manner in which such inspec-

[1] Abstract of "Quality Control System Requirements" (MIL-Q-9858, April, 1959). Number references are those specified in the referenced specification; only those paragraphs which are considered most important have been included. See Chap. 15 for discussion of this abstract.

tion and test procedures are to be used. This description may be a compilation of existing shop travelers, routing cards, inspection methods sheets, test procedures, route sheets, or other documents normally used by the contractor to define inspection operations. The description of the quality control system and all applicable inspection and test procedures shall be available to the Government representative.

3.3 Drawing and Change Control. A procedure shall be maintained by the contractor to assure that the latest applicable drawing, technical requirement and contract change information will be available at the time and place of contractor inspection. Concurrently with the effectivity of revised drawings or changes, the contractor's drawing and change control shall assure that obsolete information is removed from all points of issue and use. All changes shall be processed in a manner which will assure accomplishment on the affected supplies at the specified effective points. The contractor shall maintain a record of the point of effectivity of changes. This record shall be available for ready reference by the Government representative.

3.4 Measuring and Testing Equipment. Unless otherwise specified in the contract, the contractor shall provide and maintain gages and other measuring and testing devices necessary to assure that supplies conform to contract requirements. These devices shall be calibrated against measurement standards or designated measuring equipment at established periods to assure continued accuracy. The contractor shall prepare and maintain a written schedule for the maintenance and calibration of such equipment based on type, purpose and degree of usage.

3.4.1 Production Tooling Used as Media of Inspection. When production jigs, fixtures, tool masters and other such devices are used as media of inspection, they shall be initially inspected or, by other suitable means, proved for accuracy prior to release for production use. These devices shall be reinspected or proved at established intervals.

3.4.2 Use of Contractor's Inspection Equipment. The contractor's gages, measuring and testing devices shall be made available for reasonable use by the Government when required to determine conformance with contract requirements. If conditions warrant, contractor's personnel shall be made available for operation of such devices and for verification of their accuracy and condition.

3.5 Control of Subcontracted Supplies

3.5.1 Responsibility. The contractor is responsible for assuring that all supplies and services to apply on Government contracts conform to the contract requirements whether manufactured or processed by the contractor or procured from subcontractors. The selection of sources and the nature and extent of control, including both contractor incoming inspection and surveillance, if any, at the subcontractor's plant, shall be based on and adjusted according to the nature of the supplies, the quality evidence furnished by the subcontractor and his demonstrated capability to perform in the specialized field involved. To assure an adequate and economical system for the control of purchased material, the contractor shall utilize to the fullest extent practicable, objective evidence of quality furnished by his subcontractor.

3.5.2 Subcontract Data. The contractor shall assure that applicable requirements are properly included or referenced in all subcontracts for supplies ultimately to apply on a Government contract. These subcontracts shall contain at least the following information:

(*a*) The applicable Government contract number, name and address of the subcontractor and the consignee.

(*b*) A clear description of the supplies ordered, including as applicable: (1) Specifications, drawings, process requirements, preservation and packaging requirements, classifications of defects, inspection of defects, inspection instructions and other necessary data. (2) Requirements for qualification or other Government or contractor approvals.

(*c*) Data necessary when provision is made for direct shipment from the subcontractor to Government activities.

3.5.3 Government Inspection of Subcontracts. The Government reserves the right to inspect, at source, supplies or services not manufactured or performed within the contractor's facility. Government subcontract inspection shall not constitute acceptance, nor shall it relieve the contractor of his responsibility to furnish an acceptable end item. The purpose of this inspection is to assist the Government representative at the contractor's facility to determine the conformance of supplies or services with contract requirements. Such inspection can only be requested by or under authorization of the Government representative.

3.5.3.1 When Government subcontract inspection is required, the contractor shall add to his subcontract the following statement:

"Government inspection is required prior to shipment from your plant. When material is ready for inspection or, if practical, ten (10) days in advance thereof, notify the Government representative who normally services your plant."

3.5.3.2 When, under authorization of the Government representative, copies of the subcontract are to be furnished directly by the subcontractor to the Government representative at his facility rather than through Government channels, the contractor shall add to his subcontract a statement substantially as follows:

"On receipt of this order, promptly furnish a copy to the Government representative who normally services your plant, or, if none, to the nearest Army, Navy or Air Force inspection office in your locality. In the event the representative or office cannot be located, our purchasing agent should be notified immediately."

3.5.4 Review and Processing of Subcontracts. All subcontracts and referenced data for supplies applying to a Government contract shall be available for review by the Government representative to determine compliance with the requirements for the control of such purchases. Copies of subcontracts required for Government purposes shall be furnished in accordance with the instructions of the Government representative.

3.5.5 Receiving Inspection. Subcontracted supplies shall be subjected to inspection after receipt, as necessary, to assure conformance to contract requirements. In adjusting such inspection consideration shall be given to the controls exercised by the subcontractor at source and evidence of sustained quality conformance. The contractor shall provide procedures for withholding from use all incoming supplies pending completion of required tests or receipt of necessary test reports, except that supplies may be released when under positive control. The contractor shall initiate corrective action with his subcontractors upon receipt of nonconforming supplies, whether or not Government source inspected, as indicated by the nature and frequency of the nonconformance. The contractor shall report to the Government representative any nonconformance found on Government source inspected supplies, and shall require the subcontractor to coordinate with his Government representative on corrective action.

3.6 Inspection during Manufacture. The contractor shall establish and maintain inspection at appropriately located points in the manufacturing process to assure continuous control of quality of parts, components and assemblies.

3.7 Special Processes. When Government approval or certification of processes, equipment or personnel is required under the contract, the contractor shall assure that he and his subcontractors are fully qualified prior to requesting Government approval.

3.8 Inspection of Completed Supplies. The contractor shall inspect completed supplies as necessary to assure that contract requirements have been met.

3.9 Sampling Inspection. Any sampling procedures, in addition to those required by contract, used by the contractor to determine the acceptability of supplies, shall afford reliable assurance of the maintenance of acceptable quality levels.

3.10 Indication of Inspection Status. The contractor shall maintain a system for identifying the inspection status of supplies. Identification may be accomplished by means of stamps, tags, routing cards, move tickets, tote box cards or other normal control devices. Such controls shall be of a design distinctly different from Government inspection identification.

3.11 Nonconforming Supplies. Procedures shall be provided for control of nonconforming supplies, including procedures for the identification, presentation and disposition of reworked, repaired or waived supplies. The acceptance of nonconforming supplies is a prerogative of and shall be as prescribed by the Government. All nonconforming supplies shall, when practicable, be diverted from normal material movement channels. The nonconforming supplies shall be positively identified to prevent use until disposition is made. Holding areas mutually agreeable to the contractor and Government representative shall be provided.

3.12 Government Property

3.12.1 Government-furnished Material. When material is furnished by the Government, the contractor's procedures shall include at least the following:

(*a*) Examination upon receipt, consistent with practicability to detect damage in transit.
(*b*) Inspection for completeness and proper type.
(*c*) Periodic inspection and precautions to insure adequate storage conditions and to guard against damage from handling and deterioration during storage.
(*d*) Functional testing, either prior to or after installation, or both, as required by contract to determine satisfactory operation.

3.12.1.1 Damaged Government-furnished Material. The contractor shall report to the Government representative any Government-furnished material found damaged, malfunctioning, or otherwise unsuitable for use. In the event of damage or malfunction during or after installation, the contractor shall determine and record probable cause and necessity for withholding material from use.

3.16 Quality Control Records. The contractor shall maintain adequate records throughout all stages of contract performance of inspections and tests, including checks made to assure accuracy of inspection and testing equipment and other control media. All quality control records shall be available for review by the Government representative, and copies of individual records shall be furnished him upon request.

3.17 Corrective Action. The contractor shall take prompt action to correct conditions which might result in defective supplies or services. Use shall be made of feedback data generated and furnished by using activities as well as that generated in the contractor's facility.

APPENDIX 5

RELIABILITY MONITORING PROGRAM[1]

2. Scope of the Program

The reliability monitoring program begins when the contract is awarded and continues through the phases of design, development, production and major product improvement in the life cycle of the weapon system. (It is essential that an adequate reliability program be incorporated in the proposal forming one of the bases for the contract award.)

The program consists of a number of fixed points at which reliability is objectively reviewed. If the nature of the project does not permit an assessment at each of the eight designated review or monitoring points, the program may be adjusted by either varying the points or omitting those that are not applicable.

The monitoring program is intended to supplement, not to replace, the contractor's regular reliability monitoring activities and surveillance by the appropriate Air Force activities. Continuous monitoring by the contractor provides basic data for use at each review point in the reliability monitoring program. It also provides indications of reliability difficulties that must be promptly considered by the designers and production engineers, thereby paving the way for early remedial action to avoid costly delays and rework.

The implementation of this program is based on an expression of quantitative contractual reliability requirements that is acceptable to both the Government and the contractor. The contract must also include mutually acceptable definitions of success and failure and a definitive description of test conditions and procedures to be used for evaluating and demonstrating reliability at the proper points.

The reliability monitoring program is designed to permit the contractor a maximum degree of freedom and to be compatible with Air Force approved specifications and policies and practices in progress reporting.

At least during the initial phases of the program, the best available techniques for estimating reliability must be used. It is generally recognized that the accu-

[1] Abstract of "Reliability Monitoring Program for Use in the Design, Development, and Production of Air Weapon Systems and Support Systems" (USAF Spec. Bull. 506, May 11, 1959). Number references are those specified in referenced bulletin; only those paragraphs which are considered most important have been included. See Chap. 15 for discussion of this abstract.

racy for predicting reliability must be increased. Nevertheless, reliability predictions are essential if a reasonable balance between reliability and other system parameters is to be achieved.

The first predictions or estimates of reliability, made in a feasibility study or in the project's prehardware phase, will be derived mainly by using analytical prediction techniques, and they may not inspire a high degree of confidence. But as reliability-prediction techniques improve and as the revised estimates increasingly reflect the results of component-testing programs, they can be used with growing confidence.

3. Reliability Monitoring Points

The sequence of the program's monitoring points (described below) is generalized. It is not intended to delineate an ideal weapon system life cycle but to suggest typical points at which the program may be monitored.

The points at which reliability is predicted, assessed, or measured and reviewed are at the end of each of the following phases of the weapon system's life cycle (see Appendix A for amplification of the points):

1. Detailed Design Study: This phase begins with the contract award and ends with a design report that includes studies of system and subsystem reliability that encompass the entire weapon system design and includes an assessment of reliability, using prediction techniques wherever feasible.

2. Preprototype: Here the initial system design is nearly complete, and many component parts and assemblies have undergone some development testing. This point may be identified by some such phrase as "95 per cent of engineering released," "design engineering inspection" or "the time at which initial design is essentially complete."

3. Prototype: The first complete sets of weapon system hardware or major weapon system subsystem hardware items are available at this point and can be assembled into the general physical configuration that they will have when used by the Air Force. Laboratory testing has been conducted to demonstrate the compatibility of weapon system and subsystems. Special test-vehicle flights to obtain data for design improvement of airborne elements of the system are performed. During this phase, all necessary research and engineering data are obtained and the basic design firmly established.

4. Preproduction Demonstration: At this point, the production design of the weapon system is essentially complete and the system is considered ready for production. A demonstration of the reliability achieved during this stage provides one of the bases for assessing the system's readiness for full-scale production.

5. Demonstration of Service Readiness: Here the contractor is required to show that the weapon system which is usually built under the limited or pilot production program has reached the reliability objectives—that the system can be produced in quantity without significant loss in performance or reliability.

6. Service Evaluation: During this phase, the Air Force uses its own personnel to perform its own weapon system evaluation tests. If the weapon system is found to be operationally acceptable and is capable of being produced in quantity without significant loss in reliability or performance, approval of production for operational use is usually given at this monitoring point.

7. *Full-scale Production:* The primary aim at this state is to ensure that the level of reliability designed into the system is maintained during production.

8. *Demonstration of Major Product Improvement:* At this point, the reliability and overall value of major product improvements are demonstrated and may be approved for incorporation into the weapon system.

ABSTRACTS FROM APPENDIX A

1. Detailed Design Study

The Air Force will provide the contractor an analysis of the intended operational uses of the weapon system which includes numerical reliability requirements for service use. Concepts of operational maintenance and overall logistics will also be furnished. If these are not provided by the Air Force, the contractor should explicitly state his assumptions and conclusions relating to the concepts and indicate how they relate to his design proposal. If accepted by the Air Force, the assumptions and conclusions must be considered mutually binding and initial logistic plans should be based on them. The report at this monitoring point should describe how the specified reliability is to be achieved and should include at least the information described below, which is the minimum required for an adequate assessment of reliability:

1. Design maturity and experience background on major portions of the proposed system should be evaluated. Wherever possible, a quantitative analysis of the reliability of the weapon system should be provided. Analyses of several aspects of the program may be needed to define the different aspects of the overall reliability problem. The following are examples of factors which, as applicable, should be studied and estimated: storage, readiness, checkout, captive or training flight, launching, and in-flight reliability. The analyses should achieve a synthesis of the known or estimated reliabilities which will serve as reliability allowances and objectives for these systems and subsystems.

The basis or means used for arriving at the reliability estimates should be described.

2. The administration and improvement activities of the reliability program during design, development and production should be described in detail with respect to (*a*) optimizing operational reliability; (*b*) a survey of reliability personnel and the organization, facilities, and procedures to be used; (*c*) the acquisition and analysis of data, e.g., failure and success, performance limits, and time; and (*d*) the design of special requirements for production quality control. This description should also delineate the laboratory program for testing component parts, subsystems and systems as well as the plan for reliability control and improvement. The basis for determining inspection and test criteria including sample size and test conditions should be presented. Sequential sampling and analysis procedures should be used wherever applicable, and confidence levels should be established for laboratory and production testing.

2. Preprototype

Here the initial system design is nearly complete, and many component parts and assemblies have undergone some developmental testing. This point may be identified by some such phrase as "95 per cent of engineering released," "design engineering inspection," or "the time at which initial design is essentially complete."

2.1 Assessment Procedure to be Used. A review should be made of the division of system and subsystem reliability requirements, i.e., an allocation of reliability allowances for portions of the system developed under Section 1. System reliability estimates must be reconsidered in accordance with results of established reliability activities for which contracting and budgeting have been completed, using calculations and test or performance data from other sources. Reliability predictions and estimates should also be reported.

Analyses should be performed to identify the detailed reliability requirements for design and evaluation purposes and, wherever possible, to determine their compatibility with the concepts of military usage. The maintenance, logistic and human factors aspects of the reliability program must be included, and the analyses should encompass the entire life of all parts of the system. A reliability requirement should then be established and reported for various logistic phases.

2.2 Environmental Tests. The conditions of environments derived for the analyses should be measured, where possible, or estimated and should be used as criteria for design and environmental testing of the complete weapon system and for subsystem items. These criteria are likely to differ for various parts of the system, e.g., ground based equipment, equipment located in different areas of an air vehicle, et cetera. At this time in the program, environmental testing of at least the most critical subsystems or parts should have begun. It is not expected that full tests on new parts will be accomplished in the preprototype phase. The contractor is responsible for the performance of all tests during this phase.

2.3 Design Changes. Design changes should be formally reviewed by the contractor. Any changes submitted to the Air Force for approval should include estimates of their effect on weapon system reliability and logistic factors, as well as project cost and schedule. This procedure should be followed throughout the program.

2.4 Documentation. The documentation of necessary design changes and tests of reliability should be described in the reports, which should also include a complete description of the data and test conditions.

2.5 Specifications. All applicable military and contractor specifications should be reviewed in the light of requirements established. These may be revised or superseded by more appropriate specifications. Specification changes should be submitted to the Air Force for review and approval when required by, and in accordance with, established policy and procedures.

3. Prototype

The first complete sets of weapon system hardware or major subsystem hardware items are available at this point and can be assembled into the general physical configuration that they will have when used by the Air Force. Laboratory testing has been conducted to demonstrate the compatibility of weapon system and subsystems, and special test-vehicle flights are performed to obtain data for design improvement. During this phase, all necessary research and engineering data are obtained and the basic design firmly established.

3.1 Analysis and Measurement. By this time, qualification testing of electronic, electrical, electromechanical and mechanical component parts and subsystems should be well under way. This should include tests to determine compliance with Government and contractor specifications and tests to measure whatever additional requirements have resulted from reliability studies and engineering analyses of other data. These tests, combined with reliability studies, should provide information on performance limits, operating time to failure[1] and mode of failure. Testing that cannot be accomplished by stimulating air vehicle environments in the laboratory should be included in an adequately instrumented test-vehicle flight program.

Ground testing should be conducted to assess reliability and measure operating time to failure. The results of these tests should serve to verify or modify the requirements determined as the result of the analyses.

3.3 Documentation. A progress report on the system's development through this phase should contain the engineering (R and D) data obtained from qualification testing and the results of tests to determine reliability. It should also include a comparison of present reliability levels of the weapon system and subsystems with the objectives established in the design proposal phase and a predicted reliability growth curve justified by experience and facts.

A positive program to improve equipment that fails to meet the reliability requirement established in the design proposal shall be included in the report made at this point. The established procedure for changes should be followed.

4. Preproduction Demonstration

At this point, the production design of the weapon system is essentially complete and is considered ready for production. A demonstration of the reliability achieved during this stage provides one of the bases for assessing the system's readiness for full-scale production.

5. Demonstration of Service Readiness

Here the contractor is required to show that the weapon system built under the limited or pilot production program has reached the reliability objectives—that the system can be produced in quantity without significant loss in performance or reliability.

An adequate and uniform production sample for service-readiness tests should be available from this production program. The design and development of the weapon system should be completed by this time, although improvements may be contemplated.

5.1 Tests. These will be in accordance with Air Force policy and procedures. Contractor or Air Force personnel will demonstrate the operation of the weapon system equipment by means of laboratory, captive-firing, flight and field tests and show that the weapon system is sufficiently mature for release to the Air Force for independent evaluation.

5.2 Analysis and Measurement. An analysis should be made of the potential operational reliability of the weapon system, using techniques that will permit a

[1] This applies particularly to ground and airborne equipment for weapon systems and, where feasible, to the air vehicle itself.

comparison of the ground and flight-test measurement with the reliability goals defined in Section 1 and as required by the contract.

5.3 Documentation. In the report made at this monitoring point, a comparison of the measured reliability of the weapon system and its subsystems with the objectives given in Section 1 will be given. The report should also include descriptions of the analysis techniques and test procedures used. The predicted reliability growth curve and how it was derived should be included.

A positive program should be presented in the report, with details of the improvements to be made on the parts of the system that failed to meet the reliability objectives given in Section 1. It will include details of proposed evaluation procedures, experimental design and data reduction methods for consideration by the Air Force in connection with planning for service evaluation.

All applicable military specifications should be reviewed to determine whether the results of the foregoing sections indicate that some specification change is needed. Specification changes shall be submitted to the Air Force for review and approval when required by, and in accordance with, established policy and procedures.

6. Service Evaluation

During this phase, the Air Force uses its own personnel to perform its own weapon system evaluation tests. If the weapon system is found to be operationally acceptable and is capable of being produced in quantity without significant loss in reliability of performance, approval of production for service use is usually given at this monitoring point.

6.1 Tests. These consist of operational tests at test ranges or proving grounds. In addition, they may involve technical evaluations and flight or laboratory tests.

Throughout this evaluation, a major item of interest to the Air Force is the measurement of system reliability while it is being employed and maintained under conditions that simulate combat operations. The entire weapon system should be evaluated, using an experiment designed to permit factoring out the effect of other systems and equipment on the performance of the system under evaluation. Although cost and time factors must be considered, every reasonable effort should be made to design, conduct, and analyze tests and the results so as to detect any remaining major design unbalances between performance and reliability.

6.2 Analysis and Measurement. The Air Force is responsible for the analysis of the data which will be used as one means of determining whether the terms of the reliability section of the contract have been met. Any such test analyses should follow general procedures agreed upon by the Air Force and the contractor with respect to analysis and data presentation. The data should be provided to the contractor if independent analyses are to be performed. An extensive analysis of Service evaluation results is encouraged for maximum benefit from these costly data. From a reliability standpoint, the analyses might include the derivation of reliability values for subsystems, e.g., launchers, fire control, illuminating radar, beacons, et cetera.

6.3 Documentation. The Air Force activity conducting the operational evaluation is responsible for documenting the test procedures used and the conclusions reached. The Air Force will use the results of its analyses, as well as those that may be obtained from the weapon system contractor, to document the reliability value achieved, including a description of the techniques used and the analyses completed. The results should be presented in a form that permits determining whether the reliability requirement stated in the contract has been met.

6.4 Determination of the Corrective Program. If results of Service evaluation tests indicate that the required reliability value has not been achieved, a corrective program may be initiated. If further action by the contractor is warranted, the Air Force and the contractor should prepare a mutually acceptable program to effect the correction. Agreements should be reached regarding specific areas to be remedied and the methods and criteria to be used in evaluating the effects of the changes. If major changes are to be made in the system, the program would enter the phase of major product improvement demonstration (described in Section 8). Relatively minor changes would require that the Air Force and the contractor agree on a plan similar in its aim and content to the previously described phases of the proposed reliability monitoring program.

7. Full-scale Production

The primary aim at this stage is to ensure that the level of reliability designed into the system is maintained during production.

7.1 Analysis and Measurement. Manufacturing results shall be continuously monitored using approved quality control techniques and including any special requirements for production quality control. [See Section 1, Paragraph (2).] Requirements for quality control programs are contained in the appropriate military specifications for quality control applicable to Air Force contractors. Both factory and field results shall be analyzed so as to detect or warn of reliability problems.

8. Demonstration of Major Product Improvement

At this point, the performance, reliability and overall value of product improvements are demonstrated, and they may thereafter be approved for incorporation into the weapon system.

The contractor should demonstrate the value of major changes in the weapon system by design analysis and by laboratory and system operation tests. The nature and extent of the demonstration would depend on the scope of the proposed change. The contractor should propose, and the Air Force will determine, the reliability monitoring points to be used and the way in which the improvements are to be demonstrated.

8.1 Design Analysis. The design analysis should specify the reliability and confidence levels to be achieved on applicable component, subsystem or system levels at each stage of major product improvement, and this should include a comparison with the previously achieved level of reliability. The analysis should be complete enough to permit the Air Force to assess the effect of the proposed improvement on the performance and reliability of the weapon system, to the end that this design will at least meet the reliability requirement for the system. The proposal for improvement should include a detailed description of the analysis performed and a definite program for achieving the specified operational reliability with an acceptable confidence level.

8.3 Documentation. The contractor will present the results of all tests and descriptions of any design changes in periodic progress reports, which also will contain the analysis of the demonstration's results, including a comparison of the measured reliability levels of the components, subsystems and systems with the reliability objectives. In addition, these reports will show the current performance of the weapon system in relation to the contractual requirement.

Any special requirements (equipment and its location, placement, environmental requirements, etc.) that may require special consideration or a deviation from the specifications given in the contract will be supported. Proposed revisions to contractual specifications will be submitted in accordance with established practice to the Air Force.

A positive program should be presented detailing the improvements to be made on the equipments that fail to meet reliability objectives.

APPENDIX 6

GUIDES FOR RELIABILITY ORGANIZATION[1]

PART I. RELIABILITY PROGRAM CRITERIA

1. Concept and Approach

The overall program should reveal an appreciation of the importance of reliability and of the necessity for an organized approach to the attainment. There should be recognition of the fact that reliability must be inherent in the basic design, that improvement is best accomplished in the early stages of development and test, and that the maintenance of reliability extends through the life-cycle of a product. The program should indicate an awareness of what others are doing in the reliability area ("State-of-the-Art," as it were) and the extent to which such practices are applicable and/or utilized. Specifically, the degree of reliability to be attained should be stated in quantitative terms and the considerations of environmental and operational conditions should be clearly defined. There should be a statement of existing or potential "short-comings" and an explanation of basis for "trade-off's" with other design parameters such as performance, accuracy, weight, cost, etc. The progressive steps to be taken to arrive at the reliability objective should be outlined. These steps or phases should lend themselves to periodic reporting and progress analysis.

2. Organization

The reliability organization within the overall management organization should be such that it reflects an important position in the management effort and provides not only for the establishment of policy but for the complete implementation of the established policy. Policy can be established effectively by a policy board or group composed of responsible representatives from the executive office and the operating divisions. Implementation can be accomplished by granting the policy board organization assignment authority or by creating a line organization specifically for reliability and with assignment authority. In either case, authority for assignment of specific responsibilities must be established and it must cover the Engineering, Manufacturing, Quality Control and Purchasing functions and

[1] Abstract of "Guides for Reliability Organization" (USAF Spec. Bull. 510, June 30, 1959). Number references are those specified in referenced bulletin; only those paragraphs which are considered most important have been included. See Chaps. 15 and 16 for discussion of this abstract and for other material on this subject.

must be able to integrate these functions into an effective program. There should be a Director or Manager to provide the necessary coordination, not only at the policy setting stage, but, more importantly, in the operating stage so that the program is kept on the policy course and not allowed to drift off to the detriment of any one aspect. The Director, or Manager, should be responsible, and have authority, for making sure that the necessary services (data processing, drafting, training, etc.) are provided as needed by the program and for seeing that the executive offices are informed of, and in accord with, all aspects of the reliability program.

3. Programming

There should be a time-phased plan covering each point of the reliability policy established by the policy board and all activity comprising the reliability program. The authority and responsibility of the head of each working group should be clearly specified so that specific reliability activities can be directed and their accomplishment measured, both as to progress and completion. All reliability activities should be scheduled so that they are properly phased, where phasing is required, and so that they may be accomplished when required. The program should be so well defined, and broken down to such detail, that the people, skills, facilities and elapsed time requirements for each specified activity can be estimated with sufficient effectiveness to establish program cost.

4. Quality Control

There should be recognition that, while the function of Quality Control is concerned primarily with determining compliance with existing specifications and acceptance criteria, this activity can also contribute substantially to reliability by feed-back to and close liaison with Engineering, Manufacturing, and Purchasing elements of the organization. This requires that the organizational stature of Quality Control be such that it is not subordinated to these other three or, at least, that it have equal access to the ear of management at the reliability policy level and partnership participation at the working level. In operational procedure, there should be evidence that Quality Control is not only ensuring compliance with Air Force standards (MIL-Q-9858) but is provided and utilizing whatever higher standards and criteria experience and "state-of-the-art" have indicated will contribute to improved reliability. With respect to vendors and subcontractors, there should be reference to the procedures used in performing the Quality Control function at both the vendor and sub-contractor's installations and the prime contractor's receiving and acceptance end. Such procedures could include pre-award surveys, contractual requirements, qualification testing, acceptance criteria, assistance rendered by prime contractor, feed-back data, and corrective action resulting from such data.

5. Reliability Requirement Studies

There should be full appreciation that reliability demands quantitative treatment and that such treatment involves advance study—that it is not sufficient to wait for failure reporting to set the pattern. Such studies should be aimed at establishing quantitative requirements to include specification of required function (performance limits), operating time, and environment as well as the required reliability. The requirements should be realistic, determined as early as possible, and revised as necessary. Environmental conditions specified should cover all phases from factory to the target including shipping, storage, handling, and flight. Since the reliability of a weapons system is a function of all the elements of the system, the reliability requirements for sub-systems, equipments, and parts must conform to the overall systems requirements. The terms used to express requirements must be meaningful from the standpoint of mission success and flight safety.

6. Qualification Testing

There should be a realization of the necessity to fully qualify all systems, sub-systems, and equipment before the production in quantity is undertaken. This applies to CFE as well as GFAE, and is not limited to those items whose specifications call for qualification testing for formal inclusion on the Qualified Products List. This concept requires a time-phased scheduling of qualification testing to ensure completion by the time the first production article comes into being. The basis for qualification should be defined to the extent of indicating that tests include not only compliance with government and contractor specifications, but also whatever additional requirements have been determined as a result of reliability requirement studies, engineering data and analyses, or special environment. If based on arbitrary assumptions in lieu of accurate knowledge of requirements, this fact should be noted together with the assumptions. Availability of adequate testing facilities, both prime or sub-contract, should be referenced.

7. Acceptance Criteria

In this area, there should be outlined the measures which are taken to ensure that, once an item has been qualified, the follow-on quantity production does not fall below the established standards, particularly with respect to reliability. These measures should include procedures as applied to systems, sub-systems, equipment, and components; to CFE and GFAE; and to products of vendors and sub-contractors. There should be expression of the degree of application of these procedures (100 per cent or sample) and the corrective action that is taken to both improve the product and the procedures when articles do not comply with the acceptance criteria established.

8. Failure and Deficiency Reporting, Analysis, and Correction

There should be reference to the sources of failure and deficiency data both from within (manufacturing, quality control, test) and without (technical representatives, field service, UR's, TDR's, etc.). The procedures for collection, recording, summarizing, presenting, revising, and analyzing data should be outlined. Most importantly, the organization, facilities, procedures and follow-up that ensures that corrective action is accomplished should be clearly described.

9. Relationship with Vendors and Sub-contractors

In the selection of vendors and sub-contractors, it should be evident that there is an evaluation of the supplier's ability to meet the reliability requirements; that some sort of rating system is utilized with possible inclusion of an approved listing of sources for particular types of material. There should be reference to the completeness of engineering data, including test and inspection criteria, that is incorporated in purchase orders and sub-contracts. A statement should be made as to the extent of surveillance maintained over suppliers other than inspection and test, such as review of quality control system, drawing and engineering change control, specification review. There should be a procedure for supplying vendors and sub-contractors with malfunction and deficiency data and corresponding follow-up on corrective actions. Reference should be made to the extent of assistance that is rendered suppliers in solution of immediate problems, in improvement suggestions, in providing information on new methods and techniques in engineering and manufacturing, and the like.

10. Training (Reliability Indoctrination)

The fact that reliability is to a great extent education and that there is a need for organized training with relation to reliability should be recognized. In addition to courses which are designed to raise skill levels and hence improve reliability, there should be effort directed to making personnel reliability-conscious and to giving them knowledge of wherein their particular jobs can either contribute to or detract from the reliability of the end product. How this is accomplished (lectures, films, formal courses, etc.), and to what extent each level of employment (manager, engineer, technician, worker) participates in reliability training programs should be outlined.

PART II. DISCUSSION OF RELIABILITY ORGANIZATION

1. Organization, Personnel, and Responsibilities

The organizational position and responsibilities of contractor reliability organizations vary greatly among contractors. Some of these reliability groups have functioned effectively and others have suffered under the handicap of being mislocated organizationally from not having sufficient responsibility and authority to implement an effective reliability program. The makeup of reliability groups in their organizational positions have been the subject of intensive industry-wide surveys from time to time. The following summary represents the responsibilities of a reliability organization which appears to yield best results.

1.1 The basic question always is whether reliability should be given in-line responsibility and authority or be relegated to a pure staff or service function. The conclusion is that in order to be most effective the reliability organization must have an in-line function within the framework of the contractor's overall company or project organization. This will enable them to take action on their own and to cause others to take the necessary actions to achieve the required reliability.

1.2 By not having the people responsible for reliability reporting to the man in charge of a project, several disadvantages are incurred:

(1) A reliability man in a service group has no in-line function or authority.

(2) In the press of business, in times when reliability should be considered by the project head, he may choose to talk to the reliability people and ask their

opinion or help and he may choose not to. The result is that many activities and decisions which have a direct effect on reliability may take place within a company without the knowledge of the reliability organization, and, consequently, reliability factors are not always given the proper consideration in relation to other factors.

(3) When advice or recommendations are made to the project head by someone from an outside group, he may choose to take their advice and he may choose not to.

(4) When his help is needed on a problem on a particular project, the services of the reliability man may be available or they may not be, due to the conflicting interests of other projects.

(5) A reliability group which occupies a pure staff position can on occasion resolve certain fundamental differences between themselves and other company groups by appealing to higher management for a decision. However, this procedure can only be used when such a conflict is made a major issue. Obviously this method of handling reliability cannot be used to resolve the many day-to-day working problems which are the real difference between achieving and not achieving reliability of design, since these daily working problems cannot all be made a major issue.

1.3 By having the responsibility for reliability reporting directly to the head of a particular project, several advantages are incurred.

(1) The reliability man is already accepted as a member of the team and is not considered to be an outsider.

(2) He is in a position to be intimately familiar with the daily problems on the project.

(3) Because he is associated with a working team, his recommendations will be more readily accepted than those of an independent group.

(4) He is in a better position to follow up on his recommendations and see that the critical problems are being worked on.

(5) He has an in-line function and authority.

Based on the above considerations, a preferred description of the organizational setup, type of personnel, and responsibilities to be delegated to reliability are given in succeeding paragraphs.

2. Organization[1]

The principal reliability organization must be located in engineering, *not* in quality control or standards departments.

2.1 The person directly charged with implementation of the Reliability Program shall report to that office directly responsible for all other engineering activities on the project in question.

2.2 The reliability organization should ideally consist of three basic groups: systems, environmental, and analysis or statistical.

[1] AUTHOR'S COMMENT: Another successful method is to place all reliability and quality engineers under the supervision of a product-assurance director who will detail reliability engineers to perform technical functions under the direction of the design engineer until the research and development phase is complete. The reliability engineers can then easily be shifted to production activity.

2.3 In the beginning of a program, reliability should be located in a design engineering or systems engineering organization and may shift to product engineering when the program reaches the production phase.

3. Personnel

The backbone of the reliability organization should consist of people with the experience described below.

3.1 Design background: packaging, circuit design or analysis, test procedures.

3.2 Systems engineering: including requirements, systems specifications, analysis of flight and field test results.

3.3 Environmental specialist: test procedures, specifications, determination and interpretation of environmental conditions.

3.4 Analytical or statistical: experimental design, statistical techniques, data collection and analysis.

4. Responsibilities

A properly constituted reliability organization will have assigned to it certain in-line and advisory responsibilities which will include, but may not be limited to the following:

5. Description of Activities

A brief description of the activities to be carried out under each of these categories is given in Paragraph 5.1.

5.1 *Environmental Conditions Determination.* Missile environment measurements should be reviewed, and interactive effects of the contractor's hardware on missile environment should be estimated. Environmental conditions for assembly and parts specifications should be determined. Information on environmental conditions should be disseminated to designers, specification writers, and test planners.

5.2 *Reliability Apportionment.* Detailed reliability objectives should be established by numerically apportioning the contractor's system reliability objective among the component assemblies and parts. The objectives should be disseminated for use in design guidance, in evaluations and comparisons, in planning of test programs, and in pinpointing problem areas.

5.3 *Reliability Indoctrination.* The reliability program requirements and procedures should be explained and "sold" to company personnel and to vendors. Measures such as a training program for engineers should be instituted.

5.4 *Parts Approval Verification.* The component parts engineering effort (whether centralized or not) should be monitored. Qualification test and performance data on parts should be reviewed, coordinated, and disseminated. Information sources such as preferred parts lists, data files, and vendor ratings should be maintained.

5.5 *Specification Review.* Specification writers should be assisted with regard to reliability objectives and statistical tolerance considerations. Specifications should be reviewed to assure that reliability will not be unduly compromised in such matters as reliability-performance trade-offs.

5.6 Design Review. Designers should be assisted in such matters as environment problems, tolerances, component application, developmental marginal checking, consideration of effects of production tooling methods on reliability, and human use factors. Designs should be reviewed for such reliability factors as adequate safety margins, provision for preventive maintenance, and appropriate redundancy.

5.7 Failure Reporting Surveillance. Procedures should be established and maintained for reporting individual failures in plant and field operations involving prototype and production hardware. Individual failure reports should receive engineering analysis and be distributed to design or production activities for prompt correction of troubles. Follow-up should be instituted to assure that failures are corrected. A card file should be maintained on failure report data. Summary reports should be prepared from the card file and distributed so that problem areas and progress may be defined.

5.8 Statistical Test Planning. Test engineers should be assisted so that optimum consideration is given to environmental test conditions, reliability objectives, and statistical design of experiments (particularly with regard to sample size, stress level, and arrangement of tests). Test plans should be reviewed to assure that testing will be sufficiently comprehensive to allow detection of important modes of failure and to provide a basis for effective evaluation. "Operating characteristic" curves should be prepared to define the effectiveness of the planned tests.

5.9 Statistical Test Evaluation. All test reports should be analyzed and evaluated for information on hardware capabilities and weaknesses. Data and results should be explained to designers and also stored and cross filed for reference so that accumulated information on particular designs may be readily located at any time. Information should be included on successes as well as failures.

5.10 Human Factors. Many malfunctions and holds occurring during missile inspection, checkout, and tests are directly attributable to errors on the part of personnel and not to equipment failure. In many cases these errors are not the result of "careless workmanship," but rather represent a failure of the contractor to consider seriously the fallibility of the human element, as a factor in the design of missile hardware. Contractors must continuously strive to eliminate sources of human error through proper design of the hardware, training, improved procedures, or a combination of these.

5.11 Quality Control Coordination. The quality control effort should be coordinated with regard to such matters as reliability objectives, production process control, and inspection test procedures designed to show up incipient failures.

5.12 Program Data Evaluation. All information and data obtained in the program should be organized, analyzed, and evaluated with regard to reliability. Estimates of achieved reliability should be made and projected to operational use. Reports should be issued to define problem areas and report progress. A list of critical assemblies and parts should be maintained. A special effort should be made to "close the loop" by feeding information back into the organization promptly and at points where it is needed.

5.13 Vendor Control. The contractor shall take steps to ascertain through proper tests and surveillance that parts and devices supplied by vendors and subcontractors are adequate for their intended application in the contractor's equipment. These measures shall include tests to demonstrate design capability and to provide a continuous monitoring of the vendor's quality control and product improvement programs.

5.14 Flight Test Planning. Plans and specifications for missile flight testing should be reviewed from the point of view of obtaining as much information pertinent to reliability as is possible to do so without causing undue compromise or interference with other flight test objectives during their R and D program. Instrumentation should be carefully reviewed as to adequacy of design and ability to yield data of required quality. Special attention should be given to the question whether or not certain telemetering channels should be commutated when the reliability of a missile subsystem is being evaluated by the operating time-to-failure criteria. For flights during the latter stages of the R and D program and post-IOC[1] flights, inclusion of reliability objectives and flight test planning is even more important in the sense that flight test data on missiles of operational or near operational design are especially significant in making estimates of future reliability of operational systems.

5.15 Analysis of Test and Flight Failures. All failures resulting from environmental, factory, field, and flight test should be analyzed by people in the reliability organization to determine the significance of the failures in a reliability sense. The analysis should be on both a statistical and engineering basis. An attempt should be made to determine causes of the failures and, where feasible, whether or not the deficiency was due to the design, quality control, or human factors. Where appropriate these analyses should be fed back through the failure reporting system described in Appendix E of this document.

5.16 Determination of Corrective Action. Analysis of test results and failures should have two primary purposes: to determine the need for corrective action, and to establish at least a recommendation as to the nature of the corrective action required.

5.17 Corrective Action Follow-up. This constitutes one of the most critical of all reliability activities, and procedures should be set up which will allow for a routine follow-up of corrective actions recommended or being acted upon and should contain a system of checks and balances to assure that the required corrective actions have not been forgotten or ignored by those responsible for implementing the final action.

[The balance of USAF Specification Bulletin 510 consists of a number of appendixes up to Appendix N. These describe in minute detail the specific steps which should be taken to assure devices of high reliability and maintainability.]

[1] Initial operational capability.

APPENDIX 7

TABLES

TABLE 1. SUMMATION OF TERMS OF POISSON'S EXPONENTIAL BINOMIAL LIMIT*
(1000 × Probability of c or f or Less Occurrences of Event That Has Average Number of Occurrences Equal to a or d)

a or d \ c or f	0	1	2	3	4	5	6	7	8	9
0.02	980	1,000								
0.04	961	999	1,000							
0.06	942	998	1,000							
0.08	923	997	1,000							
0.10	905	995	1,000							
0.15	861	990	999	1,000						
0.20	819	982	999	1,000						
0.25	779	974	998	1,000						
0.30	741	963	996	1,000						
0.35	705	951	994	1,000						
0.40	670	938	992	999	1,000					
0.45	638	925	989	999	1,000					
0.50	607	910	986	998	1,000					
0.55	577	894	982	998	1,000					
0.60	549	878	977	997	1,000					
0.65	522	861	972	996	999	1,000				
0.70	497	844	966	994	999	1,000				
0.75	472	827	959	993	999	1,000				
0.80	449	809	953	991	999	1,000				
0.85	427	791	945	989	998	1,000				
0.90	407	772	937	987	998	1,000				
0.95	387	754	929	984	997	1,000				
1.00	368	736	920	981	996	999	1,000			
1.1	333	699	900	974	995	999	1,000			
1.2	301	663	879	966	992	998	1,000			
1.3	273	627	857	957	989	998	1,000			
1.4	247	592	833	946	986	997	999	1,000		
1.5	223	558	809	934	981	996	999	1,000		
1.6	202	525	783	921	976	994	999	1,000		
1.7	183	493	757	907	970	992	998	1,000		
1.8	165	463	731	891	964	990	997	999	1,000	
1.9	150	434	704	875	956	987	997	999	1,000	
2.0	135	406	677	857	947	983	995	999	1,000	

* Adapted from E. L. Grant, "Statistical Quality Control," 2d ed., Table G, pp. 518–522, McGraw-Hill Book Company, Inc., New York, 1952. By permission from the publisher.

TABLE 1. SUMMATION OF TERMS OF POISSON'S EXPONENTIAL BINOMIAL LIMIT
(*Continued*)

a or *d* \ *c* or *f*	0	1	2	3	4	5	6	7	8	9
2.2	111	355	623	819	928	975	993	998	1,000	
2.4	091	308	570	779	904	964	988	997	999	1,000
2.6	074	267	518	736	877	951	983	995	999	1,000
2.8	061	231	469	692	848	935	976	992	998	999
3.0	050	199	423	647	815	916	966	988	996	999
3.2	041	171	380	603	781	895	955	983	994	998
3.4	033	147	340	558	744	871	942	977	992	997
3.6	027	126	303	515	706	844	927	969	988	996
3.8	022	107	269	473	668	816	909	960	984	994
4.0	018	092	238	433	629	785	889	949	979	992
4.2	015	078	210	395	590	753	867	936	972	989
4.4	012	066	185	359	551	720	844	921	964	985
4.6	010	056	163	326	513	686	818	905	955	980
4.8	008	048	143	294	476	651	791	887	944	975
5.0	007	040	125	265	440	616	762	867	932	968
5.2	006	034	109	238	406	581	732	845	918	960
5.4	005	029	095	213	373	546	702	822	903	951
5.6	004	024	082	191	342	512	670	797	886	941
5.8	003	021	072	170	313	478	638	771	867	929
6.0	002	017	062	151	285	446	606	744	847	916

a or *d* \ *c* or *f*	10	11	12	13	14	15	16
2.8	1,000						
3.0	1,000						
3.2	1,000						
3.4	999	1,000					
3.6	999	1,000					
3.8	998	999	1,000				
4.0	997	999	1,000				
4.2	996	999	1,000				
4.4	994	998	999	1,000			
4.6	992	997	999	1,000			
4.8	990	996	999	1,000			
5.0	986	995	998	999	1,000		
5.2	982	993	997	999	1,000		
5.4	977	990	996	999	1,000		
5.6	972	988	995	998	999	1,000	
5.8	965	984	993	997	999	1,000	
6.0	957	980	991	996	999	999	1,000

TABLE 1. SUMMATION OF TERMS OF POISSON'S EXPONENTIAL BINOMIAL LIMIT
(*Continued*)

a or *d* \ *c* or *f*	0	1	2	3	4	5	6	7	8	9
6.2	002	015	054	134	259	414	574	716	826	902
6.4	002	012	046	119	235	384	542	687	803	886
6.6	001	010	040	105	213	355	511	658	780	869
6.8	001	009	034	093	192	327	480	628	755	850
7.0	001	007	030	082	173	301	450	599	729	830
7.2	001	006	025	072	156	276	420	569	703	810
7.4	001	005	022	063	140	253	392	539	676	788
7.6	001	004	019	055	125	231	365	510	648	765
7.8	000	004	016	048	112	210	338	481	620	741
8.0	000	003	014	042	100	191	313	453	593	717
8.5	000	002	009	030	074	150	256	386	523	653
9.0	000	001	006	021	055	116	207	324	456	587
9.5	000	001	004	015	040	089	165	269	392	522
10.0	000	000	003	010	029	067	130	220	333	458

	10	11	12	13	14	15	16	17	18	19
6.2	949	975	989	995	998	999	1,000			
6.4	939	969	986	994	997	999	1,000			
6.6	927	963	982	992	997	999	999	1,000		
6.8	915	955	978	990	996	998	999	1,000		
7.0	901	947	973	987	994	998	999	1,000		
7.2	887	937	967	984	993	997	999	999	1,000	
7.4	871	926	961	980	991	996	998	999	1,000	
7.6	854	915	954	976	989	995	998	999	1,000	
7.8	835	902	945	971	986	993	997	999	1,000	
8.0	816	888	936	966	983	992	996	998	999	1,000
8.5	763	849	909	949	973	986	993	997	999	999
9.0	706	803	876	926	959	978	989	995	998	999
9.5	645	752	836	898	940	967	982	991	996	998
10.0	583	697	792	864	917	951	973	986	993	997

	20	21	22
8.5	1,000		
9.0	1,000		
9.5	999	1,000	
10.0	998	999	1,000

TABLE 1. SUMMATION OF TERMS OF POISSON'S EXPONENTIAL BINOMIAL LIMIT
(*Continued*)

a or *d* \ *c* or *f*	0	1	2	3	4	5	6	7	8	9
10.5	000	000	002	007	021	050	102	179	279	397
11.0	000	000	001	005	015	038	079	143	232	341
11.5	000	000	001	003	011	028	060	114	191	289
12.0	000	000	001	002	008	020	046	090	155	242
12.5	000	000	000	002	005	015	035	070	125	201
13.0	000	000	000	001	004	011	026	054	100	166
13.5	000	000	000	001	003	008	019	041	079	135
14.0	000	000	000	000	002	006	014	032	062	109
14.5	000	000	000	000	001	004	010	024	048	088
15.0	000	000	000	000	001	003	008	018	037	070
	10	11	12	13	14	15	16	17	18	19
10.5	521	639	742	825	888	932	960	978	988	994
11.0	460	579	689	781	854	907	944	968	982	991
11.5	402	520	633	733	815	878	924	954	974	986
12.0	347	462	576	682	772	844	899	937	963	979
12.5	297	406	519	628	725	806	869	916	948	969
13.0	252	353	463	573	675	764	835	890	930	957
13.5	211	304	409	518	623	718	798	861	908	942
14.0	176	260	358	464	570	669	756	827	883	923
14.5	145	220	311	413	518	619	711	790	853	901
15.0	118	185	268	363	466	568	664	749	819	875
	20	21	22	23	24	25	26	27	28	29
10.5	997	999	999	1,000						
11.0	995	998	999	1,000						
11.5	992	996	998	999	1,000					
12.0	988	994	997	999	999	1,000				
12.5	983	991	995	998	999	999	1,000			
13.0	975	986	992	996	998	999	1,000			
13.5	965	980	989	994	997	998	999	1,000		
14.0	952	971	983	991	995	997	999	999	1,000	
14.5	936	960	976	986	992	996	998	999	999	1,000
15.0	917	947	967	981	989	994	997	998	999	1,000

TABLE 1. SUMMATION OF TERMS OF POISSON'S EXPONENTIAL BINOMIAL LIMIT
(*Continued*)

c or *f* / *a* or *d*	4	5	6	7	8	9	10	11	12	13
16	000	001	004	010	022	043	077	127	193	275
17	000	001	002	005	013	026	049	085	135	201
18	000	000	001	003	007	015	030	055	092	143
19	000	000	001	002	004	009	018	035	061	098
20	000	000	000	001	002	005	011	021	039	066
21	000	000	000	000	001	003	006	013	025	043
22	000	000	000	000	001	002	004	008	015	028
23	000	000	000	000	000	001	002	004	009	017
24	000	000	000	000	000	000	001	003	005	011
25	000	000	000	000	000	000	001	001	003	006
	14	15	16	17	18	19	20	21	22	23
16	368	467	566	659	742	812	868	911	942	963
17	281	371	468	564	655	736	805	861	905	937
18	208	287	375	469	562	651	731	799	855	899
19	150	215	292	378	469	561	647	725	793	849
20	105	157	221	297	381	470	559	644	721	787
21	072	111	163	227	302	384	471	558	640	716
22	048	077	117	169	232	306	387	472	556	637
23	031	052	082	123	175	238	310	389	472	555
24	020	034	056	087	128	180	243	314	392	473
25	012	022	038	060	092	134	185	247	318	394
	24	25	26	27	28	29	30	31	32	33
16	978	987	993	996	998	999	999	1,000		
17	959	975	985	991	995	997	999	999	1,000	
18	932	955	972	983	990	994	997	998	999	1,000
19	893	927	951	969	980	988	993	996	998	999
20	843	888	922	948	966	978	987	992	995	997
21	782	838	883	917	944	963	976	985	991	994
22	712	777	832	877	913	940	959	973	983	989
23	635	708	772	827	873	908	936	956	971	981
24	554	632	704	768	823	868	904	932	953	969
25	473	553	629	700	763	818	863	900	929	950
	34	35	36	37	38	39	40	41	42	43
19	999	1,000								
20	999	999	1,000							
21	997	998	999	999	1,000					
22	994	996	998	999	999	1,000				
23	988	993	996	997	999	999	1,000			
24	979	987	992	995	997	998	999	999	1,000	
25	966	978	985	991	994	997	998	999	999	1,000

TABLE 2. TABLE OF CHI SQUARE*

(For larger values of ν, the expression $\sqrt{2\chi^2} - \sqrt{2\nu - 1}$ may be used as a normal deviate with unit variance, remembering that the probability for χ^2 corresponds with that of a single tail of the normal curve.)

ν	Probability										
	0.99	0.98	0.95	0.90	0.80	0.20	0.10	0.05	0.02	0.01	0.001
1	$0.0^{3}157$	$0.0^{3}628$	0.00393	0.0158	0.0642	1.642	2.706	3.841	5.412	6.635	10.827
2	0.0201	0.0404	0.103	0.211	0.446	3.219	4.605	5.991	7.824	9.210	13.815
3	0.115	0.185	0.352	0.584	1.005	4.642	6.251	7.815	9.837	11.341	16.268
4	0.297	0.429	0.711	1.064	1.649	5.989	7.779	9.488	11.668	13.277	18.465
5	0.554	0.752	1.145	1.610	2.343	7.289	9.236	11.070	13.388	15.086	20.517
6	0.872	1.134	1.635	2.204	3.070	8.558	10.645	12.592	15.033	16.812	22.457
7	1.239	1.564	2.167	2.833	3.822	9.803	12.017	14.067	16.622	18.475	24.322
8	1.646	2.032	2.733	3.490	4.594	11.030	13.362	15.507	18.168	20.090	26.125
9	2.088	2.532	3.325	4.168	5.380	12.242	14.684	16.919	19.679	21.666	27.877
10	2.558	3.059	3.940	4.865	6.179	13.442	15.987	18.307	21.161	23.209	29.588
11	3.053	3.609	4.575	5.578	6.989	14.631	17.275	19.675	22.618	24.725	31.264
12	3.571	4.178	5.226	6.304	7.807	15.812	18.549	21.026	24.054	26.217	32.909
13	4.107	4.765	5.892	7.042	8.634	16.985	19.812	22.362	25.472	27.688	34.528
14	4.660	5.368	6.571	7.790	9.467	18.151	21.064	23.685	26.873	29.141	36.123
15	5.229	5.985	7.261	8.547	10.307	19.311	22.307	24.996	28.259	30.578	37.697
16	5.812	6.614	7.962	9.312	11.152	20.465	23.542	26.296	29.633	32.000	39.252
17	6.408	7.255	8.672	10.085	12.002	21.615	24.769	27.587	30.995	33.409	40.790
18	7.015	7.906	9.390	10.865	12.857	22.760	25.989	28.869	32.346	34.805	42.312
19	7.633	8.567	10.117	11.651	13.716	23.900	27.204	30.144	33.687	36.191	43.820
20	8.260	9.237	10.851	12.443	14.578	25.038	28.412	31.410	35.020	37.566	45.315
21	8.897	9.915	11.591	13.240	15.445	26.171	29.615	32.671	36.343	38.932	46.797
22	9.542	10.600	12.338	14.041	16.314	27.301	30.813	33.924	37.659	40.289	48.268
23	10.196	11.293	13.091	14.848	17.187	28.429	32.007	35.172	38.968	41.638	49.728
24	10.856	11.992	13.848	15.659	18.062	29.553	33.196	36.415	40.270	42.980	51.179
25	11.524	12.697	14.611	16.473	18.940	30.675	34.382	37.652	41.566	44.314	52.620
26	12.198	13.409	15.379	17.292	19.820	31.795	35.563	38.885	42.856	45.642	54.052
27	12.879	14.125	16.151	18.114	20.703	32.912	36.741	40.113	44.140	46.963	55.476
28	13.565	14.847	16.928	18.939	21.588	34.027	37.916	41.337	45.419	48.278	56.893
29	14.256	15.574	17.708	19.768	22.475	35.139	39.087	42.557	46.693	49.588	58.302
30	14.953	16.306	18.493	20.599	23.364	36.250	40.256	43.773	47.962	50.892	59.703

* This table is reproduced in abridged form form Table IV of Fisher and Yates, "Statistical Tables for Biological, Agricultural, and Medical Research," published by Oliver & Boyd, Ltd., Edinburgh, by permission of the authors and publishers.

TABLE 3. AREAS UNDER THE NORMAL CURVE*
(Proportion of total area under the curve from $-\infty$ to designated Z value)

Z	0.09	0.08	0.07	0.06	0.05	0.04	0.03	0.02	0.01	0.00
−3.5	0.00017	0.00017	0.00018	0.00019	0.00019	0.00020	0.00021	0.00022	0.00022	0.00023
−3.4	0.00024	0.00025	0.00026	0.00027	0.00028	0.00029	0.00030	0.00031	0.00033	0.00034
−3.3	0.00035	0.00036	0.00038	0.00039	0.00040	0.00042	0.00043	0.00045	0.00047	0.00048
−3.2	0.00050	0.00052	0.00054	0.00056	0.00058	0.00060	0.00062	0.00064	0.00066	0.00069
−3.1	0.00071	0.00074	0.00076	0.00079	0.00082	0.00085	0.00087	0.00090	0.00094	0.00097
−3.0	0.00100	0.00104	0.00107	0.00111	0.00114	0.00118	0.00122	0.00126	0.00131	0.00135
−2.9	0.0014	0.0014	0.0015	0.0015	0.0016	0.0016	0.0017	0.0017	0.0018	0.0019
−2.8	0.0019	0.0020	0.0021	0.0021	0.0022	0.0023	0.0023	0.0024	0.0025	0.0026
−2.7	0.0026	0.0027	0.0028	0.0029	0.0030	0.0031	0.0032	0.0033	0.0034	0.0035
−2.6	0.0036	0.0037	0.0038	0.0039	0.0040	0.0041	0.0043	0.0044	0.0045	0.0047
−2.5	0.0048	0.0049	0.0051	0.0052	0.0054	0.0055	0.0057	0.0059	0.0060	0.0062
−2.4	0.0064	0.0066	0.0068	0.0069	0.0071	0.0073	0.0075	0.0078	0.0080	0.0082
−2.3	0.0084	0.0087	0.0089	0.0091	0.0094	0.0096	0.0099	0.0102	0.0104	0.0107
−2.2	0.0110	0.0113	0.0116	0.0119	0.0122	0.0125	0.0129	0.0132	0.0136	0.0139
−2.1	0.0143	0.0146	0.0150	0.0154	0.0158	0.0162	0.0166	0.0170	0.0174	0.0179
−2.0	0.0183	0.0188	0.0192	0.0197	0.0202	0.0207	0.0212	0.0217	0.0222	0.0228
−1.9	0.0233	0.0239	0.0244	0.0250	0.0256	0.0262	0.0268	0.0274	0.0281	0.0287
−1.8	0.0294	0.0301	0.0307	0.0314	0.0322	0.0329	0.0336	0.0344	0.0351	0.0359
−1.7	0.0367	0.0375	0.0384	0.0392	0.0401	0.0409	0.0418	0.0427	0.0436	0.0446
−1.6	0.0455	0.0465	0.0475	0.0485	0.0495	0.0505	0.0516	0.0526	0.0537	0.0548
−1.5	0.0559	0.0571	0.0582	0.0594	0.0606	0.0618	0.0630	0.0643	0.0655	0.0668
−1.4	0.0681	0.0694	0.0708	0.0721	0.0735	0.0749	0.0764	0.0778	0.0793	0.0808
−1.3	0.0823	0.0838	0.0853	0.0869	0.0885	0.0901	0.0918	0.0934	0.0951	0.0968
−1.2	0.0985	0.1003	0.1020	0.1038	0.1057	0.1075	0.1093	0.1112	0.1131	0.1151
−1.1	0.1170	0.1190	0.1210	0.1230	0.1251	0.1271	0.1292	0.1314	0.1335	0.1357
−1.0	0.1379	0.1401	0.1423	0.1446	0.1469	0.1492	0.1515	0.1539	0.1562	0.1587
−0.9	0.1611	0.1635	0.1660	0.1685	0.1711	0.1736	0.1762	0.1788	0.1814	0.1841
−0.8	0.1867	0.1894	0.1922	0.1949	0.1977	0.2005	0.2033	0.2061	0.2090	0.2119
−0.7	0.2148	0.2177	0.2207	0.2236	0.2266	0.2297	0.2327	0.2358	0.2389	0.2420
−0.6	0.2451	0.2483	0.2514	0.2546	0.2578	0.2611	0.2643	0.2676	0.2709	0.2743
−0.5	0.2776	0.2810	0.2843	0.2877	0.2912	0.2946	0.2981	0.3015	0.3050	0.3085
−0.4	0.3121	0.3156	0.3192	0.3228	0.3264	0.3300	0.3336	0.3372	0.3409	0.3446
−0.3	0.3483	0.3520	0.3557	0.3594	0.3632	0.3669	0.3707	0.3745	0.3783	0.3821
−0.2	0.3859	0.3897	0.3936	0.3974	0.4013	0.4052	0.4090	0.4129	0.4168	0.4207
−0.1	0.4247	0.4286	0.4325	0.4364	0.4404	0.4443	0.4483	0.4522	0.4562	0.4602
−0.0	0.4641	0.4681	0.4721	0.4761	0.4801	0.4840	0.4880	0.4920	0.4960	0.5000

Adapted from E. L. Grant, "Statistical Quality Control," 2d ed., Table A, pp. 510–511, McGraw-Hill Book Company, Inc., New York, 1952. By permission from the publisher.

$Z = (X_i - \bar{X})/\sigma$ [see Eq. (4-2)].

TABLE 3. AREAS UNDER THE NORMAL CURVE (*Continued*)
(Proportion of total area under the curve from $-\infty$ to designated Z value)

Z	0.00	0.01	0.02	0.03	0.04	0.05	0.06	0.07	0.08	0.09
+0.0	0.5000	0.5040	0.5080	0.5120	0.5160	0.5199	0.5239	0.5279	0.5319	0.5359
+0.1	0.5398	0.5438	0.5478	0.5517	0.5557	0.5596	0.5636	0.5675	0.5714	0.5753
+0.2	0.5793	0.5832	0.5871	0.5910	0.5948	0.5987	0.6026	0.6064	0.6103	0.6141
+0.3	0.6179	0.6217	0.6255	0.6293	0.6331	0.6368	0.6406	0.6443	0.6480	0.6517
+0.4	0.6554	0.6591	0.6628	0.6664	0.6700	0.6736	0.6772	0.6808	0.6844	0.6879
+0.5	0.6915	0.6950	0.6985	0.7019	0.7054	0.7088	0.7123	0.7157	0.7190	0.7224
+0.6	0.7257	0.7291	0.7324	0.7357	0.7389	0.7422	0.7454	0.7486	0.7517	0.7549
+0.7	0.7580	0.7611	0.7642	0.7673	0.7704	0.7734	0.7764	0.7794	0.7823	0.7852
+0.8	0.7881	0.7910	0.7939	0.7967	0.7995	0.8023	0.8051	0.8079	0.8106	0.8133
+0.9	0.8159	0.8186	0.8212	0.8238	0.8264	0.8289	0.8315	0.8340	0.8365	0.8389
+1.0	0.8413	0.8438	0.8461	0.8485	0.8508	0.8531	0.8554	0.8577	0.8599	0.8621
+1.1	0.8643	0.8665	0.8686	0.8708	0.8729	0.8749	0.8770	0.8790	0.8810	0.8830
+1.2	0.8849	0.8869	0.8888	0.8907	0.8925	0.8944	0.8962	0.8980	0.8997	0.9015
+1.3	0.9032	0.9049	0.9066	0.9082	0.9099	0.9115	0.9131	0.9147	0.9162	0.9177
+1.4	0.9192	0.9207	0.9222	0.9236	0.9251	0.9265	0.9279	0.9292	0.9306	0.9319
+1.5	0.9332	0.9345	0.9357	0.9370	0.9382	0.9394	0.9406	0.9418	0.9429	0.9441
+1.6	0.9452	0.9463	0.9474	0.9484	0.9495	0.9505	0.9515	0.9525	0.9535	0.9545
+1.7	0.9554	0.9564	0.9573	0.9582	0.9591	0.9599	0.9608	0.9616	0.9625	0.9633
+1.8	0.9641	0.9649	0.9656	0.9664	0.9671	0.9678	0.9686	0.9693	0.9699	0.9706
+1.9	0.9713	0.9719	0.9726	0.9732	0.9738	0.9744	0.9750	0.9756	0.9761	0.9767
+2.0	0.9773	0.9778	0.9783	0.9788	0.9793	0.9798	0.9803	0.9808	0.9812	0.9817
+2.1	0.9821	0.9826	0.9830	0.9834	0.9838	0.9842	0.9846	0.9850	0.9854	0.9857
+2.2	0.9861	0.9864	0.9868	0.9871	0.9875	0.9878	0.9881	0.9884	0.9887	0.9890
+2.3	0.9893	0.9896	0.9898	0.9901	0.9904	0.9906	0.9909	0.9911	0.9913	0.9916
+2.4	0.9918	0.9920	0.9922	0.9925	0.9927	0.9929	0.9931	0.9932	0.9934	0.9936
+2.5	0.9938	0.9940	0.9941	0.9943	0.9945	0.9946	0.9948	0.9949	0.9951	0.9952
+2.6	0.9953	0.9955	0.9956	0.9957	0.9959	0.9960	0.9961	0.9962	0.9963	0.9964
+2.7	0.9965	0.9966	0.9967	0.9968	0.9969	0.9970	0.9971	0.9972	0.9973	0.9974
+2.8	0.9974	0.9975	0.9976	0.9977	0.9977	0.9978	0.9979	0.9979	0.9980	0.9981
+2.9	0.9981	0.9982	0.9983	0.9983	0.9984	0.9984	0.9985	0.9985	0.9986	0.9986
+3.0	0.99865	0.99869	0.99874	0.99878	0.99882	0.99886	0.99889	0.99893	0.99896	0.99900
+3.1	0.99903	0.99906	0.99910	0.99913	0.99915	0.99918	0.99921	0.99924	0.99926	0.99929
+3.2	0.99931	0.99934	0.99936	0.99938	0.99940	0.99942	0.99944	0.99946	0.99948	0.99950
+3.3	0.99952	0.99953	0.99955	0.99957	0.99958	0.99960	0.99961	0.99962	0.99964	0.99965
+3.4	0.99966	0.99967	0.99969	0.99970	0.99971	0.99972	0.99973	0.99974	0.99975	0.99976
+3.5	0.99977	0.99978	0.99978	0.99979	0.99980	0.99981	0.99981	0.99982	0.99983	0.99983

TABLE 4. NORMAL-CURVE ORDINATES*

Ordinates (heights) of the unit normal curve. The height (Y) at any number of standard deviations Z from the mean is

$$Y = \frac{1}{\sqrt{2\pi}} e^{-Z^2/2} = 0.3989e^{-Z^2/2}$$

To obtain answers in units of particular problems, multiply these ordinates by $N\theta/\sigma$, where N is the number of cases, θ the class interval, and σ the standard deviation.
Each figure in the body of the table is preceded by a decimal point.

Z	0.00	0.01	0.02	0.03	0.04	0.05	0.06	0.07	0.08	0.09
0.0	39894	39892	39886	39876	39862	39844	39822	39797	39767	39733
0.1	39695	39654	39608	39559	39505	39448	39387	39322	39253	39181
0.2	39104	39024	38940	38853	38762	38667	38568	38466	38361	38251
0.3	38139	38023	37903	37780	37654	37524	37391	37255	37115	36973
0.4	36827	36678	36526	36371	36213	36053	35889	35723	35553	35381
0.5	35207	35029	34849	34667	34482	34294	34105	33912	33718	33521
0.6	33322	33121	32918	32713	32506	32297	32086	31874	31659	31443
0.7	31225	31006	30785	30563	30339	30114	29887	29658	29430	29200
0.8	28969	28737	28504	28269	28034	27798	27562	27324	27086	26848
0.9	26609	26369	26129	25888	25647	25406	25164	24923	24681	24439
1.0	24197	23955	23713	23471	23230	22988	22747	22506	22265	22025
1.1	21785	21546	21307	21069	20831	20594	20357	20121	19886	19652
1.2	19419	19186	18954	18724	18494	18265	18037	17810	17585	17360
1.3	17137	16915	16694	16474	16256	16038	15822	15608	15395	15183
1.4	14973	14764	14556	14350	14146	13943	13742	13542	13344	13147
1.5	12952	12758	12566	12376	12188	12001	11816	11632	11450	11270
1.6	11092	10915	10741	10567	10396	10226	10059	09893	09728	09566
1.7	09405	09246	09089	08933	08780	08628	08478	08329	08183	08038
1.8	07895	07754	07614	07477	07341	07206	07074	06943	06814	06687
1.9	06562	06438	06316	06195	06077	05959	05844	05730	05618	05508
2.0	05399	05292	05186	05082	04980	04879	04780	04682	04586	04491
2.1	04398	04307	04217	04128	04041	03955	03871	03788	03706	03626
2.2	03547	03470	03394	03319	03246	03174	03103	03034	02965	02898
2.3	02833	02768	02705	02643	02582	02522	02463	02406	02349	02294
2.4	02239	02186	02134	02083	02033	01984	01936	01888	01842	01797
2.5	01753	01709	01667	01625	01585	01545	01506	01468	01431	01394
2.6	01358	01323	01289	01256	01223	01191	01160	01130	01100	01071
2.7	01042	01014	00987	00961	00935	00909	00885	00861	00837	00814
2.8	00792	00770	00748	00727	00707	00687	00668	00649	00631	00613
2.9	00595	00578	00562	00545	00530	00514	00499	00485	00470	00457
3.0	00443									
3.5	0008727									
4.0	0001338									
4.5	0000160									
5.0	000001487									

* This table was adapted, by permission, from F. C. Kent, *Elements of Statistics*, McGraw-Hill, New York, 1924.

$Z = (X_i - \bar{X})/\sigma$ [see Eq. (4-2)].

TABLE 5. FACTORS FOR ESTIMATING σ FROM $\bar{R}$*

Number of observations in subgroup, n	Factor for estimate from $\bar{R}$, d_2	Number of observations in subgroup, n	Factor for estimate from $\bar{R}$, d_2
2	1.128	21	3.778
3	1.693	22	3.819
4	2.059	23	3.858
5	2.326	24	3.895
6	2.534	25	3.931
7	2.704	30	4.086
8	2.847	35	4.213
9	2.970	40	4.322
10	3.078	45	4.415
11	3.173	50	4.498
12	3.258	55	4.572
13	3.336	60	4.639
14	3.407	65	4.699
15	3.472	70	4.755
16	3.532	75	4.806
17	3.588	80	4.854
18	3.640	85	4.898
19	3.689	90	4.939
20	3.735	95	4.978
		100	5.015

* Adapted from E. L. Grant, "Statistical Quality Control," 2d ed., Table B, p. 512, McGraw-Hill Book Company, Inc., New York, 1952. By permission from the publisher.

TABLE 6. NATURAL, NAPIERIAN, OR HYPERBOLIC LOGARITHMS*

N	0	1	2	3	4	5	6	7	8	9
0	$-\infty$	0.0000	0.6931	1.0986	1.3863	1.6094	1.7918	1.9459	2.0794	2.1972
10	2.3026	2.3979	2.4849	2.5649	2.6391	2.7081	2.7726	2.8332	2.8904	2.9444
20	2.9957	3.0445	3.0910	3.1355	3.1781	3.2189	3.2581	3.2958	3.3322	3.3673
30	3.4012	3.4340	3.4657	3.4965	3.5264	3.5553	3.5835	3.6109	3.6376	3.6636
40	3.6889	3.7136	3.7377	3.7612	3.7842	3.8067	3.8286	3.8501	3.8712	3.8918
50	3.9120	3.9318	3.9512	3.9703	3.9890	4.0073	4.0254	4.0431	4.0604	4.0775
60	4.0943	4.1109	4.1271	4.1431	4.1589	4.1744	4.1897	4.2047	4.2195	4.2341
70	4.2485	4.2627	4.2767	4.2905	4.3041	4.3175	4.3307	4.3438	4.3567	4.3694
80	4.3820	4.3944	4.4067	4.4188	4.4308	4.4427	4.4543	4.4659	4.4773	4.4886
90	4.4998	4.5109	4.5218	4.5326	4.5433	4.5539	4.5643	4.5747	4.5850	4.5951
100	4.6052	4.6151	4.6250	4.6347	4.6444	4.6540	4.6634	4.6728	4.6821	4.6913
110	4.7005	4.7095	4.7185	4.7274	4.7362	4.7449	4.7536	4.7622	4.7707	4.7791
120	4.7875	4.7958	4.8040	4.8122	4.8203	4.8283	4.8363	4.8442	4.8520	4.8598
130	4.8675	4.8752	4.8828	4.8903	4.8978	4.9053	4.9127	4.9200	4.9273	4.9345
140	4.9416	4.9488	4.9558	4.9628	4.9698	4.9767	4.9836	4.9904	4.9972	5.0039
150	5.0106	5.0173	5.0239	5.0304	5.0370	5.0434	5.0499	5.0562	5.0626	5.0689
160	5.0752	5.0814	5.0876	5.0938	5.0999	5.1059	5.1120	5.1180	5.1240	5.1299
170	5.1358	5.1417	5.1475	5.1533	5.1591	5.1648	5.1705	5.1761	5.1818	5.1874
180	5.1930	5.1985	5.2040	5.2095	5.2149	5.2204	5.2257	5.2311	5.2364	5.2417
190	5.2470	5.2523	5.2575	5.2627	5.2679	5.2730	5.2781	5.2832	5.2883	5.2933
200	5.2983	5.3033	5.3083	5.3132	5.3181	5.3230	5.3279	5.3327	5.3375	5.3423
210	5.3471	5.3519	5.3566	5.3613	5.3660	5.3706	5.3753	5.3799	5.3845	5.3891
220	5.3936	5.3982	5.4027	5.4072	5.4116	5.4161	5.4205	5.4250	5.4293	5.4337
230	5.4381	5.4424	5.4467	5.4510	5.4553	5.4596	5.4638	5.4681	5.4723	5.4765
240	5.4806	5.4848	5.4889	5.4931	5.4972	5.5013	5.5053	5.5094	5.5134	5.5175
250	5.5215	5.5255	5.5294	5.5334	5.5373	5.5413	5.5452	5.5491	5.5530	5.5568
260	5.5607	5.5645	5.5683	5.5722	5.5759	5.5797	5.5835	5.5872	5.5910	5.5947
270	5.5984	5.6021	5.6058	5.6095	5.6131	5.6168	5.6204	5.6240	5.6276	5.6312
280	5.6348	5.6384	5.6419	5.6454	5.6490	5.6525	5.6560	5.6595	5.6630	5.6664
290	5.6699	5.6733	5.6768	5.6802	5.6836	5.6870	5.6904	5.6937	5.6971	5.7004
300	5.7038	5.7071	5.7104	5.7137	5.7170	5.7203	5.7236	5.7268	5.7301	5.7333
310	5.7366	5.7398	5.7430	5.7462	5.7494	5.7526	5.7557	5.7589	5.7621	5.7652
320	5.7683	5.7714	5.7746	5.7777	5.7807	5.7838	5.7869	5.7900	5.7930	5.7961
330	5.7991	5.8021	5.8051	5.8081	5.8111	5.8141	5.8171	5.8201	5.8230	5.8260
340	5.8289	5.8319	5.8348	5.8377	5.8406	5.8435	5.8464	5.8493	5.8522	5.8551
350	5.8579	5.8608	5.8636	5.8665	5.8693	5.8721	5.8749	5.8777	5.8805	5.8833
360	5.8861	5.8889	5.8916	5.8944	5.8972	5.8999	5.9026	5.9054	5.9081	5.9108
370	5.9135	5.9162	5.9189	5.9216	5.9243	5.9269	5.9296	5.9322	5.9349	5.9375
380	5.9402	5.9428	5.9454	5.9480	5.9506	5.9532	5.9558	5.9584	5.9610	5.9636
390	5.9661	5.9687	5.9713	5.9738	5.9764	5.9789	5.9814	5.9839	5.9865	5.9890
400	5.9915	5.9940	5.9965	5.9989	6.0014	6.0039	6.0064	6.0088	6.0113	6.0137
410	6.0162	6.0186	6.0210	6.0234	6.0259	6.0283	6.0307	6.0331	6.0355	6.0379
420	6.0403	6.0426	6.0450	6.0474	6.0497	6.0521	6.0544	6.0568	6.0591	6.0615
430	6.0638	6.0661	6.0684	6.0707	6.0730	6.0753	6.0776	6.0799	6.0822	6.0845
440	6.0868	6.0890	6.0913	6.0936	6.0958	6.0981	6.1003	6.1026	6.1048	6.1070
450	6.1092	6.1115	6.1137	6.1159	6.1181	6.1203	6.1225	6.1247	6.1269	6.1291
460	6.1312	6.1334	6.1356	6.1377	6.1399	6.1420	6.1442	6.1463	6.1485	6.1506
470	6.1527	6.1549	6.1570	6.1591	6.1612	6.1633	6.1654	6.1675	6.1696	6.1717
480	6.1738	6.1759	6.1779	6.1800	6.1821	6.1841	6.1862	6.1883	6.1903	6.1924
490	6.1944	6.1964	6.1985	6.2005	6.2025	6.2046	6.2066	6.2086	6.2106	6.2126

NOTE 1: Moving the decimal point n places to the right (or left) in the number is equivalent to adding (or subtracting) n times 2.3026.

NOTE 2:

$$\log_e x = 2.3026 \log_{10} x$$
$$\log_{10} x = 0.4343 \log_e x$$
$$\log_e 10 = 2.3026$$
$$\log_{10} e = 0.4343$$

n	$n \times 2.3026$
1	2.3026 = 0.6974–3
2	4.6052 = 0.3948–5
3	6.9078 = 0.0922–7
4	9.2103 = 0.7897–10
5	11.5129 = 0.4871–12
6	13.8155 = 0.1845–14
7	16.1181 = 0.8819–17
8	18.4207 = 0.5793–19
9	20.7233 = 0.2767–21

* From A. E. Knowlton, *Standard Handbook for Electrical Engineers*, 9th ed., McGraw-Hill, New York, 1957.

TABLE 6. NATURAL, NAPIERIAN, OR HYPERBOLIC LOGARITHMS (*Continued*)

N	0	1	2	3	4	5	6	7	8	9
500	6.2146	6.2166	6.2186	6.2206	6.2226	6.2246	6.2265	6.2285	6.2305	6.2324
510	6.2344	6.2364	6.2383	6.2403	6.2422	6.2442	6.2461	6.2480	6.2500	6.2519
520	6.2538	6.2558	6.2577	6.2596	6.2615	6.2634	6.2653	6.2672	6.2691	6.2710
530	6.2729	6.2748	6.2766	6.2785	6.2804	6.2823	6.2841	6.2860	6.2879	6.2897
540	6.2916	6.2934	6.2953	6.2971	6.2989	6.3008	6.3026	6.3044	6.3063	6.3081
550	6.3099	6.3117	6.3135	6.3154	6.3172	6.3190	6.3208	6.3226	6.3244	6.3261
560	6.3279	6.3297	6.3315	6.3333	6.3351	6.3368	6.3386	6.3404	6.3421	6.3439
570	6.3456	6.3474	6.3491	6.3509	6.3256	6.3544	6.3561	6.3578	6.3596	6.3613
580	6.3630	6.3648	6.3665	6.3682	6.3699	6.3716	6.3733	6.3750	6.3767	6.3784
590	6.3801	6.3818	6.3835	6.3852	6.3869	6.3886	6.3902	6.3919	6.3936	6.3953
600	6.3969	6.3986	6.4003	6.4019	6.4036	6.4052	6.4069	6.4085	6.4102	6.4118
610	6.4135	6.4151	6.4167	6.4184	6.4200	6.4216	6.4232	6.4249	6.4265	6.4281
620	6.4297	6.4313	6.4329	6.4345	6.4362	6.4378	6.4394	6.4409	6.4425	6.4441
630	6.4457	6.4473	6.4489	6.4505	6.4520	6.4536	6.4552	6.4568	6.4583	6.4599
640	6.4615	6.4630	6.4646	6.4661	6.4677	6.4693	6.4708	6.4723	6.4739	6.4754
650	6.4770	6.4785	6.4800	6.4816	6.4831	6.4846	6.4862	6.4877	6.4892	6.4907
660	6.4922	6.4938	6.4953	6.4968	6.4983	6.4998	6.5013	6.5028	6.5043	6.5058
670	6.5073	6.5088	6.5103	6.5117	6.5132	6.5147	6.5162	6.5177	6.5191	6.5206
680	6.5221	6.5236	6.5250	6.5265	6.5280	6.5294	6.5309	6.5323	6.5338	6.5352
690	6.5367	6.5381	6.5396	6.5410	6.5425	6.5439	6.5453	6.5468	6.5482	6.5497
700	6.5511	6.5525	6.5539	6.5554	6.5568	6.5582	6.5596	6.5610	6.5624	6.5639
710	6.5653	6.5667	6.5681	6.5695	6.5709	6.5723	6.5737	6.5751	6.5765	6.5779
720	6.5793	6.5806	6.5820	6.5834	6.5848	6.5862	6.5876	6.5889	6.5903	6.5917
730	6.5930	6.5944	6.5958	6.5971	6.5985	6.5999	6.6012	6.6026	6.6039	6.6053
740	6.6067	6.6080	6.6093	6.6107	6.6120	6.6134	6.6147	6.6161	6.6174	6.6187
750	6.6201	6.6214	6.6227	6.6241	6.6254	6.6267	6.6280	6.6294	6.6307	6.6320
760	6.6333	6.6346	6.6350	6.6373	6.6386	6.6399	6.6412	6.6425	6.6438	6.6451
770	6.6464	6.6477	6.6490	6.6503	6.6516	6.6529	6.6542	6.6554	6.6567	6.6580
780	6.6593	6.6606	6.6619	6.6631	6.6644	6.6657	6.6670	6.6682	6.6695	6.6708
790	6.6720	6.6733	6.6746	6.6758	6.6771	6.6783	6.6796	6.6809	6.6821	6.6834
800	6.6846	6.6859	6.6871	6.6884	6.6896	6.6908	6.6921	6.6933	6.6946	6.6958
810	6.6970	6.6983	6.6995	6.7007	6.7020	6.7032	6.7044	6.7056	6.7069	6.7081
820	6.7093	6.7105	6.7117	6.7130	6.7142	6.7154	6.7166	6.7178	6.7190	6.7202
830	6.7214	6.7226	6.7238	6.7250	6.7262	6.7274	6.7286	6.7298	6.7310	6.7322
840	6.7334	6.7346	6.7358	6.7370	6.7382	6.7393	6.7405	6.7417	6.6429	6.7441
850	6.7452	6.7464	6.7476	6.7488	6.7499	6.7511	6.7523	6.7534	6.7546	6.7558
860	6.7569	6.7581	6.7593	6.7604	6.7616	6.7627	6.7639	6.7650	6.7662	6.7673
870	6.7685	6.7696	6.7708	6.7719	6.7731	6.7742	6.7754	6.7765	6.7776	6.7788
880	6.7799	6.7811	6.7822	6.7833	6.7845	6.7856	6.7867	5.7878	6.7890	6.7901
890	6.7912	6.7923	6.7935	6.7946	6.7957	6.7968	6.7979	6.7991	6.8002	6.8013
900	6.8024	6.8035	6.8046	6.8057	6.8068	6.8079	6.8090	6.8101	6.8112	6.8123
910	6.8134	6.8145	6.8156	6.8167	6.8178	6.8189	6.8200	6.8211	6.8222	6.8233
920	6.8244	6.8255	6.8265	6.8276	6.8287	6.8298	6.8309	6.8320	6.8330	6.8341
930	6.8352	6.8363	6.8373	6.8384	6.8395	6.8405	6.8416	6.8427	6.8437	6.8448
940	6.8459	6.8469	6.8480	6.8491	6.8501	6.8512	6.8522	6.8533	6.8544	6.8554
950	6.8565	6.8575	6.8586	6.8596	6.8607	6.8617	6.8628	6.8638	6.8648	6.8659
960	6.8669	6.8680	6.8690	6.8701	6.8711	6.8721	6.8732	6.8742	6.8752	6.8763
970	6.8773	6.8783	6.8794	6.8804	6.8814	6.8824	6.8835	6.8845	6.8855	6.8865
980	6.8876	6.8886	6.8896	6.8906	6.8916	6.8926	6.8937	6.8947	6.8957	6.8967
990	6.8977	6.8987	6.8997	6.9007	6.9017	6.9027	6.9037	6.9047	6.9057	6.9068

TABLE 7. VALUES AND LOGARITHMS OF EXPONENTIAL FUNCTIONS*

x	e^{x}		e^{-x}	x	e^{x}		e^{-x}
	Value	$\log_{10}$	(value)		Value	$\log_{10}$	(value)
0.00	1.0000	0.00000	1.00000	0.50	1.6487	0.21715	0.60653
0.01	1.0101	.00434	0.99005	0.51	1.6653	.22149	.60050
0.02	1.0202	.00869	.98020	0.52	1.6820	.22583	.59452
0.03	1.0305	.01303	.97045	0.53	1.6989	.23018	.58860
0.04	1.0408	.01737	.96079	0.54	1.7160	.23452	.58275
0.05	1.0513	.02171	.95123	0.55	1.7333	.23886	.57695
0.06	1.0618	.02606	.94176	0.56	1.7507	.24320	.57121
0.07	1.0725	.03040	.93239	0.57	1.7683	.24755	.56553
0.08	1.0833	.03474	.92312	0.58	1.7860	.25189	.55990
0.09	1.0942	.03909	.91393	0.59	1.8040	.25623	.55433
0.10	1.1052	.04343	.90484	0.60	1.8221	.26058	.54881
0.11	1.1163	.04777	.89583	0.61	1.8404	.26492	.54335
0.12	1.1275	.05212	.88692	0.62	1.8589	.26926	.53794
0.13	1.1388	.05646	.87809	0.63	1.8776	.27361	.53259
0.14	1.1503	.06080	.86936	0.64	1.8965	.27795	.52729
0.15	1.1618	.06514	.86071	0.65	1.9155	.28229	.52205
0.16	1.1735	.06949	.85214	0.66	1.9348	.28664	.51685
0.17	1.1853	.07383	.84366	0.67	1.9542	.29098	.51171
0.18	1.1972	.07817	.83527	0.68	1.9739	.29532	.50662
0.19	1.2092	.08252	.82696	0.69	1.9937	.29966	.50158
0.20	1.2214	.08686	.81873	0.70	2.0138	.30401	.49659
0.21	1.2337	.09120	.81058	0.71	2.0340	.30835	.49164
0.22	1.2461	.09554	.80252	0.72	2.0544	.31269	.48675
0.23	1.2586	.09989	.79453	0.73	2.0751	.31703	.48191
0.24	1.2712	.10423	.78663	0.74	2.0959	.32138	.47711
0.25	1.2840	.10857	.77880	0.75	2.1170	.32572	.47237
0.26	1.2969	.11292	.77105	0.76	2.1383	.33006	.46767
0.27	1.3100	.11726	.76338	0.77	2.1598	.33441	.46301
0.28	1.3231	.12160	.75578	0.78	2.1815	.33875	.45841
0.29	1.3364	.12595	.74826	0.79	2.2034	.34309	.45384
0.30	1.3499	.13029	.74082	0.80	2.2255	.34744	.44933
0.31	1.3634	.13463	.73345	0.81	2.2479	.35178	.44486
0.32	1.3771	.13897	.72615	0.82	2.2705	.35612	.44043
0.33	1.3910	.14332	.71892	0.83	2.2933	.36046	.43605
0.34	1.4049	.14766	.71177	0.84	2.3164	.36481	.43171
0.35	1.4191	.15200	.70469	0.85	2.3396	.36915	.42741
0.36	1.4333	.15635	.69768	0.86	2.3632	.37349	.42316
0.37	1.4477	.16069	.69073	0.87	2.3869	.37784	.41895
0.38	1.4623	.16503	.68386	0.88	2.4109	.38218	.41478
0.39	1.4770	.16937	.67706	0.89	2.4351	.38652	.41066
0.40	1.4918	.17372	.67032	0.90	2.4596	.39087	.40657
0.41	1.5068	.17806	.66365	0.91	2.4843	.39521	.40252
0.42	1.5220	.18240	.65705	0.92	2.5093	.39955	.39852
0.43	1.5373	.18675	.65051	0.93	2.5345	.40389	.39455
0.44	1.5527	.19109	.64404	0.94	2.5600	.40824	.39063
0.45	1.5683	.19543	.63763	0.95	2.5857	.41258	.38674
0.46	1.5841	.19978	.63128	0.96	2.6117	.41692	.38289
0.47	1.6000	.20412	.62500	0.97	2.6379	.42127	.37908
0.48	1.6161	.20846	.61878	0.98	2.6645	.42561	.37531
0.49	1.6323	.21280	.61263	0.99	2.6912	.42995	.37158
0.50	1.6487	.21715	.60653	1.00	2.7183	.43429	.36788

* From R. H. Perry, "Engineering Manual," McGraw-Hill Book Company, Inc., New York, 1959.

TABLE 7. VALUES AND LOGARITHMS OF EXPONENTIAL FUNCTIONS (*Continued*)

x	e^x		e^{-x} (value)	x	e^x		e^{-x} (value)
	Value	$\log_{10}$			Value	$\log_{10}$	
1.00	2.7183	0.43429	0.36788	1.50	4.4817	0.65144	0.22313
1.01	2.7456	.43864	.36422	1.51	4.5267	.65578	.22091
1.02	2.7732	.44298	.36060	1.52	4.5722	.66013	.21871
1.03	2.8011	.44732	.35701	1.53	4.6182	.66447	.21654
1.04	2.8292	.45167	.35345	1.54	4.6646	.66881	.21438
1.05	2.8577	.45601	.34994	1.55	4.7115	.67316	.21225
1.06	2.8864	.46035	.34646	1.56	4.7588	.67750	.21014
1.07	2.9154	.46470	.34301	1.57	4.8066	.68184	.20805
1.08	2.9447	.46904	.33960	1.58	4.8550	.68619	.20598
1.09	2.9743	.47338	.33622	1.59	4.9037	.69053	.20393
1.10	3.0042	.47772	.33287	1.60	4.9530	.69487	.20190
1.11	3.0344	.48207	.32956	1.61	5.0028	.69921	.19989
1.12	3.0649	.48641	.32628	1.62	5.0531	.70356	.19790
1.13	3.0957	.49075	.32303	1.63	5.1039	.70790	.19593
1.14	3.1268	.49510	.31982	1.64	5.1552	.71224	.19398
1.15	3.1582	.49944	.31664	1.65	5.2070	.71659	.19205
1.16	3.1899	.50378	.31349	1.66	5.2593	.72093	.19014
1.17	3.2220	.50812	.31037	1.67	5.3122	.72527	.18825
1.18	3.2544	.51247	.30728	1.68	5.3656	.72961	.18637
1.19	3.2871	.51681	.30422	1.69	5.4195	.73396	.18452
1.20	3.3201	.52115	.30119	1.70	5.4739	.73830	.18268
1.21	3.3535	.52550	.29820	1.71	5.5290	.74264	.18087
1.22	3.3872	.52984	.29523	1.72	5.5845	.74699	.17907
1.23	3.4212	.53418	.29229	1.73	5.6407	.75133	.17728
1.24	3.4556	.53853	.28938	1.74	5.6973	.75567	.17552
1.25	3.4903	.54287	.28650	1.75	5.7546	.76002	.17377
1.26	3.5254	.54721	.28365	1.76	5.8124	.76436	.17204
1.27	3.5609	.55155	.28083	1.77	5.8709	.76870	.17033
1.28	3.5966	.55590	.27804	1.78	5.9299	.77304	.16864
1.29	3.6328	.56024	.27527	1.79	5.9895	.77739	.16696
1.30	3.6693	.56458	.27253	1.80	6.0496	.78173	.16530
1.31	3.7062	.56893	.26982	1.81	6.1104	.78607	.16365
1.32	3.7434	.57327	.26714	1.82	6.1719	.79042	.16203
1.33	3.7810	.57761	.26448	1.83	6.2339	.79476	.16041
1.34	3.8190	.58195	.26185	1.84	6.2965	.79910	.15882
1.35	3.8574	.58630	.25924	1.85	6.3598	.80344	.15724
1.36	3.8962	.59064	.25666	1.86	6.4237	.80779	.15567
1.37	3.9354	.59498	.25411	1.87	6.4883	.81213	.15412
1.38	3.9749	.59933	.25158	1.88	6.5535	.81647	.15259
1.39	4.0149	.60367	.24908	1.89	6.6194	.82082	.15107
1.40	4.0552	.60801	.24660	1.90	6.6859	.82516	.14957
1.41	4.0960	.61236	.24414	1.91	6.7531	.82950	.14808
1.42	4.1371	.61670	.24171	1.92	6.8210	.83385	.14661
1.43	4.1787	.62104	.23931	1.93	6.8895	.83819	.14515
1.44	4.2207	.62538	.23693	1.94	6.9588	.84253	.14370
1.45	4.2631	.62973	.23457	1.95	7.0287	.84687	.14227
1.46	4.3060	.63407	.23224	1.96	7.0993	.85122	.14086
1.47	4.3492	.63841	.22993	1.97	7.1707	.85556	.13946
1.48	4.3929	.64276	.22764	1.98	7.2427	.85990	.13807
1.49	4.4371	.64710	.22537	1.99	7.3155	.86425	.13670
1.50	4.4817	.65144	.22313	2.00	7.3891	.86859	.13534

TABLE 7. VALUES AND LOGARITHMS OF EXPONENTIAL FUNCTIONS (*Continued*)

x	e^x		e^{-x} (value)	x	e^x		e^{-x} (value)
	Value	$\log_{10}$			Value	$\log_{10}$	
2.00	7.3891	0.86859	0.13534	2.50	12.182	1.08574	0.08208
2.01	7.4633	0.87293	.13399	2.51	12.305	1.09008	.08127
2.02	7.5383	0.87727	.13266	2.52	12.429	1.09442	.08046
2.03	7.6141	0.88162	.13134	2.53	12.554	1.09877	.07966
2.04	7.6906	0.88596	.13003	2.54	12.680	1.10311	.07887
2.05	7.7679	0.89030	.12873	2.55	12.807	1.10745	.07808
2.06	7.8460	0.89465	.12745	2.56	12.936	1.11179	.07730
2.07	7.9248	0.89899	.12619	2.57	13.066	1.11614	.07654
2.08	8.0045	0.90333	.12493	2.58	13.197	1.12048	.07577
2.09	8.0849	0.90768	.12369	2.59	13.330	1.12482	.07502
2.10	8.1662	0.91202	.12246	2.60	13.464	1.12917	.07427
2.11	8.2482	0.91636	.12124	2.61	13.599	1.13351	.07353
2.12	8.3311	0.92070	.12003	2.62	13.736	1.13785	.07280
2.13	8.4149	0.92505	.11884	2.63	13.874	1.14219	.07208
2.14	8.4994	0.92939	.11765	2.64	14.013	1.14654	.07136
2.15	8.5849	0.93373	.11648	2.65	14.154	1.15088	.07065
2.16	8.6711	0.93808	.11533	2.66	14.296	1.15522	.06995
2.17	8.7583	0.94242	.11418	2.67	14.440	1.15957	.06925
2.18	8.8463	0.94676	.11304	2.68	14.585	1.16391	.06856
2.19	8.9352	0.95110	.11192	2.69	14.732	1.16825	.06788
2.20	9.0250	0.95545	.11080	2.70	14.880	1.17260	.06721
2.21	9.1157	0.95979	.10970	2.71	15.029	1.17694	.06654
2.22	9.2073	0.96413	.10861	2.72	15.180	1.18128	.06587
2.23	9.2999	0.96848	.10753	2.73	15.333	1.18562	.06522
2.24	9.3933	0.97282	.10646	2.74	15.487	1.18997	.06457
2.25	9.4877	0.97716	.10540	2.75	15.643	1.19431	.06393
2.26	9.5831	0.98151	.10435	2.76	15.800	1.19865	.06329
2.27	9.6794	0.98585	.10331	2.77	15.959	1.20300	.06266
2.28	9.7767	0.99019	.10228	2.78	16.119	1.20734	.06204
2.29	9.8749	0.99453	.10127	2.79	16.281	1.21168	.06142
2.30	9.9742	0.99888	.10026	2.80	16.445	1.21602	.06081
2.31	10.074	1.00322	.09926	2.81	16.610	1.22037	.06020
2.32	10.176	1.00756	.09827	2.82	16.777	1.22471	.05961
2.33	10.278	1.01191	.09730	2.83	16.945	1.22905	.05901
2.34	10.381	1.01625	.09633	2.84	17.116	1.23340	.05843
2.35	10.486	1.02059	.09537	2.85	17.288	1.23774	.05784
2.36	10.591	1.02493	.09442	2.86	17.462	1.24208	.05727
2.37	10.697	1.02928	.09348	2.87	17.637	1.24643	.05670
2.38	10.805	1.03362	.09255	2.88	17.814	1.25077	.05613
2.39	10.913	1.03796	.09163	2.89	17.993	1.25511	.05558
2.40	11.023	1.04231	.09072	2.90	18.174	1.25945	.05502
2.41	11.134	1.04665	.08982	2.91	18.357	1.26380	.05448
2.42	11.246	1.05099	.08892	2.92	18.541	1.26814	.05393
2.43	11.359	1.05534	.08804	2.93	18.728	1.27248	.05340
2.44	11.473	1.05968	.08716	2.94	18.916	1.27683	.05287
2.45	11.588	1.06402	.08629	2.95	19.106	1.28117	.05234
2.46	11.705	1.06836	.08543	2.96	19.298	1.28551	.05182
2.47	11.822	1.07271	.08458	2.97	19.492	1.28985	.05130
2.48	11.941	1.07705	.08374	2.98	19.688	1.29420	.05079
2.49	12.061	1.08139	.08291	2.99	19.886	1.29854	.05029
2.50	12.182	1.08574	.08208	3.00	20.086	1.30288	.04979

TABLE 7. VALUES AND LOGARITHMS OF EXPONENTIAL FUNCTIONS (*Continued*)

x	e^x Value	e^x $\log_{10}$	e^{-x} (value)
3.00	20.086	1.30288	0.04979
3.05	21.115	1.32460	.04736
3.10	22.198	1.34631	.04505
3.15	23.336	1.36803	.04285
3.20	24.533	1.38974	.04076
3.25	25.790	1.41146	.03877
3.30	27.113	1.43317	.03688
3.35	28.503	1.45489	.03508
3.40	29.964	1.47660	.03337
3.45	31.500	1.49832	.03175
3.50	33.115	1.52003	.03020
3.55	34.813	1.54175	.02872
3.60	36.598	1.56346	.02732
3.65	38.475	1.58517	.02599
3.70	40.447	1.60689	.02472
3.75	42.521	1.62860	.02352
3.80	44.701	1.65032	.02237
3.85	46.993	1.67203	.02128
3.90	49.402	1.69375	.02024
3.95	51.935	1.71546	.01925
4.00	54.598	1.73718	.01832
4.10	60.340	1.78061	.01657
4.20	66.686	1.82404	.01500
4.30	73.700	1.86747	.01357
4.40	81.451	1.91090	.01227
4.50	90.017	1.95433	.01111
4.60	99.484	1.99775	.01005
4.70	109.95	2.04118	.00910
4.80	121.51	2.08461	.00823
4.90	134.29	2.12804	.00745
5.00	148.41	2.17147	.00674
5.10	164.02	2.21490	.00610
5.20	181.27	2.25833	.00552
5.30	200.34	2.30176	.00499
5.40	221.41	2.34519	.00452
5.50	244.69	2.38862	.00409
5.60	270.43	2.43205	.00370
5.70	298.87	2.47548	.00335
5.80	330.30	2.51891	.00303
5.90	365.04	2.56234	.00274
6.00	403.43	2.60577	.00248
6.25	518.01	2.71434	.00193
6.50	665.14	2.82291	.00150
6.75	854.06	2.93149	.00117
7.00	1096.6	3.04006	.00091
7.50	1808.0	3.25721	.00055
8.00	2981.0	3.47436	.00034
8.50	4914.8	3.69150	.00020
9.00	8103.1	3.90865	.00012
9.50	13360.	4.12580	.00007
10.00	22026.	4.34294	.00005

TABLE 8. CRITICAL VALUES OF KAPPA SQUARE (K^2)
(Confidence level, %)

k \ γ	99.96%	99.73%	97.5%	95.0%	90.0%	85.0%	80.0%
2	75.12	60.06	29.24	22.39	15.68	12.09	9.55
3	24.04	19.23	9.36	7.17	5.02	3.87	3.06
4	12.89	10.30	5.02	3.84	2.69	2.07	1.64
5	8.44	6.74	3.28	2.51	1.76	1.36	1.07
6	6.13	4.90	2.39	1.83	1.28	0.99	0.78
7	4,76	3.80	1.85	1.42	0.99	0.77	0.60
8	3.85	3.08	1.50	1.15	0.80	0.62	0.49
9	3.22	2.58	1.25	0.96	0.67	0.52	0.41
10	2.76	2.20	1.07	0.82	0.57	0.44	0.35

For other values not shown, refer to Fig. 11-1.

BIBLIOGRAPHY

Calabro, S. R.: "Availability," presented before IRE, Willkie Memorial Building, Mar. 15, 1960.

Calabro, S. R., M. Barov, and V. Selman: "Availability: A New Approach to System Reliability," Fifth National Symposium on Global Communications, May 22, 1961.

Campbell, C. C.: "Safety Margins Established by Combined Environmental Tests Increase Atlas Missile Component Reliability," *Proc. IRE Intern. Conv.*, Mar. 21–24, 1960.

Carhart, R. R.: "A Survey of the Current Status of the Electronic Reliability Program," *Rand Corporation Research Memorandum RM-1131*, Aug. 14, 1953.

Chin, J. H. S.: "Optimum Design for Reliability the Group Redundancy Approach," *Proc. IRE Wescon Conv.*, Aug. 19–22, 1958.

Cochran, W. G.: "Sampling Techniques," John Wiley & Sons, Inc., New York, 1953.

Conner, J. W.: "Measurement Engineering: A Key Reliability Tool," *Proc. 5th Natl. Symposium on Reliability and Quality Control*, 1959.

Cramer, H: "Mathematical Methods of Statistics," Princeton University Press, Princeton, N.J., 1946.

Deming, W. E.: "Some Theories of Sampling," John Wiley & Sons, Inc., New York, 1950.

"DOD Proposed Reliability Monitoring Program for Guided Missiles," Ad Hoc Group on Guided Missile Reliability (ACGMR), April, 1958.

Dodge, H. F., and H. G. Romig: "Sampling Inspection Tables: Single and Double Sampling, 2d ed., John Wiley & Sons, Inc., New York, 1959.

Eisenhart, C., M. W. Hastay, and W. A. Wallis: "Techniques of Statistical Analysis," McGraw-Hill Book Company, Inc., New York, 1947.

"Electronic Equipment Reliability Program," *Tech. Rept. No. 45*, Vitro Corporation of America, Contract No. 52215, Mar. 31, 1953.

"Electronic Reliability in Military Applications," *Gen. Rept. No. 2*, Aeronautical Radio, Inc., Washington, July 1, 1957.

Epstein, B.: "Estimates of Mean Life Based on the *r*th Smallest Value in a Sample of Size *n* Drawn from an Exponential Population," *Tech. Rept. No. 2*, July, 1952; research done under ONR Contract Nonr-451(00), Wayne University.

Epstein, B., and M. Sobel: "Some Tests Based on the First Ordered Observations Drawn from an Exponential Distribution," *Tech. Rept. No. 1*, Mar. 1, 1952; ONR Contract Nonr-451(00), Wayne University.

Epstein, B., and M. Sobel: "Life Testing," *J. Amer. Statist. Assoc.*, vol. 48, pp. 486–502, 1953.

Epstein, B., and M. Sobel: "Some Theorems Relevant to Life Testing from an Exponential Distribution," *Ann. Mathematical Statistics*, vol. 25, pp. 373–381, 1954.

Flehinger, B. J.: "Reliability Improvement through Redundancy at Various System Levels," *Proc. IRE Conv. Record*, part 6, pp. 137–151, 1958.

Folley, J. D., Jr., and J. W. Altman: "Guide to Design of Electronic Equipment for Maintainability," *WADC Tech. Rept. 46-218*, 1956.

"A General Guide for Technical Reporting of Electronic Systems Reliability Measurement," Electronic Industries Association (formerly RETMA), December, 1956.

Gordon, R.: "Optimum Component Redundancy for Maximum System Reliability," *Operations Research*, vol. 5, no. 2, pp. 229–243, April, 1957.

Hadden, F. A., and L. W. Sepmeyer: "Techniques in Putting Failure Data to Work for Management," *Proc. 1st Nat. Symposium on Reliability and Quality Control*, November, 1954.

Henney, K. (ed.): "Reliability Factors for Ground Electronic Equipment," McGraw-Hill Book Company, Inc., New York, 1956.

Herd, G. R.: "Estimation of Reliability Functions," *ARINC Monograph No. 3*, ARINC Research Corporation, May, 1956.

Heyne, Jay B.: "Integrating Reliability Considerations into Systems Analysis," *Proc. IRE Wescon Conv.*, Aug. 19–22, 1958.

Heyne, Jay B.: "On an Analytical Design Technique," *Proc. IRE Natl. Conv.*, Mar. 24–27, 1958.

Hill, D. A., and R. H. Myers: "Procedures and Tables Based on Acceptable Reliability Levels," *Proc. 12th Nat. Conv. Am. Soc. for Quality Control*, May, 1958.

Kuehn, R. E.: "Some Results of an Early Reliability Program," *Proc. IRE Intern. Conv.*, Mar. 21–24, 1960.

Landers, R. R.: "Systems Reliability Measurement and Analysis," *Proc. 4th Natl. Symposium on Reliability and Quality Control*, January, 1958.

Luebbert, W. F.: "Principles and Concepts of Reliability for Electronic Equipment and Systems," *Tech. Repts. Nos. 90 and 91*, Stanford University Electronics Research Laboratory, Aug. 18, 1955.

McLaughlin, R. L., and H. D. Voegtlen: "Electronic Equipment Support Cost vs. Reliability Maintainability," *Proc. 5th Natl. Symposium on Reliability and Quality Control*, January, 1959.

"Methods of Field Data Acquisition, Reduction, and Analysis for Ground Electronic Equipment Reliability Measurement," Rome Air Development Center TN-58-183 (ASTIA AD-148801).

Moore, E. F., and C. E. Shannon: "Reliable Circuits Using Less Reliable Relays," *J. Franklin Inst.*, vol. 262, pp. 191–208, September, 1956; pp. 281–297, October, 1956.

Moskowitz, F.: "The Statistical Analysis of Redundant Systems," *Proc. IRE Intern. Conv.*, Mar. 21–24, 1960.

Moskowitz, F., and J. B. McLean: "Some Reliability Aspects of Systems Design," *IRE Trans. on Reliability and Quality Control PGRQC-8*, pp. 7–35, September, 1956.

"Multi-level Continuous Sampling Procedures and Tables for Inspection by Attributes," *Inspection and Quality Control Handbook (Interim) 11106*, U.S. Department of Defense, Oct. 31, 1958.

Naresky, Joseph J.: "Reliability Prediction and Test Results on USAF Ground Electronic Equipment," *Proc. IRE Natl. Conv.*, Mar. 24–27, 1958.

Nucci, E. J.: "Progress Report: Ad Hoc Study on Parts Specifications Management for Reliability," *Proc. IRE Natl., Conv.*, March, 1959.

Patterson, Mel: "Design Techniques for Upgrading the Reliability of Weapon Systems during Flight-readiness Check Out," *Proc. IRE Wescon Conv.*, Aug. 19–22, 1958.

"Reliability and Components Handbook," secs. 3.2.1 and 3.2.2, Western Military Electronics Center, Motorola, Inc., Phoenix, Ariz., Jan. 30, 1959.

"Reliability of Electronic Equipment," MIL-STD-441.

"Reliability of Military Electronic Equipment," *AGREE Task Group 9 Rept.*, U.S. Government Printing Office, June, 1957.

"Reliability Stress Analysis for Electronic Equipment," *Tech. Rept. TR 1100*, Nov. 28, 1956. (Available under publication number PB 131678 through Office of Technical Services, Department of Commerce.)

Rosenblatt, Joan Raup: "Statistical Models for Component Aging Experiments," *Proc. IRE Intern. Conv.*, Mar. 21–24, 1960.

Ryerson, C. M.: "The Confidence That Can Be Placed on Various Reliability Tests," *Proc. IRE Wescon Con.*, Aug. 19–22, 1958.

Ryerson, C. M.: "Numerical Assessment of Reliability," *Proc. 2d Natl. Symposium on Reliability and Quality Control*, January, 1958.

Sharp, D. W.: "Data Collection and Evaluation," *Proc. 5th Natl. Symposium on Reliability and Quality Control*, Jan. 12–14, 1959.

Sternberg, Alexander, and John S. Youtcheff: "Reliability Trade-off Analysis," *Proc. IRE Intern. Conv.*, Mar. 2–23, 1961.

"A Study for Reliability Data Analysis and Processing Program," (U)OR319, The Martin Company, November, 1958.

"Suggestions for Designers of Electronic Equipment," U.S. Navy Electronics Laboratory, San Diego, 1955.

Sumerlin, W. T., "Quantitative Reliability Acceptance Testing," *Proceedings of the Third National Symposium on Reliability and Quality Control*, January, 1957.

"A Summary of Component Failure Rate and Weighting Function Data and Their Use in Systems Preliminary Design," WADC TR 57-668 (ASTIA AD-142120), December, 1957.

"A Summary of Reliability Prediction and Measurement Guidelines for Shipboard Electronics Equipment," *Vitro Lab. Rept. No. 98*, Apr. 15, 1957.

"System Reliability Considerations," *Proc. 4th Natl. Symposium on Reliability and Quality Control*, January, 1958.

"Techniques for Reliability Measurement and Prediction Based on Field Failure Data," *Vitro Lab. Rept. No. 80*.

"Techniques of System Reliability Measurement," vols. 1 and 2, ARINC Research Corporation, December, 1958.

Thorndike, F.: "Standard Cumulative Poisson Probability Charts," *Bell System Tech. J.*, October, 1926.

Wald, A.: "Sequential Analysis," John Wiley & Sons, Inc., New York, 1947.

Walker, R. M.: "Reliability Improvement by the Use of Multiple Element Circuits," unpublished IBM Watson Laboratory memorandum, April, 1957.

INDEX